Amazon Expeditions

AMAZON
EXPEDITIONS

My Quest for the Ice-Age Equator

Paul Colinvaux

Yale University Press
New Haven & London

Library of Congress Cataloging-in-Publication Data

Colinvaux, Paul A., 1930–
Amazon expeditions : my quest for the ice-age equator / Paul Colinvaux.
p.   cm.
Includes bibliographical references.
ISBN 978-0-300-11544-4 (cloth : alk. paper)
1. Natural history—Amazon River Region.   2. Climatic changes—Amazon River
Region.   3. Geology, Stratigraphic—Pleistocene.   4. Colinvaux, Paul A., 1930–
—Travel—Amazon River Region.   I. Title.
QH112.C65 2007
560′.453098—dc22
2007021088

A catalogue record for this book is available from the British Library.
The paper in this book meets the guidelines for permanence and durability
of the Committee on Production Guidelines for Book Longevity
of the Council on Library Resources.

10   9   8   7   6   5   4   3   2   1

For Llewellya Hillis, Ph.D., Hon. D.Sc.
My beloved, and talented, scientist wife.
She searched the coral reefs of the world for reef-building algae
of the genus *Halimeda*, working the oceans while I searched the land
for ancient lakes. And we shared the rearing of our children
as we staggered our expeditions.

CONTENTS

PREFACE

The Amazon forest is a big place. It occupies the larger part of the area drained by the Amazon River system, an area roughly the same size and shape as the continental United States. The basin defined by this river system spreads into the territory of seven sovereign states, Brazil holding the largest part. It is bounded on three sides by mountains, particularly the giant Andes to the west. The fourth side, on the eastern seaboard of Brazil, has the modest and interrupted hills of the coastal range but essentially meets the open Atlantic Ocean, the source of much of its high rainfall.

This book describes a forty-year effort to discover the effect of the climate changes of the ice ages on the Amazon forest. Were the trees still there in the ice age, the forest still intact? Or had the forest been radically altered by some cataclysmic change in climate associated with the huge ice sheets that covered so much of the northern hemisphere? Or, far more exciting, did the forest meet climate change with a reshuffle of species, a response that could also have served to meet the long succession of ice ages that came before as well as this, the last of the line. This would be a dynamic forest always with us, bending to climate change, the individual species always surviving. The resulting forest might look slightly different with every stage of a glacial cycle but it should still be the forest home for all the plants and animals of the forest.

As late as 1964, when this tale began, there was not one radiocarbon date from the Amazon basin; not a single ice-age site or deposit had been identified. Had a cartographer undertaken to make a vegetation map of the ice-age Amazon he could have done no more than draw in the traceries of the river system with all the interfluves left empty and white. Terra incognita.

Terra incognita is an explorer's dream, and this one we knew how to explore. Ice-age deposits must be there if you looked. Find them, date them by radiocarbon, seek out fossil evidence of the past, principally the tell-tale pollen of trees. And for good measure take in the whole of the American equator, not just the Amazon. Survey across the Andes Mountains, and over to the Galapagos Islands 600 miles out to sea. Pencil in, as it were, the ice-age climate and vegetation of the entire American equator from the Pacific Ocean to the Atlantic. 'Twas a worthy dream. I still think so after squandering forty prime years on it. We pulled it off, too; at least as far as a single lifetime allows, and now the next generation carries on.

The ice age matters; not because it was terrible (it wasn't) but because it was recent. For the natural history of the earth it was an instant ago, its last gasp now reliably dated by radiocarbon to about 10,000 years BP (before present). Ten thousand years is just forty lifetimes for trees that live 250 years each. Compare this with the span of forty human life-times of fifty years each. This takes us back 2,000 years, say roughly to Julius Caesar's crossing of the Rubicon, the fall of the Roman Republic, and the birth of the Roman Empire; events so recent that our generation still reads memoirs of the principal actors.

Life systems of the earth are still as influenced by the zeitgeist of an ice age as are we by the legacy of Roman civilization. Many of us still speak versions of the Roman language. Life on earth is quite as much in thrall to the large and recent upheaval of the ice age, an event for the lives of species as close to the twenty-first century as the Roman Empire is to our own polity.

More than being recent, an ice age has been our normality. Each glaciation has lasted roughly a hundred thousand years, and they came in a long succession each separated by 15–20,000 years or so of warm interglacial times like those in which we lived before our dangerous experiment of greenhouse warming. We humans, together with all other living things that move or grow on the surface of the earth, are ice-age animals and plants, adapted to ice-age climates, whatever those might have been. The paltry 10,000 years of comparative warmth since the last glacial has been far too short for most of our current species to have evolved. Where ice-age climates were hostile, the life we know migrated, found strange neighbors, lived in different ecosystems. Where environmental hostility was extreme, the migrations and accommodations southwards should have been extreme also. Our working hypothesis was that much less happened in the Amazon than elsewhere, but accommodations to climate change should have been of the same kind: migration, strange neighbors, revised ecosystems, though all such accommodations would be muted by northern standards.

It is in northern latitudes that the zeitgeist of ice ages is still most apparent. Consider the absurdly impoverished forests of the British Isles, with their miserable total of a dozen species of native trees. Glaciers scraped Britain bare to the valley of the Thames, while what was to be the south British coast and northern France remained too cold for trees. Modern Britain got its trees back by migration from the south when the warming came, but the migrant flow lasted only until the sea rose to flood the English Channel early in postglacial time. By then Britain had its dozen and could get no more.

The gardeners of England will tell you that far richer forests can grow in England. It was the ice age that said they should not, because Britain was an island with but a shallow moat around it. But the Amazon was neither small, nor an island, nor subjected to an annihilating scourge of bulldozing ice.

So the Amazon should have had a more subtle history. Reading that history might require techniques more refined than those that had been designed to work in Europe. Europeans plotted the advance of glaciers, hence destruction of forests, by marks on the land, but the replacement of ice by modern forests through the analysis of fossil plant pollen. We Amazonians had no glacial marks to help us, rather all forest changes should be marked by fossil pollen. And the pollen of the Amazon was known to be as different in kind to the pollen of Europe as the forests and landscapes were different.

Pollen analysis was invented nearly a century ago in Sweden, a land well supplied with peat bogs to store pollen. In Europe, and other northern latitudes where the glaciers came, nearly all the trees are wind pollinated. Instead of the gorgeous flowers of many a tropical tree they have numerous bunches of anthers shaking like small rattles in the wind, casting out copious pollen as they shake. In a like manner, lonely stigmas spread their sticky surfaces, waiting in hope for a pollen grain to fall their way. In Europe too, most open land is grassland, and the grasses too, are pollinated by wind. Pollen brought by the wind is the name of the European game.

Only a tiny few of these wind-blown pollen grains make it to a stigma, the remainder float free where their concentration in the air is attested to by sufferers from hay fever. Remaining pollen clouds drift above the forests until a rainstorm washes them out to the ground beneath. When the "ground" happens to be a peat bog, the tiny pollen husks, or outer coats, are preserved virtually forever, after the contents have rotted away. Peat has lots of them, it is quite usual to find many thousand pollen shapes in a single cubic centimeter of rotted peat. Pollen grains are typically 30 microns across, too small to be seen by the naked eye but beautiful and big to a research microscope. It is slightly shocking to think of a micron as a millionth of a meter and yet be able to see an object just 30 microns across clearly with an ordinary microscope, but you can. This is old technology. The measurement also explains why there can be so many in samples as tiny as a cubic centimeter. To avoid clutter we usually use 1/4 cc samples.

The toughness of the pollen husks lets them survive in rotting peat for millennia preserving in minute detail the shape of the living pollen grain. This toughness also makes them resistant to chemical digestion of the peat, as well as chemical removal of their own protective wax, making it simple, though time-consuming, to clean and extract the pollen. The end result is a preparation of thousands of nice clean grains on a microscope slide. The shapes are distinctive enough for recognition, at least to genus (say as an oak tree rather than a particular species of oak). At this level they are easy to learn. In Europe you need learn less than fifty types to do a satisfactory pollen analysis and twenty types is usually more than you will find in significant numbers. These few will tell you the local forest trees or grasslands at the time the peat was laid down. Do it every 10 cms of a 10 m cliff in Swedish peat diggings, and you have an outline history of the migration that led to the modern forest.

The dream of plotting the fates of the Amazon forests through glacial cycles would have to use the pollen method, but the rules of the game would be different. There were too many species, for a start. And not only was this number almost beyond understanding, but they belonged to a huge array of genera and families, most of them unknown in Europe. Good enough to call pollen "oak" in a European land where only one or two oak species were likely, but Amazon trees might have hundreds of species in a family, all with similar pollen. It is common even for a single genus to have scores of species. The fig (*Ficus*) genus in the Amazon has in excess of three hundred tree species, all with the same pollen type.

Worse than this, most Amazon tree pollen has nothing to do with wind, having real flowers, pollinated by insects, birds, or even bats. Could there even be pollen clouds in the tropics to be washed by rainstorms to the ground? Faegri and Iverson, the leading manual on pollen analysis in the 1950s, predicted that pollen analysis would be impossible in the wet tropics because most trees would be "silent" in the pollen rain. Grasses would still be wind pollinated, as well as their allies the bamboos, but if you wanted a history of forest, forget it.

Also those wonderful pollen traps, the raised peat bogs of Europe, rarely exist in warmer latitudes. Thus were tropical pollen projects damned from the start: a huge pollen vocabulary (80,000 species in the Amazon), wind pollination rare, and no peat bogs. Add in a certain difficulty of access and the terra incognita on the hypothetical ice-age forest map of the Amazon in 1964 is no surprise.

The difficulty of no peat bogs was soon solved: use lake sediments instead. Lakes collect and mix pollen falling all over their surfaces, the pollen becomes waterlogged, sinks, and settles to the mud where it is beautifully preserved. Instead of stomping about in peat bogs, time is more graciously spent messing about in boats. Lakes beat peat bogs for pollen analysis technically, statistically, and aesthetically: no contest.

Exploration of the ice-age forests of the Amazon was to be a search for lakes ancient enough to hold ice-age mud. A lake is as old as the hole it fills (or as young) and the age of the hole was decided by geological process. Most of the lakes you see across the soggy Amazon plains are far too young for our purpose; I know, I have drilled. But the land is big, the lakes many, sooner or later you hit an old one. Our best little Amazon lake, many years down the searching road, gave strong evidence of spanning one hundred and seventy thousand years in just six meters of mud, with pollen all the way.

Because of that little lake, and others like it, we do have the story of the ice-age Amazon to tell. But three different stories entwine and plait together on the long road that took us there: the tale of lake searching, the tale of pollen meaning, and the tale of the opposing paradigm.

The tale of lake searching is a simple tale, though the most frustrating, the one with sleepless nights. Ours were the days before modern technology. Of course we thought our technology was modern, that was our ignorance of what was to come. We used air photographs, when we could find them. Mapping from space was still a military experiment, and not applied to the jungles of the equator. I did find two lakes by looking at photographs taken by the Gemini astronauts with their Hasselbads. Many published maps in our collections were plain white sheets with just the skeletons of rivers. We found lakes by talking to locals, to missionaries, to oil men, to bush pilots. That was not modern technology, that was nineteenth century. We were among the last to be privileged to go exploring as they used to do. Now my successors do their lake searches on their office computers, letting the remotest portions of the earth drift across their screens until a promising lake appears. Because so old fashioned, the lake search dominated my work and is the connecting theme of this story, with after each search the struggle to reach and core what was often a remote target. The usual result was sediments that were too young to be of use. I borrowed a useful term from oil men for this, "A dry hole."

I was afraid of the tale of "pollen meaning" from the start. Real mastery of their meaning, identity, and method of distribution from forest to lake mud was the work of Mark Bush. I was saved by comrades who did. They built our vocabulary from plant collections in the great herbaria: Eileen Schofield for the Galapagos Islands, Paulo de Oliveira and Enrique Moreno for the Amazon. Thirty-five years into the project we published our Amazon Pollen Manual and Atlas.[1] This atlas reduced that enormous flora of the Amazon to about 400 useful and identifiable kinds of pollen. This, thank goodness, was the limit of what could be done with a light microscope. But it was enough. Most pollen were identified to family rank only, each pollen type thus representing perhaps hundreds of relatives. No good if we were to have delusions of making species counts, but adequate, indeed excellent, for our purpose of testing whether the forest was still there, and whether it had changed much. And we solved that problem of Faegri and Iversen's "silent trees" which, being pollinated by animals, should leave no trace. Or rather Mark Bush and Paulo de Oliveira solved it through an inspired program of pollen trapping. They showed that flowers and pollen are beaten to the ground by rain and washed into our lakes, not of course in the huge quantities of the wind-blown types, but enough to find. So we had two independent pollen records: one of wind-blown types dumped on the lake surface as in classic European sites, and one of the trees actually growing in the watershed, whose pollen came with soil runoff water, a result vastly richer than the experience of the founders of pollen analysis in Europe (as seems fitting for the largest, richest forest on earth).

The third tale, that of the opposing paradigm, started as a hypothesis that begged all questions. The Amazon rains failed in an ice age, so it stated, failed so badly that the forest largely withered away. Drought, not cold, was the ice-age affliction. Trees survived only on places like hillsides that received extra rain, all the rest of the Amazon lowlands turning into something like savanna the while. Jürgen Haffer invented this scenario to explain some peculiar distributions of forest birds, suggesting that they still lived where they had survived and evolved through the ice age, in those remaining wet patches where the forest "took refuge." In these patches of Haffer's imagination both forest and birds had survived the terrors of the ice age. A thrilling tale. Many a biologist called it "Beautiful."

The refuge hypothesis was published in 1969, five years after I sat down to write the grant proposal to begin my ambitious transect, and three years after my first field season. The laboratory was beginning to learn the pollen. Despite the five years start, it would take years of work before anyone could test with real data the claim that drought destroyed the great forest.

Yet the refuge hypothesis seemed on everybody's lips. When Sherlock Holmes said that Professor Moriarty's treatise on the binomial theorem "had a European vogue," he was giving rather less of a puff than was given to the refuge hypothesis. In a very few years it was treated with the reverence of a paradigm. Drought-with-refuges was the ice-age Amazon in the average biologist's mind. Many went to the Amazon, especially from Brazil, specifically to look for evidence to support the paradigm.

To go there as a possible skeptic, as one who would test hypotheses with actual data, did not seem so popular. An old military maxim says that any natural obstacle is to be preferred to human resistance. The role of the "paradigm" in my tale was to provide the human resistance, showing the military maxim to operate in exploration science as well.

ACKNOWLEDGMENTS

Daniel A. Livingstone of Duke University trained me as a corer of lakes, passing to me his Alaskan torch as he set out to core lakes in East Africa. Edward S. Deevey of Yale University, by a superb piece of social engineering, then diverted me to core lakes at the American equator. G. Evelyn Hutchinson, mentor to both men, aided and abetted their cause to me directly in critical conversations in the old Osborne Laboratory at Yale. These three men drew me into their mighty company, thus making possible the near half century of exploration in mind and body that has been my happy lot.

The United States National Science Foundation (NSF) was my principal funder from start to finish, albeit with lacunae when my colleagues, the reviewers, punished substandard proposals. The NSF is one of the most enlightened organizations ever to emerge from human societies. The phrase "without fear or favor" might have been coined for it. Hail to thee.

THE GALAPAGOS EXPEDITIONS OF THE 1960s

Roger Perry, then director of the infant Charles Darwin Station, was my essential advisor and friend on the islands. My coring companions were Curt McGinnis and Erno Bonebaker (Yale), David Greegor (Ohio State), and William Horn (Duke). Further data were collected in subsequent years by my Ph.D. students, Daniel Goodman (chemical history of Genovesa lake), P. Dee Boersma (Galapagos penguins) and Dwayne Maxwell (Galapagos ocean productivity), as they performed their separate studies. These three, together with my colleagues Jerry F. Downhower (Darwin's finches) and Charles Racine (giant cactus), who developed their own Galapagos studies, formed a daily lunch

seminar in my laboratory on Galapagos ecosystems that I remember with gratitude and happiness as a high point in my research career. Eileen Schofield prepared the pollen reference collection, curated the Galapagos herbarium, and
shared with me the pollen analyses.

THE SEARCH OF ECUADOR FOR ANCIENT LAKES, 1972–1985

Miriam Steinitz-Kannan, at first a graduate student in my laboratory at Ohio
State, now a Regents' Professor at Northern Kentucky University, was the guiding spirit and planner of all our operations in Ecuador. Some of her remarkable
achievements are described in the stories that follow. Her parents, Hans and
Trude Steinitz, took me into their home in Quito and made me feel I have a sec-
ond home for life. Our companions on the many coring expeditions in Ecuador
were drawn from M. C. Miller, K-b. Liu, M. B. Bush, M. Riedinger, A. Anton, K.
Olson, I. Frost, M. Frost, M. Scheutzow, P. de Oliveira, and Ecuadorian students
E. R. Asanza, A. C. Sosa-Asanza, and C. Vazquez. We were the beneficiaries of
the extraordinary hospitality of Sister Rosa Vargas and the nuns of the Salesian
Mission at Yaupi and of the muscular help of their flock from the Shuar people.
We were also given friendly assistance by a unit of a very different organization,
the army of Ecuador, when we arrived unannounced at a remote jungle station.
In general, we thank the people and officials of Ecuador for the friendship or
hospitality we invariably received.

THE DECISIVE CORE COLLECTION FROM BRAZIL, 1986–1998

The organizer of our Brazilian expeditions was Paulo de Oliveira, a native of
São Paulo and graduate of the University of São Paulo, who had earned his master's degree from the University of Cincinnati with my friend M. C. Miller. He
completed his doctoral studies in my laboratory and was a postdoctoral fellow for
the latter half of our Brazilian operations. Without his organizational skills in
Brazil it is doubtful if our field work could have been completed. Paulo also re-
searched the critical sites, something vanishingly hard from outside Brazil. In addition to de Oliveira, my field companions were drawn from M. C. Miller, M. B.
Bush, N. Carter, S. Weber, S. Carter, and J. Curtis. Various Brazilian students
and teachers accompanied our field parties. Dra. E. Franzinelli arranged a vital
collaboration that made possible the decisive work at Six Lakes Hill and the earlier find of the lake of the floating forest. We thank particularly the safe hands of
the helicopter pilots of Kovacs. The mayor and people of the village of Recreio
allowed the use of their soccer field as a helicopter base. Six splendid woodsmen

from Recreio, strong, willing, and rich with jungle lore, flew with us to be our guides and safekeepers on the remote inselberg of Maicuru.

Simon Haberle contributed the critical pollen analyses of the Amazon fan sediments. Dolores Piperno guided our exploration of Panama, taking us to La Yeguada, and to the Darien airstrip from whence we discovered Lake Wodehouse. Piperno was also responsible for the phytolith studies that so enriched our pollen data. Enrique Moreno curated the Amazon pollen reference collection, analyzed the pollen in the L. Dragão core, and made the photomicrographs and plates for our pollen manual. The Field Museum of Natural History in Chicago provided space and facilities to base the building of the reference collection, as well as the use of its superb herbarium. The herbaria of the New York and Missouri Botanical Gardens both gave access to their outstanding collections of Amazon plants.

I am immensely grateful to my literary agents, Margaret Hanbury in London and Robin Straus in New York, for help with many a difficulty in the publication process, and especially for critically important boosts to my morale. The final shape of both text and illustrations has the imprint of my editor, Jean Thomson Black.

PROLOGUE: TRIAL IN ALASKA

Small lakes, generally no more than thirty meters (say a hundred feet) deep, can be cored by hand, using a device called a "Livingstone sampler," named after its designer, Daniel Livingstone. Dan had earned his doctorate, and designed his sampler, for pioneering work in lakes of the Arctic coastal plain north of the Brooks Range in Alaska, in his day before oil (1950s) an exceedingly remote region. His lake cores had yielded the first long pollen history of remote Arctic tundra since the last ice age.

I learned to core from the master himself in the Bay lakes of North Carolina in 1959. Livingstone was then a young professor at Duke University, and I was his graduate student. That I should learn to core lakes was inevitable, as was the fact that we should talk about Alaska.

Just before Christmas 1959, Dan handed me a reprint, "You should read this." A "royal" command. It was bound with brown paper covers, it had the title *History of Imuruk Lake, Seward Peninsula, Alaska*, by David Hopkins.[1] Just a report of a government geologist, reconstructing the history of one of those holes in the ground that become lakes, a fairly big one eight miles long by five wide, but dreadfully shallow (flat bottom at three meters), and so remote as to be reachable only by float plane or a hike through a mountain range. Yet Hopkins sent the reprint to lure Dan with his "Livingstone sampler," because he thought Imuruk Lake might occupy a very old hole, perhaps one or two hundred thousand years old. His paper suggested it should be well worth exploring by core drilling and pollen analysis. Dan's pioneering histories of Alaskan pollen spanned only the last 12,000 years or so. Now this.

The position of the lake on Seward Peninsula magnified the importance of the find. Seward Peninsula is not a household name, yet I dare say most literate people know its shape on the map. It is the pointing finger of Alaska that just fails to reach Siberia across Bering Strait. This meant that Imuruk Lake lay squarely in the middle of what was once the Bering land bridge, the ancient causeway between Asia and the Americas.

Wonderful opportunity, but Dan couldn't do it. His plans were far advanced to go coring in Africa, a dilemma to be solved in the classical professorial manner, "Send a student." He sent me. Imuruk Lake was to be my taskmaster in the art of coring lakes, a nice harsh one, as it turned out, suited to fit me for anything the Amazon might be able to throw. In that Christmas of 1959 I had no idea of ever going to the Amazon, but I can now see the Amazon spectre behind every line of the grant proposal I wrote at Duke during Christmas break.

The proposal was easy to write (in Livingstone's name, of course, to reviewers). Two fat plums might be plucked. The first was to use pollen analysis for a direct measure of the environment of the Bering Land Bridge, the homeland of the first Americans. The second was an outside chance of getting a climate record of glacial times certain enough to test whether the adjacent seas had permanent ice cover or none, a subject of strong controversy in those far off days for oceanography.[2]

But it is as an exercise in the theory and practice of coring a remote and ornery lake by hand that Imuruk served as a prologue to the Amazon adventure.

The only skills needed for coring are those of watermanship; the rest is technique and the right equipment. Imuruk needed all three more than any lake I was to find in the coming decades. It was the Imuruk technique and equipment, worked out under duress of an eight-mile shallow lake, with a hard clay bottom, and subject to storm that was to make possible our prowess at coring in the Amazon. To follow the plot it helps to know how the trick of coring is done.

First anchor your raft to three long ropes set out like the star on the hood of a Mercedes. This is where the call for watermanship comes in. The anchors, typically rocks of twenty-five pounds or so (mud anchors), must be planted by a man in a dinghy while he who is placing the anchors stands on the raft paying out line, directing the dinghy, shouting the moment to let go. The raft, made of rubber boats, drifts in the wind the while, making nice the decisions of where to direct the dinghy for the second and third anchors so that the raft ends up in the center of the star when the three ropes are pulled tight and made fast. After this the raft must not be able to move in any direction, no matter what the wind does. A rough rule is that each anchor rope must be at least three times the depth of water; I am happier with six times, and in a shallow, stormy varmint of the Imuruk kind have the ropes pulling in as close to the horizontal as your rope supply allows.

Next lower a pipe, in sections, through a hole in the raft to the bottom of the lake and make the top fast to the raft. This we call our "casing." We drill inside it, using it to find the old hole as we core in sections. The casing must remain perfectly vertical for coring. If it does not, this is evidence that one of the anchors is dragging and you need to do something about it.

The actual drilling is like coring an apple, a straight pushing operation, no twisting. A geologist's drill will not do, it would make a hole all right but the mud would be reduced to soup. We need a perfectly undisturbed sample, liking to say that the mud must not know it has left the lake. For this, the simplest apple coring technique will not do. If you push an empty long tube (apple corer) into mud it will not cut a core. Instead the end of the tube will be bunged up with mud, and further pushing merely makes a hole, collecting nothing. So ours must be an apple corer with a difference.

The enemy that stops mud entering your long apple-corer tube is friction. Civil engineers have long known how to overcome this friction when the corer is under water. Close the open end of the tube with a captive piston suspended on a cable, then push the tube down past the piston. Because nature abhors a vacuum, mud is drawn into the tube. The effect is like a bicycle pump in reverse. To pump a tire you push the piston down the pump tube, driving out air. To sample sediment under water you hold the piston stationary and push the core tube down past the piston, pulling in sediment. The force coming from the weight of water on the mud surface is adequate to overcome friction of wet mud against the inside of your tube. The result is a tube full of mud that "doesn't know it has left the lake." This was the key principle that Dan Livingstone introduced into lake coring.

But there is a catch. From our raft of rubber boats we only have room, to say nothing of physical strength, to manipulate short cores. Our tradition is to work with sample tubes with space for just one meter of mud. Since we always want more, we must go down the old hole to its very bottom for the next meter, and do this again and again, until we reach the bottom. Easy to find the old hole because of the casing. But we must get the piston to the bottom of the hole as well as the end of the empty tube. And there is the rub: water pressure in the narrow hole will tend to push the piston up the tube.

In the Livingstone sampler, this problem was solved by holding the piston down on the end of a long metal rod reaching through the whole length of the empty tube. At the top, this rod was attached to the rods of the drill string, thus keeping the piston captive throughout the descent of the old hole. Once at the bottom, the piston could be released by pulling up the drill string the one meter length of the sample tube until a half turn of the drill string locked the bottom of the sliding rod to the tube holder. Clamp the piston cable to the raft, and all is now set to take the next meter of core.

This system works beautifully on most lakes small or shallow enough to be worked by hand and has been for forty years the workhorse of those who reconstruct the climate and vegetation of the continents of the last thirty thousand years or so. But until modified, it did not work at Imuruk Lake. The sliding rod and its lock lacked the rigidity needed to transfer sufficient force to the sample tube, in short, it could not be driven efficiently with a hammer.

In the second half of June 1960, I flew out of Kotzebue, in Arctic Alaska north of the Arctic Circle, south to Imuruk Lake, with John Cross in a Cessna 180 mounted on floats. John Cross was an old pilot, "old" in every generous sense of the word. He had been a bush pilot in Alaska so long that his past was legend, someone telling me that he had flown a fighter in France during the first World War. Wise, careful, and full of knowledge, no matter the thick glasses on the aged eyes, there was peace and safety in his cockpit. He had passed over Imuruk a week before, then still covered with ice, but thought it should have opened up by now.

Packed into the thin, dragonfly body of the little Cessna were the longer bits of the Livingstone sampler, the rest of the aircraft stuffed with equipment and stores. John Cross would bring my undergraduate assistant and the rest of the stores in a second trip. Across Norton Sound, over the tiny village of Deering, up the valley of the Inmachuk River we went and over the Asses Ears range, passing close by the ears themselves. Ahead, the expanse of Imuruk Lake spread away from us, its surface the dull gray of ice in the morning light.

A grand circuit of the shoreline skimming low above the ice, and there was water, a quarter-mile patch along one shore. "I can put you down there?" "OK." The rushing whisper of a float plane coming down with engine throttled back, a splash, and with scarcely a jolt from the glass-smooth water we crept up to a low promontory of lava rock. John Cross, already in hip boots, was out in a moment, hauling us to the bank. I joined him on the tundra.

We unloaded. "Got anything to keep the bears off?" I pointed at the shotgun. John climbed back into the cockpit, I pushed off, the motor roared, and he was gone. And I felt the glorious silence of a spring morning in the Arctic. On one horizon the jagged Bendeleben Mountains, rising 5,000 feet between me and Nome, on the other the provocative Asses Ears above the less jagged 2,000 foot ridge over which the Cessna had vanished, and the great lake motionless under its melting ice. Maybe the first Americans saw this sight. My home for the summer. Bliss.

In a few days of bright sun the ice went, the lake surface turned into modest waves, and my undergraduate assistant and I anchored the raft half a mile from shore for our first shot at coring, towing it out with a small rubber boat fitted with a 3 hp outboard. Down casing, and insert the sampler to feel for the bottom. No difficulty in finding bottom in this lake, it was a hard mixture of clay and silt. Raise the drill string to pull up the holing rod and free the piston, twist and lock. Clamp the piston cable and push. After about half a meter we were stuck, the combined shoving of two young men could do no more. Reach for the primitive hammer I had designed (badly) and bang our way down: slow, but it worked. Pulling up was harder: visualize two young men with strong backs straining until the raft started to sink under the pressure of their feet before the mud finally lost its grip and the tube slowly withdrew. One meter secured. Fit a fresh sample tube, load the piston, move the piston cable through the vise by precisely one meter so that the strain would not come on until the piston was at the bottom of the old meter-deep hole. Down the casing once more, into the old hole, and force the tube down against the resistance of the hole's sticky sides until the tautness of the cable registered that the piston had reached the bottom of the hole. Raise the drill string one meter, twist and lock. Then push and hammer as before.

Meter 3 was the end of that try. A sudden squall of wind out of the mountains raised whitecaps that swept over the raft, finally throwing the tender onto it. The flation tube of the tender struck a sharp piece of metal, causing a 2 inch tear and instant deflation. We hauled the wreck aboard to save our precious outboard motor, collected everything onto the raft, and glumly rowed the whole thing the half mile back to camp. Imuruk had begun its long training of the finer points of coring.

Days of repair and redesign, then try again in the middle of the lake, miles from shore. We got stuck at the fourth or fifth meter and were utterly unable to pull the sampler out again. We were lucky in that our struggles slipped the holder out of the tube so that we retrieved everything except the tube full of mud. The hole was blocked by the shed sample tube, though. Back to camp and invent plan B.

Plan B was to core in very short sections that we could pull out, say, 20–30 cms at a time. But my tubes were designed to hold a meter of mud, plus the length of the piston, plus the length of the holder. My tool kit was primitive, but I had a hacksaw, hand drill, and set of files; all I needed. I sawed the tubes in half and drilled the holes for the lugs of the holder. Then I walked thirty miles across the Asses Ears range to the Fink Creek gold mine on the Inmachuk to get an engineer to weld a striking plate to a steel tube that would slide over the top of the drill string to be pounded with a sledge hammer. And walked back with it.

More than half the summer had gone by the time we had the raft, now strengthened with timbers found on shore, anchored for plan B. But at last the weather was perfect, a flat calm that lasted nearly two weeks. Plan B worked. We hammered even our short tubes until we found resistance building up, then stopped and pulled out. In two weeks of steady hammering we reached something over six meters in about twenty sections. The calm spell had ended and now Imuruk was getting choppy. Perhaps it was this that let us hurry. Whatever, we beat a tube in to yield its entire half meter and got stuck again, stuck beyond the limits of our strength. So try to let the waves work the tube loose.

We filled the boats of which the raft was made with water, then pulled our mightiest to sink the raft. We had bolstered the rubber boats with heavy timbers left behind by gold miners long ago, boosting the carrying power of the raft, yet we still got the whole thing sunk. We hauled in the steel piston cable and clamped it securely to a baulk of timber, and let go. Now the steel "airplane" cable was doing the pulling, a relentless steady pull made fiercer when waves tried to lift the raft. We bailed out the rubber boats to maximize "lift." Then we relieved the raft of our weight too, going ashore in our tender with remaining drill rod etc., leaving wind and water to do our work.

The wind blew to a gale that night, growing fiercer over the next three days. I call it a gale for want of a better word, 'tempest' perhaps. By the second day the wind was strong enough to make walking on the open tundra difficult. I went on hands and knees to the top of the ridge to look for the raft among the whitecaps through binoculars. The raft was still there, the six anchors we had used, on long nylon ropes, apparently doing their job. But on the third day I could not see it.

All storms end, this one after five days. The raft had been reduced to floating debris. The drill string had been broken off under water, and the steel piston cable had snapped, breaking our last connection to the sampler still stuck deep in the Imuruk mud. Reestablish contact, or give up? Stark choice.

I swam to the bottom of the lake with the insane idea of uncoupling the broken length of rod and making a connection to a new rod under water, but the bitter cold stopped me. Numbed hands in water close to $5°$ C failed to move the aluminum "sleeve-and-bullet" coupling. Naked, I sat on the mud at the bottom of Imuruk Lake, feet in the entry to the hole, enlarged as it was by the battering given the casing and drill string by the storm waves. Clutching the broken drill rod as it protruded from the hole, and with faint light swirling in the icy water above my head, I surrendered. Round one to Imuruk.

I walked out again over the Asses Ears to the Fink Creek mine on the Inmachuk, bummed a truck ride to Deering, and used the radio connection of Deering schoolhouse to call the air charter company to come and get us. And I resolved to come back in winter when the frozen lake would keep still. How I wanted it to keep still. Ice half a meter thick would also be the perfect drilling platform, strong, horizontal, an unsinkable fulcrum for a winch to pull out those stuck samplers. So south to Seattle to reequip.

Reequipping meant more than a winch, a tripod, and a mechanical hammer. It also meant fitting sample tubes with steel cutting shoes, and abandoning (for good, as it turned out) the Livingstone principle of holding the piston in place with a sliding rod.

The steel shoe is a device for reaming out a tubular hole of slightly wider diameter than the sample tube. If the shoe is only about 5 cm long, as mine were, friction should be no more than is applied to the first 5 cm of the sample tube. But being twice as thick as the tube, it would ream out a hole that much wider. In the best of theoretical worlds, the sample tube could follow down in the wider hole almost without friction. The real world is, of course, different. The tube rubs one side or the other and mud tends to squeeze back into the reamed-out hole. Still the reamer shoe ought to help a bit. It did; it does.

I designed those shoes in moments of despair at Imuruk. Having no training as an engineer, they were my own idea, indeed had to be. Only when I was back in contact with Dan did he refer me to the standard reference on the design of such shoes, I rather think published in the late nineteenth century. But I have to say it feels great to think of yourself as clever enough to redesign the wheel from first principles.

But my shoes did more than merely ream out the hole, because they also held the piston and did away with the sliding rod. This feature was to be the most important contribution to the Amazon project, where I was to encounter mud as stiff and sticky as at Imuruk. The shoes held the piston on the wine cork principle. The sharp leading edge of my shoes projected half a centimeter beyond the sample tube to make a little ledge on which the tube rested (necessary to transfer the driving force to the shoe). But I had the ledge machined wider still, by about a thirty-second of an inch (I was working with a Boeing subcontractor in Seattle, and machinists did not measure in metric in those days). I then made the leading rubber seal of the piston the right size to be rammed into this narrow orifice like a cork into the neck of a wine bottle. And there it would stay until jerked out by a blow on the drill string after the piston cable was taut and clamped to the raft (or ice!). Can I claim the world's longest corkscrew?

And so back to Alaska in late November, when the ice was already thick but the last hours of daylight still gave time to work. The ice operation worked. Using a fifty-pound drop hammer suspended on a fifteen-foot tripod driven by a 4 hp Briggs and Stratton engine bolted to the ice. The giant hammer drove the sample tubes with a few strokes, like a skilled carpenter driving tenpenny nails. To get them out again I mounted a 2,000 lb come-along (ratchet hoist) on the tripod. With this rig two Inuit men from Deering village and I did in one short Arctic winter day what I and my assistant could not do in six weeks in summer. We still did not get all the mud that was there because at the eighth meter the guide rod for the drop hammer snapped in two. The rod was solid steel one and a half inches thick, but it was a victim of low-temperature metal fatigue. A day's hammering at about twenty below F ($-30°C$) did it in as surely as similar fatigue had brought down the Comet airliners. But I had what I needed. The 8 m core turned out to span the glacial, and with it the complete duration of the Bering land bridge of the last ice age.

Adventure came at the last, as ice fog settled over the lake, apparently spreading to the mountains on either side. We heard aircraft in the distance but saw none. Then, when gathering willow twigs one morning for a fire, I looked up toward the outlet of the lake to see a Cessna on skis, sandwiched between ice and fog. It came in straight and smooth to where we ran waving. John Cross had found the lake through the ice fog by following the outlet river from the coast, his skis down in the river valley all the way. The door opened, and we heard John Cross say, "Just the men and their weapons," and then to his radio, "I've got the men and they are all safe." The Cessna climbed up through the ice clouds to the light. I went back for the core in good weather on December 4.

An author's apologies for this digression into the esoteric, doubtless boring, trivia of coring lakes. But coring lakes across the Amazon, the Andes, and beyond was the technique that was to undo the Amazon paradigm. As in so much in science, technique makes results possible. No mechanical drop hammers or winches in the rubber-boat world we should inhabit in the Amazon, but otherwise the same rig, complete with steel shoes to ream out holes and grip the piston wine-cork-style. Our lot would be to hump loads of duralumin pipe, drill rod and sample tubes, ten-pound drop hammers, inflatable yacht tenders (Avon Redstarts mostly, weighing in at more than 30 lbs each dry), wooden raft platforms, and hundreds of meters of anchor rope, as well as assorted equipment for lake surveys.

Perhaps less boring, though equally esoteric, is the pollen analysis of the Imuruk Lake core. This is a game of search and count. Look the slide over at a low magnification of 100× for an idea of what is there, how much pollen, how much trash. Switch to 400× and the pollen grains show up sharp and clear. You can go to 1,000× when a pollen grain will fill the entire field of view with very fine details resolved, but this means an oil immersion lens and assorted bother that slows the whole process. Stick with the 400×, all you need to identify the wind-blown pollen of the Arctic, and indeed of the whole north-temperate belt too.

In arctic Alaska the wind-blown pollen story is told in just six pollen shapes, the sedges of the sedge tussocks, true grasses, *Artemisia* (wormwood genus that includes sagebrush), birch, alder, and spruce. The basic vegetation of Seward Peninsula is tussock tundra, seemingly endless fields of cotton grass (a sedge) tussocks, mounds about a foot across, with an ice core and a fringe of living sedge leaves and cotton balls like hair on a head. Dreadful stuff to walk over, even when only a foot high but I encountered some crotch-high, making walking uncomfortable indeed. Search for a caribou trail or a stream bottom and follow that instead. Tussock pollen is distinctive but a mess. It is pear-shaped, with a granular texture with side patches where the granules disappear, serving as pores from which the pollen tubes can emerge. We see them as flattened three-dimensional objects, letting you look through a "pore" to the granules of the other side. Also they crumple and fold, resulting in the mess. But you quickly learn that the mess is cotton grass pollen from a tussock: lots of them, easy to count.

The other pollen are easier still. Grasses, many of which grow on the sedge tussocks, are beautiful spheres like tiny ping-pong balls with a single pore like a porthole with an annulus of thickening around it. The only other wind-pollinated herb is *Artemisia*, not sagebrush, which does not make it to the cold Arctic, but small and inconspicuous members of the same genus that live in rocky outcrops, or even among the tussocks. The pollen is a delight to see, covered in short spines, walls immensely thick, built of two stout layers separated by struts, a miniature engineering feat; three grooves with pores as a way out for pollen tubes past the bastions of the walls between.

The three woody plants are just as easy. Birch has a shiny sphere with three thickened pores equally spaced to distort the sphere into a triangle. Alder pollen has a similar structure but with five pores and the pentagonal shape amplified by thickenings, miniature curved girders between the pores. And spruce, the largest grain of the arctic suite, looks like Mickey Mouse's head, an elliptical head with two enormous bladders like Mickey's ears.

Birch on Seward Peninsula is not a tree, or even a respectable shrub, but a true dwarf, *Betula nana*. Its common habitat near Imuruk Lake is in what should be space to place a boot between the cotton grass tussocks, space which it fills like a vegetable barbed wire. A true plant of the open tundra. Alder is a true shrub on Seward Peninsula, taller than a man, but the nearest alder bush to Imuruk Lake is ten miles away, down slope toward the sea. Spruce is the forest tree of the interior, the edge of the modern forest being fifty miles away at the base of the peninsula. With the massive greenhouse warming of Alaska now in progress I suspect the spruce will make short work of colonizing the peninsula; hard to think of Imuruk Lake in the forest.

Neither spruce nor alder had colonized the peninsula in the ice age, the pollen made this quite clear (there wasn't any). The surface mud of the last few decades did have significant spruce and alder pollen (more than ten percent), showing that the pollen clouds had blown at least fifty miles for spruce to show up and ten miles for alder. Earlier than two or three thousand years ago there was none of either. Birch was there throughout, but much less of it early on. To me, all this meant cooling. Imuruk Lake was probably recording the near destruction of the spruce forest of the interior in ice-age time as well as the impoverishment of the local tundra.

I had enough to lay claim to having tested both the warmer, nicer Bering land bridge hypothesis, and the warmer, unfrozen ocean hypothesis, finding them both wanting.

Nature published the land bridge results in 1963 and my monograph followed.[3] After that my postdoctoral years were virtually predetermined. Opportunity and funding said, "Back to the Arctic." And it was so. I cored in the Pribilof Islands to get the southern (and presumably warmest) coast of the Bering land bridge. Jerry Brown, then working for the army's cold regions engineering laboratory, sent me the North Coast in the form of frozen peat dug out of the permafrost at Barrow, already radiocarbon dated.

I now had a complete north-to-south span of the old Bering land bridge. The Pribilof pollen record backed up Imuruk Lake, more extreme arctic tundra in glacial times, suggesting that any human migrants could not escape many of the rigors of that cold artic place by keeping to the southern parts. But the big news was from Jerry Brown's northern peats tunneled from the permafrost. Glacial-age pollen from those peats revealed a tundra more extreme than anything on the coast today, bolstering my deduction that the sea ice was always present. *Science* published that one.[4]

I was then working at Yale under Edward S. Deevey and G. Evelyn Hutchinson. I was thus at the center of the new emphasis in ecology on efforts to explain diversity, or as Hutchinson put it in a celebrated address and paper, "Why are there so many kinds of animals?," going on to say, for that matter, "Why are there not more kinds?" He conceded that perhaps the underlying question was, "Why are there so many kinds of plants?," but as a zoologist he preferred to stick with animals. His students spread these titillating questions as they joined the faculties of great universities: Robert MacArthur, Peter Klopfer, Lawrence Slobodkin. It was the subject of the day.

Deevey, long before, had been one of Hutchinson's first students, and they had then worked together at Yale for the best part of a lifetime. As early as 1949 Deevey had published a hundred-page monograph on the "Biogeography of the Pleistocene" (i.e. ice ages). He went on to be a pioneer of the use of lake sediments as historical tools, and had published a paper on "Coaxing History to conduct Experiments." He had been Dan Livingstone's doctoral advisor when Dan invented the Livingstone sampler, the better to secure the histories in sediments that could be coaxed into doing the long-term experiments that ecologists could not do in a human lifetime.

One day Ed Deevey called me into his office for a talk. Ed always had something of the old-time professor about him, tweed jacket or cardigan, soft-spoken, kindly, habitually puffing at a pipe, glorious intellect masked by an unassuming demeanor. These traits were particularly in evidence that day. "Paul, you have done all this work in the Arctic!," a pause to puff, "and you have shown us," another puff, "that when the glaciers came, it got colder." He looked me steadily in the eye in silence.

In this space should be a cartoon drawn in the older traditions of *Punch*, with the elderly professor saying those words to the brash young man, and the caption reading, "Collapse of young man."

This cartoon would not have been unfair. I had been feeling pleased with my arctic successes; the land bridge a cold, inhospitable place, the Arctic Ocean always frozen, findings good enough for *Nature* and *Science*. Yes; pleased with myself, not necessarily a nice trait in the young. Ed punctured that bubble in the kindest of ways: gently pointing out that all I had done was spend more than four years of my life showing that proximity to ice meant being cold.

The burning question of my specialty was the state of the tropics far away from where the ice was spreading. Ed was saying, "Go south, young man." Thus began what was to turn out to be near forty years of work: and adventure. Dan was already doing Africa; I chose the Amazon.

We cannot see the future. When I took the fatal decision to undertake the ice-age history of the Amazon, I foresaw some massive obstacles to be overcome: those 80,000 pollen-bearing species, tree pollen mostly traveling on animals not the wind, the region poorly mapped, no known ancient lakes, logistics often difficult. I did not foresee that it would take me forty years, nor that the hardest part would be the obstacles and objections of those who became devoted to what I think of as a fairy-tale hypothesis (widely known as refuge theory), and whose devotion raised this fairy tale to the status of a paradigm.

1

THE REASON WHY

The reason why was diversity. Why are there so many species in the tropics? This has long been one of the knottiest problems of ecological theory. Life gets richer as you go from temperate regions toward the equator. In warmer climes there are more kinds of living things than in the colder north; many, many more kinds. But why should this be?

The temptation is to say, "Obvious! It is nicer in the tropics; more productive; wet and warm; no winter; living is good and lots of species take advantage of it. Next question please." But that answer is no answer. Lots of living things in Europe and North America, thousands of kinds of animals and plants. The problem is that the wet tropics have more kinds still, many more.

In the tropics, the Amazon being the most spectacular example, there are perhaps eighty times as many species of plant as in a northern region. I get this absurd statistic by comparing the Amazon with Alaska, big places both. Accepting the commonly touted (especially by me) figure of 80,000 plant species known from the Amazon basin and comparing it with the known species list of Alaskan plants which is something over 1,000, I get that ratio of 80 to 1. As a statistical exercise this calculation is absurd, but like the generality of lying statistics, it adds drama to a mundane observation. The plant species list of the Amazon basin is hugely diverse.

More honest statistics can make the same point. The island of Britain is credited with about a dozen kinds of native tree (taking "tree" to mean anything that grows at least 30 feet tall): oak, beech, ash, elm, willow, alder, birch, sycamore, pine, larch, and yew; allow two species each of oak and birch and that is a generous dozen. A good splitter, perhaps armed with DNA technology, can break out a few more species but not many. The island of Britain, for all that it is rich in agriculture and well worth fighting for, is a poor place for botany, though good for gardening.

A portion of the great Amazon forest the same size as the British island would have possibly a thousand distinct kinds of tree. This seems grossly unfair. Why should a British naturalist get away with learning just a dozen or so tree names while to be truly competent a Brazilian should learn a thousand?

Even this statistical comparison, however, has its components of falsehood. The most immediate untruth is hidden in comparing an island with a continental expanse because it is as hard for trees to invade islands as it is for armies. Islands always have fewer species than continental bits of the same size. But however dud the statistics, the wet and warm places are always more richly diverse than the lovely temperate lands of the "West." Why should an eternity of natural selection have left more species in wet, tropical places like the Amazon than anywhere else?

The sublime explorer, Alfred Russel Wallace, found this problem awaiting him when he thrust himself into the Amazon jungle in 1848. Wallace was to find fame as the co-discoverer with Darwin of evolution by natural selection. But he started out as a restless young man in England, with little formal education, in straightened circumstances but with a thirst for reading, and, by his own admission, a passion for collecting beetles. He worked, saved, and went with his friend Henry Bates to the Amazon by the cheapest route.

Wallace existed on practically nothing at all for four years along the rivers and in the forest; collecting, always collecting. His younger brother came out to help him but died of yellow fever within a year. Wallace collected with gun, net, or forceps, whatever it took; every living thing was his target. In four years he had amassed thousands of specimens. He lost the lot when the old tub of a ship taking him home to England burned and sank off Bermuda. But the memory of what he had seen and learned did not burn. He had lived four years in a world so rich in life and species that England was a barren rock by comparison.

Two years in England were enough for young explorer Wallace before he was off again to another spell of penury and illness in the jungle, this time to the Malay Archipelago. This second self-imposed sentence was for eight years. All the way down the archipelago, through the islands that are now Indonesia, and on to New Guinea was that same incredible richness of life he had seen in the Amazon, so different from England. Again he collected with dedicated vigor. But the catch was not the catch made familiar in the Amazon. Animals might look the same, come in as many kinds, and do the same sorts of things, but the species were different. Wallace had shot and trapped them on two continents, and he knew the differences when he saw them. From this observation came the two great contributions to learning for which Wallace is known: biogeography and the discovery of natural selection. But he also wondered why England was different.[1]

Wallace's self-education by reading had included Charles Lyell's *Principles of Geology*, the stem treatise of modern geology. Wallace knew about the ice age. He knew the havoc the ice must have wrought in Europe. No sign of ice works in the Amazon or the Malay Archipelago, of course. Here was a prime hypothesis to explain the relatively barren life of England: the glaciers had killed off so much of what ought to have been there. In the tropics, there were no glaciers, thus no extinction.

A reasonable idea, this; and one that has partly survived the test of time, though only partly. As explained in the preface it accounts for the poverty of trees in Britain compared with continental Europe, and the poverty of both compared with the American Midwest.

On continental Europe the southern reach of the ice sheets was made more effective as tree destroyers when they stretched southwards to meet other glaciers coming north or west from the Alps or Carpathians. Thus the vegetation of central Europe was caught in an icy pincer movement: ice sheets to the north, glaciers to the south, bitter cold between. In the western bits of Europe retreat southwards from glacial cold was impeded by the Mediterranean Sea. The forest diversity, not just of Britain but of the whole of Europe, was savagely reduced by extinction as the forests were caught in this glacial trap. These events were very recent, ending just 12,000 to 10,000 years ago. People saw them happen. Pity they had not yet invented writing, their memoirs would make good reading.

In European strata older than the ice ages we find fossils of trees that still grow elsewhere in the world, China and North America mostly. That these were killed off in Europe by the ice and have been unable to get back since seems certain. These trees, not seen in Europe since the Miocene, include the long list of trees familiar to American naturalists but known only as ancient fossils to European botany; trees like sweet gum, black gum, tulip poplar, and sassafras which so delight European naturalists when first they see them. Russian visitors too. Back in the days when the USSR exerted its power to restrict travel, a distinguished Russian paleobotanist, expert in the fossil history of Russian plants, walked through a forest in the American Midwest for the first time during an international congress. In emotional awe he turned to his hosts to say, "I am walking through the Miocene of the Russian plains."

The extinction of this Miocene flora in Europe derives from the peculiar properties of European geography that set up the land for catastrophe by ice.

The principal mountains run west to east, making possible that deadly pincer movement of mountain glaciers running northwards to meet the ice sheets heading south.

But in North America the mountains run north and south. When the ice sheets reached down from the Arctic, snuffing out forests in Canada and the northern United States, escape was always possible to the south. Forests might not escape intact; we are now sure that they didn't. But most of the species did escape, to become southern colonists living in forests quite different from modern forests.

Pollen analysis was a European invention, devised essentially to plot the return of forests after the glacial retreat. Those forests were simple, with few species, from which two great ecological fallacies followed: that plants lived in obligate *associations* (after all, you always found the same ones together) and that you could use the most prominent or persistent species as "indicators" for the whole group. The inventors of pollen analysis in Europe and their well-trained successors convinced themselves that they could identify whole associations by finding "indicator" pollen. Thus were pollen analytical interpretations made simple.

This belief in obligate associations of plants took root in North America too, particularly in the first half of the 20th century. An eloquent teacher, Frederic Clements, persuaded three generations of plant ecologists that not only was the taxonomic unit of vegetation the *association* but also that this "super organism," as some called it, could grow and colonize fresh ground by plant succession. In Clements' own words, "As *an organism the formation arises, grows, matures, and dies. . . . The life-history of a formation is a complex but definite process, comparable in its chief features with the life-history of an individual plant.*"[2]

This was the European idea of plants living in obligate associations pushed to the limit of credulity. For a botanist steeped in this viewpoint, the glacial history of eastern North America was both obvious and simple. The associations migrated south as neighborly groups, some jostling, perhaps, but the whole thing rather orderly until all the old associations were hunkered down in warmer latitudes. When the glaciers melted, they all moved north again, in sequence. Simple.

Too simple, America was soon well supplied with skeptics of the whole permanent association idea that made this glacial history seem highly improbable. Trees were individuals of individual species, provided with independent traits; they did not, and could not, belong to organizations of different species. Careful plotting of long transects in mature America forests, or up mountainsides, demonstrated continual change and overlapping populations of species. These careful examinations were enough in themselves to show that real associations did not exist except as abstractions, they were merely collections of individuals with common interests in a local habitat.[3]

The definitive demolition of this idea of forests or associations migrating across landscapes was done by two pollen analysts, Margaret Davis and Tom Webb working in the eastern United States, in what I think of as the finest achievement of pollen analysis in my professional lifetime. They severally collected together all the pollen diagrams from lake sediments known from eastern North America, filling in gaps with new lake cores of their own, then mapped the relative importance of tree pollen types individually each millennium for ten thousand years. At the end of glacial time *none* of the associations familiar to modern botanists could be found, although all the species could, scattered well to the south of the ice front, even down to the Mexican border, but in eclectic array. Then the successive maps showed the northward migrations, species by species, at different rates and by different routes, until that final fateful meeting with the New Englanders and their axes. A by-product extremely useful for squashing error in Amazon interpretations was that since associations had no fixed existence, there could not be indicator species for those associations. Ignore claims for indicator species of Amazon forests or, more likely, Amazon savannas!

The Davis and Webb studies were completed only in the 1980s, when they gave immense confidence to the attitudes to Amazon pollen analysis we had already developed. This immense forest (and all other vegetation there) should not be regarded as static, but in a constant state of reshuffle. There should be no "indicator species" of anything at all beyond simple physical needs like "needs a swamp" or "needs frost to break seed dormancy" (not likely in the Amazon, but we did encounter one such species in southeastern Brazil).

This elegant demonstration of the independence of plant species was to strengthen our hand in the 1980s and 1990s when old-fashioned beliefs in indicator species and the integrity of associations incredibly were still guiding pollen analysts trained in the old European traditions. They tended still to look for indicator species. They, and their Brazilian colleagues, were also avid supporters of the dry Amazon hypothesis so that finds of pollen types dubbed as indicators of open space, or genera known to be found in savannas, were taken to allow reconstructions of Amazon vegetation that conformed to the expectations of the refuge hypothesis. The hypothesis, in its more powerful guise as a paradigm, was to come at the last to call on outmoded and falsified science to its defense.

The Davis and Webb work in America gave final understanding of why Wallace's hypothesis of extinction by northern glaciers could not account for the ubiquitous phenomenon of richer diversity as you move from a pole towards the equator. It works fairly well for his native England, for Europe, and for the Russian plains, but perhaps only for trees, the least mobile of things. That seems to be the limit. The trees of the Miocene, more than five million years ago, survived the coming of the glaciers in America and China. I wish I could claim that hearing that Russian paleobotanist talking in the 1960s of "walking through the Miocene of the Russian plains" had set me to thinking that the Amazon forests might be Miocene also, but I cannot. Forty years later, however, that is where we were to end up.

Back in 1876, when Wallace published his great treatise on zoogeography there seemed to be no alternative to the hypothesis of glacial extinction to explain low diversity in the north.[4] The corollary was high diversity in the Amazon because no glaciers and thus no extinction. In the absence of an alternative hypothesis, this was to become the textbook explanation for the best part of a century.

One objection is obvious: we did not know what the tropics were like in an ice age, particularly the Amazon. The hypothesis merely assumed that nothing happened to Amazon climates when huge areas of the earth were covered with ice, when the very oceans were so drained that continents and islands were joined together into new configurations of land. The ice-age earth was not the earth in which we live now. Did any of it have climates quite like those of modern times? It seems unlikely. How could we know that the Amazon of those days had not also changed before the mighty convulsion that had gripped the whole earth? The answer was that we couldn't know, and that we didn't know.

We still didn't know in 1964 when Ed Deevey called in his postdoc to needle him with having found out that in Alaska it got colder in the ice age. The Arctic did get colder, unsurprising; but did the Amazon get colder? Was it otherwise changed by the global climatic cataclysm? We did not know. This was the Reason Why. We did not know what the Amazon was like in an ice age. The new ecology needed to know.

Yale, in the early 1960s, was the natural place for thinking these thoughts. G. Evelyn Hutchinson had made Yale a venue of choice for young ecologists. In the 1950s Robert MacArthur ran with Hutchinson's early thoughts about species diversity to develop formal models of relative abundance, commonness, and rarity and the links between diversity and ecosystem stability. This work had much influence on the discovery of ecology by politics.[5] "Diversity begets stability": an appealing, if flawed, idea had emerged.[6] Half a century later political pundits can still be heard mouthing about "diversity" needed for "stability"; often hogwash, but sometimes not.

Hutchinson's special expertise was as a limnologist, a student of lakes. The hosts of tiny animals and plants of the plankton made of lakes microcosms to reveal process in the mega world of forest and animals. He was intrigued by the oddity of the large numbers of species of microscopic planktonic plants that manage to live together, stirred up with the water in which they all have their being, feeding together in the same way by photosynthesis, their resource needs the same and met from the same reservoir, how could they coexist without ruinous competition leading to extinction? His elegant paper setting out this problem, *The Paradox of the Plankton*, has been the starting point for many a professional discussion of diversity.[7] Trying to resolve this paradox has given ecologists much fun over the years. It still does. With the "paradox of the plankton," Hutchinson's microcosm anticipates the Amazon question of "why so many species mixed into that great green forest," but in miniature.

In the graduate schools of my day, Duke among them, we read avidly of the paradox of the plankton and the lecture given in 1959 to the American Society of Naturalists by Hutchinson, who was then the Society's president. Presidential addresses give license to think new thoughts, free from battles with pesky referees. Hutchinson, by common consent probably the most gifted ecologist of his day, used his presidential address to ask the biggest "why" question an ecologist can dream up: Why are there so many kinds of animals?

Hutchinson added to the daring of his topic a whimsy, by grafting onto his working title the name of the patron saint of Palermo in Sicily, Santa Rosalia. Little is known of Rosalia's sainthood. Hutchinson's whimsy was that she should be declared the patron saint of diversity because of the variety of corixid bugs (water boatmen or backswimmers, in the vernacular) he had found in ponds near her sanctuary. The full title of his presidential address became "Homage to Santa Rosalia or Why Are There So Many Kinds of Animals?"[8] In the trade it is known simply as the "Santa Rosalia paper."

"Santa Rosalia" explored ways in which natural selection might parcel out the resources of a place into more and more specialized packets, each occupied by a specialist species, and how many links could there be in a food chain before attrition of energy supplies became limiting. How close can specializations be without ruinous conflicts for resources? And the grandest question of all, for crowded places like the Amazon, was there a critical number of species above which a community of interacting species might become so stable that extinction should be rare?

More than half a century later "Santa Rosalia" is still read in graduate schools, but largely now for its historical importance. Like many a great paper of daring and novelty before it, "Santa Rosalia" has been dissected, scrutinized, argued, and tested, its speculations, including a necessary connection between species diversity and ecosystem stability, often found wanting and so dismissed. But, a century after Wallace and Darwin had started the chase, "Santa Rosalia" signaled that we now knew enough of ecology and the workings of natural selection to hope to ask questions like, "Why more species in equatorial regions?"

For the Amazon, a first task must be to document the magical high diversity, with its patterns of rarity or abundance. But it is not easy. Start with the trees. Counting tree species in tropical rain forest is hard. From a tortuous trail cut through the forest what you see in the pervading gloom is dark trunks of trees soaring above you to the shading canopy far overhead, and plant taxonomy is based on flowers. Rain forest trees are in flower only briefly, and the flowers are usually high — 20, 30 or even 40 meters above the ground. A complete local census of a patch of "Tall Lowland Amazon Forest" (the UN-approved term for Amazonian rain forest) was long in coming.

The first such complete count of local Amazon trees was published only in 1988. Alwyn Gentry of the Missouri Botanical Garden identified and counted every tree in the 2.4 acres of a one-hectare plot in the lowland rain forest of Amazonian Peru. This is in the wettest region of the Amazon basin, part of the long strip of forest in Ecuador and Peru where moist air rises against the flanks of the Andes Mountains. Local precipitation can be 5, or 7, or even 9 meters per year in parts of this tremendously green area. Gentry included in his count all trees with trunks more than 10 cm across at breast height, finding 580 trees in his hectare. This density, as opposed to variety, of trees was not spectacular; roughly 600 trees per hectare should be expected in temperate forests of Europe or the United States. But those 580 Peruvian trees were spread among 283 different species. Thus in just one hectare there were more tree species than in the whole of North America or 40 times as many as in the British island.

Not every hectare of Amazon forest has in the order of 300 species of tree, but some do, and most are far more diverse than are temperate forests. A second site in the Peruvian Amazon had "only" 189 species in a hectare, and near Manaus in central Brazil the number is perhaps a little less still. More than this we cannot yet say because of the difficulty of doing the counting.

Al Gentry was the best there was at this trade. Even he had to climb many of those 283 trees, with ropes and irons, to find flowers or fruits in the canopy before he could name them. A few years later Gentry undertook a forest reconnaissance for conservation planning in Amazonian Peru, using a light aircraft. This crashed into the trees he was surveying. When a rescue party reached the downed aircraft after more than a day, Al Gentry had died of his injuries.

Through such adventurous labors has the astonishing diversity of Amazon plants been confirmed. The numbers are beyond all comparison with species lists in more temperate lands; indeed, they are beyond easy understanding. How on earth can 283 different kinds of tree share so small a habitat as a hectare, competing for the light, water, and nutrients of the place as they must? It seems a miracle of coexistence, Hutchinson's paradox of the plankton written in rather larger letters.

With smaller living things, both plant and animal, even these prodigious numbers are atrociously magnified. Insects are the most vexing because we find it impossible to count them. Over a million species of insects have definitely been recorded from the Amazon in museum collections, but we know that this is no more than a sparse sampling. If you want the thrill of a "species new to science," go to the Amazon and collect insects; you will probably encounter something "new" in your first days of collecting. Of course, you must first have spent years of your life learning entomology or you will not know that what you found is "new"!

Our best count of insects is a truly beautiful lying statistic. Terry Erwin of the Smithsonian did it. He counted the beetles in the canopies of a few trees of one species, then extrapolated. To count the beetles he enclosed branches in plastic tents, inserted poison gas, and later shook out the corpses. There were several hundred species of beetle living in that one species of tree, many of them "new," a goodish number host-specific. From these counts Erwin compiled the beautiful statistic. Multiply by the number of known tree species, then factor in a multiplier for beetles that live in other ways or places than tree canopies, then multiply by a guess at the ratio of other insects to beetles, and you are on your way. The answer for all the Amazon basin came out to thirty million species, give or take a few million. As a "statistic" this result is best used to illustrate our true ignorance of Amazon diversity rather than as a statement of Amazonian insect numbers. Yet the number seems proportionate to numbers of more conspicuous life forms that are easier to count, like birds or trees; and Terry Erwin is a first-class naturalist. His statistic serves to encourage us in the conclusion that the Amazon basin holds insect species by the million.

In some mysterious way wet tropical places like the Amazon must have special properties that let more species share resources than is possible in other climes, we should say to "coexist." Or the process of speciation is so rapid that new species are mass-produced to overwhelm whatever local extinctions there might be. Or both.

A deceptively beguiling approach to the problem of excessive coexistence in the wet tropics is to suggest that unfailing wet and warm lets species get packed in tightly as every last calorie of resources can be safely used. Stability lets niches be smaller, as "ecospeak" puts it. Ways of life like symbiosis or narrowly defined lifestyle should help pack the species in. Probably all true, but investigating how the allocation of resources to species is fine-tuned in the wet tropics cannot answer why coexistence is possible. If there are more species in a place there are, by definition, more niches. If resources have to be allocated to more niches, it follows that niches must be smaller. To say that wet tropics provide for close-packed living (more niches), therefore, is a tautology that gets us nowhere.

Explaining high tropical diversity directly from niche theory, a central argument of "Santa Rosalia," thus went *kaput*. More promising was finding properties of the tropics that let, or forced, species to rotate, leading to a dynamic system of sharing. If this could be, then the species inventory of a forest patch could be multiplied like the dynamic inventory of a warehouse in just-in-time manufacturing. This seems, on the face of it, a tall order, but two properties of tropical life do in fact suggest the required dynamic.

For all its equable climate of wet and warm, the rain forest can be a violent place. Thunderstorms are amazingly violent, local, and frequent. They beat the ground into submission with rain that seems to a bemused northerner more like a fire hose than a shower. Storm winds tear the soggy, top-heavy trees down; and the tangle of woody creepers called lianas which tie to neighbors bring those down too. The forest is well sprinkled with the resulting gaps. Although each gap means death to some, it also frees space for others. In the structure of forest and climate, therefore, is a dynamic that lets species use constricted space in turn.

Ecologists call this pattern of local destruction by storm "intermediate disturbance," disturbance not of the prolonged savagery of a northern winter, or, even worse, a Pleistocene ice sheet, which can kill on a continental scale causing extinction. Intermediate disturbance opens local space not preempted by superior competitors.

At least as important as intermediate disturbance is the potency of hungry predators in the tropics, particularly those that feed on plants and are known as herbivores. Herbivore populations can be maintained at high levels year round. This means that they can devastate plants, particularly young plants, in ways unusual for herbivores in high latitudes. A devastated plant population opens space for other plants to come in.

Trees have ways to disperse seeds, by wind or by luring animals with tasty morsels to act as porters. Even so, most seedlings gamble for life close by the parent tree. This is where tropical predators are so devastating. Close by the parent tree the predators are waiting: both seed predators and straight herbivores. No winter has decimated their numbers or forced them into the immobility of resting stages. They are alert, numerous, and ready to pounce. Most of these predators are insects of specialized taste, with digestive systems adapted to handling the particular chemical defenses of the target tree. The immediate vicinity of a parent tree is a killing zone for the succulent seedlings or seeds.

Other seed predators are mobile animals, rodents particularly, for which the parent tree is a flag that says, "Here be seeds." In these circumstances "near mother" is a bad place for a baby tree to be. Getting just a little way from mother's shade may be a good enough tactic in a temperate forest but it is not good enough in the Amazon rain forest. "Near" is still in the killing zone. Predators waiting near the parent tree will soon find you. So an enormous premium is placed on getting clean away.

I like Steven M. Stanley's term for the opening up of resources by predators, "the cropping principle," because space or food is made available to others by the "cropping" of the previous tenants. Like "intermediate disturbance" it offers resources temporarily freed from debilitating competition, and hence opportunities for other species to be established.

These two processes, intermediate disturbance and cropping, certainly increase the number of local species that a collector can find. They are real world, real ecology processes. They let species survive without fancy methods for coexistence. For a certain kind of ecologist, me for instance, they are gratifying. But they scarcely answer the big questions of Amazon diversity; the numbers are just too big. Think of those millions of insects, how did they get there? One way out of the dilemma would be for the Amazon environment to have properties that speed up the process of evolution by natural selection itself. Then the huge size of the Amazon, and processes like intermediate disturbance and the cropping principle, might conserve the product as species became ever more numerous.

Speeding up evolution is a tricky concept.

Natural selection is always busy; in crowds, in the scattered, in the ubiquitous, in the isolated. It is the common experience of life that individual offspring are differently endowed with heritable properties, that because of these different endowments they have different chances of survival and, more important still, different chances of reproductive success. This is natural selection. To miswrite a sentence from the scriptures, "Natural selection is with us always."

But though natural selection is ever present, only rarely is it in the business of making new species. Mostly it removes deviants, maintaining the average. And if some deviant does well, the genes of its deviancy will be swamped in the next generations as the fortunate possessors find mates in the overwhelming number of normal individuals around them. Thus making a new species requires the rare circumstance of sufficient isolation that new types cannot be swamped at the next breeding. The classic example of how this can happen is a remote island in the oceans. Natural selection can work there without the results being swamped by out-crossing.

But the isolated population need not be small as on oceanic islands, though if you seek bizarre results, small founding populations are a great idea; being isolated is what matters. Then natural selection, working on the whole population, will shift the average to local circumstance. The role of the isolating barrier is to restrict or prohibit gene flow between the isolated population and the population with which it was once united. The trick to get new species thus is to arrange discontinuities in the geographic spread of populations. The rest can be left to natural selection and local ecology.

An outstanding consequence to this process is that, as Ernst Mayr put it in 1963, *"Regions which in any sense of the word are insular always show active speciation, whereas continental regions show speciation only where physiographic or climatic barriers produce discontinuities among populations."*[9]

That quote, from Mayr's influential *Animal Species and Evolution*, sets out most plainly a critical aspect of the Amazon diversity problem. Continental regions with minimal physical or climatic barriers are least expected to "show active speciation." But here was a river basin of continental size, mostly occupied by one great forest, with little more to be seen in the way of barriers than the rivers themselves. Yet here was the largest number of species on earth.

This was the limit to my self-briefing when in 1964 I began to plan how to reconstruct the forest history. I had no theory of diversity to proclaim, no hypothesis of evolution rates or of landscapes and vegetation in flux, no hypothesis of climate change to test. These, if any, should come later. First go to the *terra incognita* of the ice-age Amazon, explore it, know it. Perhaps then the reasons for the evolution and preservation of such diversity would show themselves.

I did not know then that I was making my plans a mere five years before the invention of a plausible history of the Amazon that for many seemed to explain everything. This was the so-called "refuge hypothesis" of Jürgen Haffer, published in 1969, with its conception of massive climate change in the ice-age Amazon after all, but through drought, not cold. This drought was supposedly so prolonged and severe as to wipe out the great forest, leaving only small patches on uncommonly wet hillsides. These hillside patches were to be the refuges. All the rest of the vast Amazon plain was to be savanna.

This conception did not come from the exploration of *terra incognita* and discovery of the past to which Ed Deevey had nudged me, rather it was little more than the pure flight of a gifted imagination to explain what seemed a curious pattern in the distribution of forest birds. If this grand scenario of the Amazon past should prove to be correct, however, it would feed directly into the strongest consensus of contemporary evolutionary thinking: vicariant populations evolving in isolation.

This fantasy had instant appeal, and no wonder. By smashing the wet forest into fragments with continental drought, it made barriers galore. And by doing it with an ice age, it allowed a hundred thousand years for the barriers against the ten thousand years of the green near-homogeneity of the modern forest.

All this was yet five years away when Ed Deevey called me in to make the ice-age history of Amazon climate and vegetation my goal. We had no fantasies in mind. But that there should have been changes we were sure. Our view was that these must be known before a realistic hypothesis and understanding of Amazon diversity should be possible. Discover what changes there were, if any. This was "The Reason Why."

The coming of the refuge hypothesis as a public property when we were but a few years into our project meant perforce that we must test it with our historical reconstructions as a necessary way station to our goal. It should be simple enough. The refuge hypothesis predicted that the pollen record of the ice-age lowlands would be savanna instead of forest. Our paths were sure to cross the refugialists. Before they could cross, however, we were to be given many a lesson in the human side of science. For the refuge theory came to be loved, to find scientific lives devoted to its elucidation and defense, and eventually to be elevated to the high intellectual rank of a paradigm. It had guerrilla forces to fight back. For all that, it was to fail the test of accurate reconstructions of the ice-age Amazon.

2

THE GALAPAGOS THAT DARWIN KNEW

I began my transect of the American equatorial lands with the Galapagos Islands. No thought here of starting at one end of the transect and working to the other, I'm not that organized. The Galapagos were simply the easiest place to begin; also the sexiest. I had to write the grant proposal first, my first as an independent investigator, it had better be neither grandiose nor mundane, but it must be attractive. The Galapagos fit that bill nicely.

The islands also fit my personal bill of being "easy," easy that is in the technical sense. Where the Amazon had 80,000 species of plants, the Galapagos had something like 600; a nice gentle introduction to the pollen of the American equator.

My going there when I did, in 1966, was timely beyond my imagining. The title for this chapter, "The Galapagos that Darwin knew," is not an unfair description of the Galapagos I was to see and search: easier to go by boat between islands than in Darwin's time because of better propulsion systems in tricky, not to say dangerous waters. Also the nuisance of an airstrip built by the U.S. Air Force in the war with Japan, and the extinction of the race of land iguanas round the airstrip connected to the temporary housing of a division of soldiers. But the tiny human population of the archipelago had had little more impact than in Darwin's day. Many more feral animals released and tortoises looted almost to extinction, but otherwise Darwin would have felt the place familiar. This state of affairs was about to end.

After swearing I should never go back, I was lured to the Galapagos in 2005, thirty-nine years after my expedition. The bait was prodigious. Young scientists with knowledge and equipment impossible to the 1960s called to say my Galapagos climate history of the eastern Pacific was still the principal record from the region and they wanted to rework my old discoveries using modern methods. We went there together, we saw, they cored, and now the new generation has taken over where I had begun. This is close to the ultimate happiness for a gray-bearded professor.

But the islands that Darwin knew are gone forever. The island of Santa Cruz, my original base, had less than eighty people then, no motor vehicles, and no roads. Now it has six thousand people and a traffic problem, complete with dangerous intersections and traffic lights. The island of San Cristóbal, where Darwin made landfall, has twelve thousand people and an airstrip built to take tourist jets.

True, large-scale and intensive conservation programs exist on paper, particularly aimed at the animals tourists come to see. The large island of Santiago is preserved entire, off limits for settlement, a gorgeous time-capsule as long as political resolve lasts. Dedicated people run these programs, subject of course to political whims as their peers are the world over. But the Galapagos that Darwin knew, it is not.

Nor in this contemporary Galapagos would it be possible to give free rein to scientific exploration on my scale. My grant from the U.S. National Science Foundation (NSF) was to search the entire archipelago for ancient lakes and to make such baseline collections as should be necessary to interpret the evidence of the sediments. This included making as complete a Galapagos herbarium collection as possible (in plain language, collect every plant in sight).

It still seems to me extraordinary how little was known of the Galapagos in 1964, when I began to plan. The name was on every biologist's lips but on virtually none of their itineraries. Galapagos learning of the preceding hundred years came from a series of elite expeditions; ship-mounted collecting forays essentially. Some of these "expeditions" were really voyages on the yachts of the wealthy, who took along cargoes of eager young naturalists. Many were on ships commissioned by zoological societies in New York or California, either converted yachts or larger, and were likewise paid for by wealthy donors. They all collected what they could reach from their floating homes, mostly in the same places. The last of them was in the late nineteen-thirties, after which certain world events put an end to the era of expedition yachts.

I had read and been thrilled in my youth by an account of the Galapagos more splendid than any reports of these yacht-mounted jaunts. This was in a book, read, if memory serves, on daily commutes on the London Underground. The book was called *Journal of Researches into the Geology and Natural History of the various Countries visited by H.M.S. Beagle, under the Command of Captain FitzRoy from 1832 to 1836*, by one Charles Darwin.[1] Darwin devotes only 40 pages to the Galapagos, but he says more of substance and excitement in those

40 pages than is in all the subsequent expedition accounts put together. If you would know the Galapagos Islands, read Darwin first.

The Galapagos Islands are the emergent tops of volcanoes, as Darwin saw,

> *"Some of the craters surmounting the larger islands are of immense size, and they rise to a height of between three and four thousand feet. Their flanks are studded by innumerable smaller orifices. I scarcely hesitate to affirm that there must be in the whole archipelago at least two thousand craters."*

That catches the Galapagos perfectly.

Captain FitzRoy's immaculate chart-making revealed that the sea between the islands was not deep.[2] He found only 70 fathoms between Santa Cruz and Floreana and as little as 22 and 45 fathoms in the passage between James and Jervis islands (those needing to think in meters, it is near enough to multiply by two). Once out in the open ocean and well clear of the archipelago, depth falls to more than 1,000 fathoms, where FitzRoy's sounding lines could not reach.

The discovery of plate tectonics in the 1960s, and the effort to map the ocean floors that followed, tell us more of Galapagos origins. The islands are protruding volcanoes of the Galapagos Rise, a submarine ridge continually forming as molten magma upwells at the place where crustal plates are drifting apart. The more easterly islands are quiet now, but those to the west like Fernandina (the *Narborough* of Darwin's account) are still forming and still violent, a dichotomy that Darwin noticed. But all are geologically young, again as Darwin guessed.

The age of the earliest habitable Galapagos rock is uncertain but cannot have been much more than 3–4 million years. This is the oldest potassium/argon dating of hard rock samples bored at the bottom-most stratum exposed along sea cliffs. As volcanoes, the islands are built by pouring lava of successive eruptions on top of each other and can be imagined as flattish, cone-shaped onions. It is possible that the larger and older islands like San Cristóbal (*Chatham* to Darwin) have a buried inner cone that these drill samples have not reached which is older, but probably not by much. These shield volcanoes form quickly (in geological terms, that is).

The oldest marine fossils associated with the cliffs are only some two million years old, bringing the origin close to the beginning of the Pleistocene ice ages.[3] It seems likely, then, that evolution on the Galapagos has largely occurred during the span of ice-age time. No Miocene origins of Galapagos spectacular species as those in the American Midwest that so delighted the Soviet paleobotanist, nor of the Miocene origins I was later to sense for the forests of the Amazon lowlands. The Galapagos story was an ice-age story.

The Galapagos lie in a desperately dry part of the climate system, borderline between the dry strip crossing the Pacific Ocean just south of the equator, now known as the "central Pacific dry zone," and the desert coasts of southern Ecuador and northern Peru. Wind directions and cold water are the culprits. From the southeast come both trade winds and currents, their directions conforming to the iron rules of Coriolis force, their origins far to the south in the cold antarctic regions.

More important for sea temperature is that the south equatorial current is skewed away from the continental land by the push of the southeast trade winds, causing an upwelling of cold water from below. The trade wind in turn is also cooled from below by this upwelled seawater over which it blows. The cooled air, bottom heavy, flows on as a stable mass. Thick stratus clouds form at the top of this body of cold air marking an inversion of warm air over cold. No air rises from the wet ocean surface to pierce this inversion, so no rain can fall.

These clouds above the inversion must be a familiar sight by now to the tourist multitudes who fly to the Galapagos out of Quito or Guayaquil. The clouds form the unbroken grey carpet far below the aircraft, hiding the sea from view for most of the hour-and-a-half flight. They are the product of trade winds blowing over a cold, upwelled sea.

Enhancing the effect of coastal upwellings is a second source of upwelling, this time from the west, as deeply buried currents from the central Pacific hit the underwater massif of the Galapagos Islands and are driven to the surface. They add to the cooling of the sea surface and the surface winds. The combined result is that most of the time the Galapagos Islands have cloud but no rain, except in the higher parts of the larger islands where the volcanoes reach the inversion to pierce the clouds and be wet with dripping fog. A dense forest, jungle almost, occupies the fog belt of these islands, cut off from the sea below by the coastal desert.

Captain FitzRoy brought the *Beagle* to the Galapagos from Chile and Peru in September 1836, riding the southeast trade wind. Darwin writes of the climate:

> "*Considering that these islands are placed directly under the equator, the climate is far from excessively hot; this seems chiefly caused by the singularly low temperature of the surrounding water, brought here by the great southern Polar current.*"

The *Beagle* had indeed traveled north with the polar current, but the cold water that cooled the islands had mostly come from the depths in upwellings, not from the polar region. Seamen of Darwin's day had not discovered upwellings. But the critical part of the observation was correct and so was what followed:

> "*Very little rain falls, and even then it is irregular; but the clouds generally hang low. Hence, whilst the lower parts of the islands are very sterile, the upper parts,*

at a height of a thousand feet and upwards, possess a damp climate and a tolerably luxuriant vegetation."

One aches for Darwin to have discovered atmospheric inversions and their causes while accurately observing their effects, but perhaps founding modern biology was enough for one voyage without founding modern meteorology too.

And Darwin tells you what to expect of these volcanic islands without rain below the cloudline when you first set foot on the Galapagos from a boat:

> *"Nothing could be less inviting than the first appearance. A broken field of black basaltic lava, thrown into the most rugged waves, and crossed by great fissures, is everywhere covered by stunted, sunburnt brushwood, which shows little signs of life. . . . We fancied even that the bushes smelt unpleasantly. . . . Such wretched-looking little weeds would have better become an arctic than an equatorial flora."*

The landfall for H.M.S. *Beagle* was on San Cristóbal (Chatham) but the description fits all the islands well enough, perhaps particularly if you have been weeks at sea with drinking water rationed. The coast probably looked even worse to the typical Galapagos visitors of the times, who came in whalers or pirate ships unlikely to have been provisioned as well as a king's ship. The Galapagos are true desert islands, places where castaways can easily die of thirst.

Yet visiting ships had been many and frequent: a corps of plunderers of the South Seas; whalers, traders, and pirates; with a sprinkling of patrolling frigates from the "powers." A U.S. commerce raider, the frigate U.S.S. *Essex*, commander David Porter, established a base there in the War of 1812 and did sore damage to British commerce.[4] These ships needed both water and meat; and they got both: meat by collecting giant tortoises by the hundred, but water on a desert island?

Water was the most important resource, particularly to me planning to reconstruct the ice-age climate of these "desert" islands from sediment cores. I must have lake mud: travelers of old, please tell me of lakes in your writings! But none of the travelers' tales talked of freshwater lakes, Darwin included. The water that made possible the victualing of ships came from coastal springs, perhaps with small catchment ponds, at just three or four places in the whole archipelago: Freshwater Bay on San Cristóbal and similar well-known sites on Santiago (James) and on Isabela (Albemarle).

There could be no doubt that the coastal springs were fed by water from the cloud forests of the interior, percolating down through the fissures with which volcanoes are well supplied. There ought in fact to be fairly copious water in the more mature forests because they had not only the cloud drip for eight months of the year but also the possibility of serious rain for the other four months.

Despite the rigorous stability of an inversion cooled from below by the up-welling of cold water, the system can be overturned, as happens annually for a few weeks in the northern winter. The trick is to divert ocean currents so that they no longer push upwellings so strongly. Then the sea surface warms, air is no longer cooled from below, the inversion is broken, and rain is possible.

In late December of every year this sequence happens when the climatic equator, where the northeast and southeast trade winds meet (called the Inter-tropical Convergence Zone, or ITCZ) is briefly pushed south of the geograph-ical equator by the passage of the seasons. Rain comes to the desert coasts of Ecuador and Peru when this happens, every year at Christmastime. The people rejoice and say the Christ child (*El Niño*) has truly come.

Every few years the Christmas rains are massively increased as another process obliterates the upwellings and inversion on a grander scale. This other process is a flood of warm water arriving from Indonesia and Australasia after crossing the whole Pacific. Huge cumulus and cumulonimbus clouds form, pouring out tropical rain, and the welcome rains of Christmas are changed to devastating floods. The climate modeling trade has appropriated the name "El Niño" to mean just these intermittent and catastrophic versions of the late December rains. But both the blessed and the cursed versions of El Niño can bring rain to the Galapagos from January to April.

So the theoretical potential for a freshwater lake in the interior of one of these islands was clear; the problem was that none of the classic accounts reported see-ing one. Instead they reported small ponds in the forests where tortoises went to drink. Darwin called these "wells" and describes the long journeys to them taken by tortoises:

> "*Broad and well-beaten paths branch off in every direction from the wells down to the sea coast. . . . When the tortoise arrives at the spring, quite regardless of any spectator, he buries his head in the water above his eyes, and greedily swal-lows great mouthfuls.*"

The tortoise trails are all long gone now, along with the tortoises themselves.

If all the early visitors from ships used tortoise trails to find water but still never mentioned a lake, my quest did seem hopeless. After all, the tortoises knew the is-lands best! I could only appeal to more recent travelers, mostly by mail. Fortu-nately, before 1965 there had been a sudden influx of scientific visitors to the Galapagos, pushed, as it were, by the fortunate circumstance that the first direc-tor-general of UNESCO was Julian Huxley, of the Huxley family of evolutionary fame. Under his prodding the Charles Darwin Foundation was created in 1959, and work was started to build a scientific station on the most unspoiled island, Santa Cruz (Indefatigable). Robert Bowman, one of the few modern biologists to have done thesis work on the Galapagos, organized a visit and symposium of likely scientists for the NSF, at once creating a bank of Galapagos expertise for the likes of beginners like me.[5]

My mail gave hints that a freshwater lake did exist, high on the summit of the easternmost, and hence oldest, island, San Cristóbal (Chatham). This was the island best known for watering ships from the springs of Freshwater Bay. Standing water on the heights above Freshwater Bay did seem plausible. True, the letters I had were conflicting. One talked of reading a report by a whaling captain of seeing a lake three miles long, a piece of hyperbole which seemed to both my correspondent and me to have the aura of myth about it, much though I wanted to believe. No Galapagos map even hinted at such a lake. But another correspondent had either seen a lake there or had spoken to someone who had. It was enough on which to hang a proposal; I could start writing.

The proposal to the NSF said we should core the lake on San Cristóbal (if such a lake did indeed exist). We also said we should look for more lakes on other islands that might have been overlooked as the San Cristóbal lake had been; and in addition make what we could of mud in tortoise ponds and peat bogs recently reported for the highlands of Santa Cruz. We should also core salt lakes by the sea, some well known and grandly beautiful. Darwin again gives the idea:

"We anchored in Bank's Cove in Albemarle Island. The next morning I went out walking. To the south of the broken tuff-crater, in which the Beagle was anchored, there was another beautifully symmetrical one of an elliptic form; its longer axis was a little less than a mile, and its depth about 500 feet. At its bottom there was a shallow lake, in the middle of which a tiny crater formed an inlet. The day was overpoweringly hot, and the lake looked clear and blue: I hurried down the cindery slope, and choked with dust eagerly tasted the water — but to my sorrow, I found it salt as brine."

Bank's Cove is now known as "Tagus Cove," on the west coast of Isabela, across the strait from Fernandina, a cozily safe anchorage well known to yachtsmen. The lake from which Darwin tried to drink is clearly in Beagle Crater, identified both by position and the "tiny crater" of the inlet. Darwin apparently overlooked (at least did not mention) the small but spectacular crater lake hidden behind the ridge at the head of Tagus Cove. For one of my correspondents this crater lake, thought to be deep, was possibly the best prospect for a long sedimentary record in all the Galapagos. Both it and the Beagle crater lake went onto the target list in the proposal.

Darwin also visited a salt-mine lake on Santiago (James) Island,

"[We walked to a] tuff-crater, at the bottom of which the salt-lake lies. The water is only three or four inches deep, and rests on a layer of beautifully crystallised

*white salt. The lake is quite circular and is fringed by a border of bright green suc-
culent plants; the almost precipitous walls of the crater are clothed with wood, so
that the scene was altogether both picturesque and curious. A few years since, the
sailors belonging to a sealing-vessel murdered their captain in this quiet spot;
and we saw his skull lying among the bushes."*

Not a promising prospect, but my correspondents suggested there was more
than one salt lake. I put the James salt-mine lakes on the target list.

Last was the spectacular great Genovesa (Tower) crater lake. Neither Darwin
nor other early travelers had any idea that this glorious salt lake existed, for it lay
in the middle of remote, flat, and waterless Genovesa Island, more than five kilometers of lava desert away from the visible coasts. William Beebe described it in
1923, having persuaded the captain of the yacht *Nova* to risk his ship by moving
into the uncharted bay on the south side of the island in quest of a place to go
ashore.[6] This Beebe did, then wandered across the island in his characteristic
way, chancing on the great hidden crater as his reward. The lake fills this central
crater, a collapsed magma chamber of the low shield volcano that forms the island, a stretch of salt water a third of a mile (500 m) across, lapping against the
vertical 200 foot (60 m) cliffs that surround it. Collapsed magma chambers tend
to be deep holes, making it likely that a thick span of mud might be present,
though this idea needed faith to back it up. No one had ever launched a boat on
the lake's waters, let alone taken soundings. Also a pollen record from a low island covered with no more than desert scrub seemed a poor possibility for a
record of climate change. But I had seen a photograph of the lake and hungered
after it. Down it went to complete the target list.

The proposal was short, derisorily short it seems to me now. It had only six
pages of narrative text when NSF allowed (and still allows) fifteen. Never again
would I engage in the folly of using fewer than the allowed number of pages for
a proposal, each packed with data and hypotheses. Moreover, reading through
that old proposal now, I am embarrassed by how little it says: how could a proposal say little more than, "We know nothing of the climatic history of this exciting place; we ought to know; if I am lucky, I might find a sedimentary record despite the odds against it: and I might even be able to use the local tropical pollen
record to reconstruct a climate history; give me the money and I shall try." But
NSF did fund it, for the full amount I asked.

Those were still the days when NSF did not let applicants see the reviewers'
comments; these became available to investigators only many years later, when
Congress insisted. All you got then was a telephone call from the program director, saying yes or no and passing on whatever reviewers' comments he thought
fit. The voice on the telephone told me that one reviewer had written something

to the effect that "there are no lakes; that young Colinvaux will most likely die of thirst looking for them." But the voice went on to say that the panel had decided that since nobody really knew if there were any lakes or not, NSF had best let me go to find out. So they did.

The way to the Galapagos in those days was the monthly boat from the Ecuadorian port of Guayaquil, and the way to Guayaquil was by air from Miami. Send the drilling equipment and stores for two months for a party of five by air freight in good time to meet the boat, and fly down yourselves. This procedure has one certain destination: a customs shed, staffed by unsympathetic personnel with arcane rules that must be followed. So it was in Guayaquil.

Equipment can be got out of customs; it must be, or South American seaports would develop to be nothing but vast, ever-expanding sheds. But oddball expedition equipment does not come out quickly or cheaply. As it happened, time was not important to us because there was no boat for the Galapagos. The monthly boat had got sick and was in dry dock for an indefinite period of repair. There was no other boat.

I went the rounds, to the U.S. Consulate (known to my informants as the "golden ghetto"), to people the consulate suggested, to people suggested by the people the consulate suggested, and to other venues of desperation now blessedly forgotten. Days passed before I had the idea. I went to the British Consul.

In those days I was still a green-card American, which meant that I traveled on my British passport. So, ho for the door with the lion and unicorn emblem. The British consul in Guayaquil turned out to be a veteran who knew the ropes. In just three days our heavy equipment was out of customs, duty free, and actually loaded onto an Ecuadorian warship bound for the Galapagos. We five went down to the docks, where a sign said "*Armada del Ecuador.*" In the history books of my English boyhood the word "Armada" had a prominent, special, and unfriendly meaning; Francis Drake, Good Queen Bess, and all that. But we boarded the warship of this Armada with gratitude and as welcome guests.

The good ship *Jambeli* was an ex-U.S. LST, or Landing Ship Tank, designed for storming beaches during World War II. Possibly the *Jambeli* had once done just that in the days when it had had a number rather than a name or had even gone to Normandy. LSTs are long narrow ships, with a tank alley down the middle and a drop ramp in front. They are designed to charge beaches, let the ramp go, and loose the alley full of tanks to join in whatever is happening beyond the beach. The *Jambeli* carried one gun of substance in an open turret forward, where I slept to escape the stuffy tier bunk below decks. The tank alley was filled with cargo, including ours; also by Galapagos residents or immigrants, who had rigged plastic or canvas shelters so that the open alley looked like a peasant street market.

After days at sea, we closed with San Cristóbal, the Chatham Island of Darwin's first landing, though his landfall was from the southeast, ours from the northeast. But there was the same "broken field of black basaltic lava" and the "stunted, sunburnt brushwood," looking just as lifeless and sterile as it had to Darwin. The rounded and worn shapes of the old volcanoes of the summit were humps, misting into the clouds. This was the easternmost, the oldest, of the Galapagos Islands. And the one that I hoped might hold a lake in its clouds.

We cruised along the north shore past the "sleeping lion," the islet formally named "Kicker Rock." This rock had fascinated Captain Colnett, R.N., who did the first survey of the Galapagos for the Royal Navy in 1793–1794 in the sloop-of-war H.M.S *Rattler.* He referred to the sleeping lion repeatedly in his logs.[7] I remember staring at the rock with equal fascination as it slowly changed shape with our passage past it: the lion came and went, and I wasted film trying to catch it napping.

The *Jamboli* anchored in Wreck Bay, as the *Beagle* once had. But now a small town was there, with a naval base and the administrative capital of what was a growing province of the Republic of Ecuador. Yet the shore still looked the same, though we knew the tortoises were gone.

For the deck passengers this was journey's end. But most of the cargo, as well as my party, was bound for Santa Cruz in the morning. We duly arrived in Academy Bay with all our equipment and checked in to Charles Darwin Station, director Roger Perry. I learned a lesson even from that simple act of arrival. Never ship expedition supplies in big boxes. I learned this while watching our precious cargo being dangled down the side of the warship on ropes to balance the crates across dugout canoes for the journey to shore.

And so to work. The first task was to learn our environment by doing something strenuous together. So the five of us left shortly before first light on our first day ashore to walk to the summit of Santa Cruz, in 1966 still the least changed of the larger, fertile islands. A road carries tourist buses across the island now, but in 1966 not a single motor vehicle was on the island, and the total population was less than one hundred. Blessed wilderness, still almost as it was in Darwin's time. We could walk a trail in single file or, at the higher elevations, on no trail at all. It was just possible for fit young men to do the round trip from coast to summit in twelve hours, all the daylight that an equatorial day ever gives you, while still allowing time for pauses to take in some natural history on the way.

This walk took us through all the classic vegetation types of an ascent of a Galapagos island, vegetation that I must learn to distinguish by its pollen: coastal desert grading into cloud forest, the bushland above, and finally the treeless uplands. Santa Cruz in 1966 had enough of the original vegetation left to show what once it had been. I started building my herbarium, grabbing every kind of plant I found as Darwin had before me.

Darwin had passed his collections to the leading plant taxonomist of his time, Joseph Hooker of the British Museum, and he summarized the then still unpublished Hooker conclusions in his account of his travels. Hooker found that the total of known Galapagos species was 225, of which Darwin had collected 193 in his forays ashore from the *Beagle*. Of the 185 flowering plants in Darwin's total about 100 were new to science, a quite incredible result. Hooker further found that all the flora, without exception, was of what Darwin called "*an undoubted Western American character.*"[8]

These "American" plants that were not the same as those in America provided one of the puzzles that led Darwin to his discovery of evolution.

"Why, on these small points of land, which within a late geological period must have been covered by the ocean, which are formed of basaltic lava, and therefore differ in geological character from the American continent, and therefore acting on each other in a different manner — why were they created on American types of organisation?"

It would be fourteen years before Darwin published his answer in the *Origin of Species*, in 1859, but the answer was obviously in his mind when he read Joseph Hooker's report on his plant collection.

Might it be possible to find traces of these evolutionary events in the pollen record? An impracticable dream: even if pollen could be read with sufficient accuracy the chance of finding old enough deposits was vanishingly small. But dreams are hard to stop; I collected all the plants I could get.

By 1966 the known list of Galapagos plants had grown to about 600. I needed to know the pollen signal of all of them, and I meant to do so. That impossible dream of plotting evolution in action should certainly require that much; besides, like the rest of my profession, I had no idea what kind of pollen signal would come from this weird tropical vegetation, where pollination by insects and birds was the norm for every plant, including the trees, and pollination by wind the exception. I should send my collection to the contemporary equivalent of Joseph Hooker, Ira Wiggins of Stanford, then engaged in writing the definitive flora of the Galapagos Islands.[9] I sent him 400 specimens, but only one of my 400 was new to science: an honor to have been scooped by Darwin.

The dawn start of our ascent of Santa Cruz was across the coastal desert belt, broad behind Academy Bay because of the wide coastal shelf there. Giant cacti define it in memory: tall organ pipes of *Jasminocereus* are no more remarkable than the saguaros of Arizona or Mexico, but the giant *Opuntia* are sights to see, towering as high as the organ pipes, their broad, spiny disks stacked improbably high above your head: nothing like that in Arizona.

Scattered between the cacti are the shrubs of which Darwin spoke, gray of stem and gray of leaf. He follows his remark about the "wretched-looking little weeds" (given above) with,

> *"The brushwood appears, from a short distance, as leafless as our trees during winter, and it was some time before I discovered that not only almost every plant was now in full leaf, but that the greater number were in flower. The commonest bush is one of the Euphorbiaceae: an acacia and a great odd-looking cactus are the only trees that afford any shade."*

Darwin was too gloomy. This dry region of "brushwood" turns out to be the most diverse vegetation in the archipelago, with many more species, novel species, most of them, than any other vegetation on the islands. The Euphorbiaceae are certainly present in force, good news that because the family tends to be wind-pollinated. But so were the families Celastraceae, Verbenaceae, Combretaceae, Boraginaceae, Compositae, Gramineae, Amaranthaceae, Rhamnaceae, Burseraceae, and Solanaceae. Darwin remarked on one legume with *Acacia*; probably he saw *Parkinsonia* but without its fine yellow flowers, which would have given it away. *Bursera* trees, scrawny and not very tall, would have been truly leafless, because it takes real rain of the El Niño events (both meanings) to make them put out leaves, let alone fruit: nothing to identify in a lifeless-looking trunk with branches.

To a collector the dry regions, at least on Santa Cruz, are a delight where something fresh keeps turning up as one prowls among the giant cacti over the cracked lava slabs. But timing is probably everything: Darwin was there around October; we were there around July. Among our privileges in the desert region were native tomato plants, draped over the naked lava, rooted in cracks, their grape-sized yellow fruits delicious.

And Darwin seems to have been spared the "shade" tree among the Euphorbiaceae, *Hippomane mancinella*, a dreaded plant with the local name manzanillo. It looks like a great, gnarled apple tree when full grown, complete with globular fruit superficially like apples. But do not eat them; better yet, do not even touch. Naïve but dedicated, I collected a specimen thoroughly; flowers, leaves, sliced twigs; and got the sap on me. Within little more than an hour I was in fever, and the world was going dim as my sight faded to black. The rest of the day on my bed in a darkened room made me right again, but I shall leave collecting *Hippomane* to others in future.

The sun was on us as we emerged from this botanical desert wonderland and started to climb a steep path winding upward. At once the effects of height in this place of peculiar climate were apparent: trees were more common among the thinning stands of giant cactus; vegetation in transition. Large but leafless trees yet had splendid red flowers as big as oriental lilies. This was *Erythrina velutina*, a leguminous tree with lovely figured hardwood, good for furniture making.

A sense of continuous change follows with ascent through this transition to forest. The cacti and apparently lifeless *Bursera* trees swiftly drop out, with fine large trees in the family Nyctaginaceae (*Pisonia floribunda*) becoming prominent. Unlike *Erythrina*, the *Pisonia* carried a proper canopy of leaves when we saw them. *Pisonia floribunda* was named by Hooker, suggesting that it was among the new species collected by Darwin himself. Soon the first specimens of *Zanthoxylum* (Rutaceae), *Piscidia* (Leguminosae), *Psidium galapageium* (Myrtaceae; another Hooker/Darwin new species of tree), and *Sapindus* (Sapindaceae) appear. Then it is apparent that the transition is over at something less than 200 m elevation. This is the forest fed by the stratus clouds.

The forest continues for the next 200 m or so of the climb, essentially staffed with the same trees that appear in the transition, though increasingly loaded down with epiphytes, mostly lichens and ferns, but a few orchids and one endemic bromeliad (*Tillandsia insularis*).

But in the higher part of the forest a Galapagos marvel appears: the composite tree *Scalesia pedunculata*. I think of it as the tree made out of a dandelion, though of course it is nothing of the sort. But it is a tree in that it has a woody stem, is tall enough to take its place in the forest canopy, and even makes closed canopies of its own in nearly pure stands. At the same time, it is clearly of the family Compositae, most of whose members, including dandelions, are herbs or low shrubs. Imagine a dandelion twenty or thirty feet tall that got to that prodigious size by inflating all its parts so that its leaves and flowers were as huge in proportion as its stem. Then imagine an attempt, not completely successful, to shrink the leaves and flowers back to more reasonable proportions. The result is *Scalesia pedunculata*, an odd-looking tree whose trunk would be scorned by any woodworker and which has flowers somewhere between a dandelion and a sunflower.

The genus *Scalesia* is now credited with eleven species, all of them found in the Galapagos and nowhere else in the world. *S. pedunculata* has the best pretensions to be a real tree, though a forest of *S. affinis* in the hidden forest at the top of Fernandina (old Narborough) achieves a good second best. Other *Scalesia* species are merely large shrubs. To a naturalist the appearance of these arborescent members of the family Compositae in this remote place makes eminent sense. They use parachute seeds, like those that make dandelions so effective at colonizing lawns from parent plants outside the garden; but seeds of the *Scalesia* ancestor crossed a thousand kilometers of open ocean before lighting down on unoccupied soil. Very few forest trees could colonize the Galapagos, competition from big plants was muted, and natural selection favored large scalesias.

Darwin saw this too, in the *Journal of Researches* calling attention to, "*Scalesia, a remarkable arborescent genus of the Compositae, is confined to the archipelago.*" He knew of only six species instead of eleven and suggested they were island-specific. But in the *Origin of Species* he harks back to this observation with,

> "*Trees would be little likely to reach distant oceanic islands; and an herbaceous plant, though it would have no chance of successfully competing in stature with a fully developed tree, when established on an island and having to compete with herbaceous plants alone, might readily gain an advantage by growing taller and taller and overtopping the other plants.*"

The upper edge of the Santa Cruz forest is far more abrupt than the lower so that none have written of a transition zone there. The forest quickly thins, and the climber is in bushland. Or rather "was" in bushland, because this is the region most favored for farming.

The bushes are another Galapagos oddity, *Miconia robinsoniana*, though clearly closely related to a species of *Miconia* from Panama, is endemic. The bushes look like laurels, the leaves glossy and dark green, but with the lattice pattern of parallel veins typical of the tropical family Melastomataceae, and they have beautiful little candelabra of tiny bells for flowers. On Santa Cruz and San Cristóbal *Miconia* grows with the bracken fern *Pteridium aquilinum*, that same bracken associated with forests and moors in Britain and Germany. The combination produces a vivid impression of novelty to the botanical mind.

Above the *Miconia* is open land, almost fit to be called a "greensward," with sedges and grass and above all ferns. Some writers have even called it a fern zone. The trade winds blow across it, but it remains dripping wet from enveloping clouds, though the very tops sometimes poke above the clouds, cool to cold; the summit ridge is a wonderful resting place after a hike through desert and forest of six hours to get there.

Two more precious Galapagos oddities adorn these uplands. The first is a tree fern endemic to the islands: *Cyathea weatherbyana*, broad and chunky compared with many of the tree ferns of the mainland forests but a massive fern dwarfing a human beside it all the same. Moreover it appears to grow only in the open, typically in gullies where the force of the trade winds are muted. Lovely thing, but possibly doomed as cattle eat and trample it. Darwin never saw a tree fern in the islands and declaimed at their apparent absence. Soon perhaps they will really be absent.

The second oddity is peat bogs, one of the lures that took me to the Galapagos. We found them on that first walk, perfect domes of bright green, raised *Sphagnum* peat, like copies of the oh-so-green Irish bogs but shaped by the circular craters in which they lie. Spores of the *Sphagnum* moss that builds them drift with the wind and so can settle in the remotest of places. Crossing 1,000 km of open sea is nothing to these plants. Yet it stirs a naturalist's soul to find this symbol of Ireland or Alaska on an equatorial desert island. The bogs would be my first coring objective.

A few days later we packed in to the *Miconia* belt and set up camp there; five little green tents almost hidden in the bushes. From there an hour's walk could get us to any of several *Sphagnum* bogs. I cored three of them with a Hiller peat borer, a gadget which is part auger and part a cylindrical box to bring deep peat samples to the surface. Disappointing exercise: two of the bogs were less than two meters deep and the best and deepest of them only five meters. I had no idea how fast a *Sphagnum* bog could grow at the equator; probably faster than in Ireland. And the rough rule of peat bog cores in Europe is one meter per millennium, which meant that I must have more than ten meters to reach the ice age. At this tropical site my best five meters probably represented only two or three thousand years. No good. But we carefully packed the cores; any part of Galapagos history should be interesting.

Talks with the station staff confirmed that there was nothing else in the sediment line to interest me on Santa Cruz except perhaps the shallow wallows that the remaining tortoises used, and they could wait. Time to try for the main hope, the reported lake high on San Cristóbal. At least Roger Perry confirmed that it really existed; his description, "Two hundred yards across in a weathered crater in the clouds" sounded good. We took ship on the station's boat, the *Beagle II*. I got seasick on the crossing but we arrived in due course at Wreck Bay where Darwin had first landed. The Ecuadorian Navy, by prior arrangement with the station, had detailed a heavy truck and driver to take the five of us, our equipment, and provisions for ten days on what was then the only road in the archipelago to the village of Progresso and up the higher mud road as far as may be. A spell of dry weather in fact allowed this to be all the way. We only had to man-haul our stuff the last one or two hundred yards to the crater rim. A truly rugged expedition.

The "remote and rugged" aspect of the undertaking was further echoed by the lake setting itself. This was a lake unknown to geography, the possible existence of which was implicitly denied by an NSF reviewer. Yet here was a barbed wire fence down to the water. Cattle and a herd of horses grazed the smiling broad pastures on the tableland around the cone of the crater. The livestock went into the crater to drink.

The lake seemed to have no name other than "La Laguna" (the lake); reasonably enough, as the only freshwater lake seen on the islands in the memory of man. Why name it more precisely? But we nitpicking scientists, playing at geography, needed a name. I liked the sound of "El Junco" (the marshy place), what people called the marshy meadow spread below the cone on the broad flank of the old volcano. And I called that lake Laguna El Junco, eventually to publish the name in 1968 in *Nature*.

We set up our two drilling rafts on this precious lake, each to its three anchors. Two 2-man teams proceeded to lower their casing and start to core. The water was six meters deep, the coring dead easy at first, as we forced our way through the top three meters of lovely organic brown lake mud. But then we hit red clay, with no clue as to how thick it might be. I had given an edict to myself and the crew that we should hammer our way through whatever we found until we hit rock, defined as any substance hard enough to deform the steel cutting shoes, the design that had handled the gray Imuruk clay five winters before. We hammered for three days. When I took breathers from hammering on my raft, I could hear the chink-chink-chink of the hammer striking the anvil on the other raft, the sound echoing from the walls of the crater. I fancy I can hear it now.

Frigate birds soared up from the sea below on their five-foot wings and great forked tails, taking the trade wind up the mountain, dipping to the lake surface, touching it at the apex of their swoop to drink while on the wing. And well the birds might drink, for the water was wonderfully pure, chemically nearly indistinguishable from rain.

I had assigned my raft the very center of the lake, with the second raft some 30–50 meters away. Not surprisingly in so symmetrical a basin, this resulted in my raft getting the longest core; something over 15 m when at last further hammering produced nothing but a sullen bounce from the anvil and the steel shoe came up not only bent but with a piece of scoria jammed in the end. But at some 13 m the second raft got the better core, because the crew thought they had hit something other than red clay at depth. Indeed they had, an ancient lake episode missed by the first core for all its greater length. Easily explained as a result of an irregular bottom in the rocky crater floor. It was the second, shorter core, core B as it were, that gave us the decisive ice-age record.

A near permanent blanket of fog hung over the lake, lifting first at one corner, then at another, but always present. We could see across the lake from the crater rim only once in the week we lived there. We rushed to photograph and sketch in that opportune half hour of revelation.

It is a wet world there in the clouds of San Cristóbal. As in the clouds of Santa Cruz, we were above the forest line in a land of grasses, sedges, ferns, and bushes. On both islands the clouds come past riding the trade wind, an endless conveyor thrusting wetness on the horizontal, forcing a leaning posture on shrubs of exposed places that botanists call "krumholtz." But only in these broad terms can the vegetation of the two islands be compared. San Cristóbal had nothing like the *Scalesia pedunculata* groves reaching out into the open as on Santa Cruz; the *Miconia* bushes were scattered; there were no *Sphagnum* bogs.

The "missing" bogs worried me: why peat bogs on Santa Cruz but not on San Cristóbal? The soggy clouds were the same; both highlands were a mass of volcanic rock. The only halfway decent hypothesis was offered forty years later by a friend who noted that Santa Cruz was 100–200 m higher, perhaps a tiny bit colder or thrust into the clouds for longer: a straw to clutch but perhaps a stout straw.

I blamed the different vegetation on people when I saw it. The cool, moist uplands of San Cristóbal, with a settled colony even in Darwin's day, had by 1966 been reduced to ranch land. The native vegetation was grazed down, and agricultural invaders had filled the gaps, particularly the guava bushes that dotted the pasture and choked the defiles, not the *Psidium galapageium* of Hooker and Darwin but the common agricultural species *Psidium guajava*, the tasty fruits of which the ranchers chewed as they rode about the range, spitting out the stones as they went. Compared with this, Santa Cruz had suffered little settlement, and this only since the 1930s.

The "people did it" explanation has now been shown to be correct, because the people have undone it. When I saw El Junco thirty-nine years later, the open rangelands in the clouds of San Cristóbal were gone; in their place a tightly packed bushland of *Miconia robinsoniana*, the finest *Miconia* stand to be seen on the Galapagos, finer even than the 1966 stand on Santa Cruz, the only other island on which it grows. The "people did" this wonder too by the simple expedient of banning farming and chopping down the guava bushes. The *Miconia* did the rest.[10]

More interesting are the real differences that make each island unique, Galapagos oddities that had set Darwin thinking. Near the end of his account in the *Journal of Researches* is this passage:

> *"I have not as yet noticed by far the most remarkable feature in the natural history of this archipelago; it is, that the different islands to a considerable extent are inhabited by a different set of beings. My attention was first called to this fact by the Vice-Governor, Mr. Lawson, declaring that the tortoises differed from the different islands, and that he could with certainty tell from which island any one was brought. I did not for some time pay sufficient attention to this statement, and I had already partially mingled together collections from two of the islands. I never dreamed that islands, about fifty or sixty miles apart, and most of them in*

*sight of each other, formed of precisely the same rocks, placed under a quite sim-
ilar climate, rising to a nearly equal height, would have been differently ten-
anted; but we shall soon see that this is the case."*

After showing that this was true not just for tortoises, but also for the iguanas
and the finches, Darwin explored his plant collections: *"If we now turn to the
Flora, we shall find the aboriginal plants of the different islands wonderfully different."* Darwin provides an analysis of the whole flora as he and Hooker knew it,
which was little more than a third of the species list now known for the islands.
There were some inaccuracies because of this, but the conclusions of startling
differences between islands were sound. When vegetation types of the different
islands look different to the eye, this is because they are, in fact, different.

When we visited El Junco, a small grove of the tree fern *Cyathea weatherbyana* grew on one side of the lake with their fronds actually drooping over the
water. As on Santa Cruz, the tree ferns are destroyed by cattle, but the El Junco
grove was in a particularly rocky part of the crater rim where footing for cattle
should be difficult; outside this rocky patch we found trampled dead stems of
tree ferns and moribund specimens, suggesting that *Cyathea* was formerly much
more prominent in the crater. In 1966 I would have predicted extinction for *Cyathea* in my lifetime; haply I should have been wrong.[11]

That the tree ferns of the two islands are the same is easily explained because
they are dispersed by tiny, wind-blown spores that are produced in amounts so
copious that some spores should reach any island nearly within sight of each
other, as the larger Galapagos islands are. For the tree ferns, hops across fifty
miles of sea are no great matter, whatever the usual winds may be. One species
fits all islands, as it were. For other plants, and even birds, a hop from one Galapagos island to another is more difficult. If these difficult crossings are sufficiently rare, then evolution should proceed in isolation in different directions on
different islands from the same common stock. Darwin explicitly noted the importance of these sea crossings in explaining differences between islands:

> *"The only light which I can throw on this remarkable difference in the inhabi-
> tants of the different islands, is, that very strong currents of the sea running in a
> westerly and W.N.W. direction must separate, as far as transportation by the sea
> is concerned, the southern islands from the northern ones; and between these
> northern islands a strong N.W. current was observed, which must effectually sep-
> arate James and Albermarle Islands. As the archipelago is free to a most remark-
> able degree from gales of wind, neither the birds, insects, nor lighter seeds, would
> be blown from island to island."*

We drilled more cores from El Junco to make sure of a sufficiency of mud for
analysis, dragged the lake for plankton, sampled the water, and collected rock

samples. We had been a week in the clouds. All five of us were wet, as were our clothes and our tents; even our sleeping bags were unpleasantly damp. The two who had the most Spanish set off on foot for the village of Progresso to telephone for the naval base truck to come and get us and the other three struck camp.

It was a Sunday, not the most active time for a provincial government. But the commandant of the islands, bless him, sent two trucks. It had rained below the clouds as well as above, and the first truck got stuck in the resulting mud, so he sent the second. We arrived at the Wreck Bay base in the blessed tropical sunlight at four o'clock that Sunday afternoon, still in wet clothes and soaking rainwear, embellished with mud and unshaven. I called on El Commandante in this condition. I remember him as immaculate in his white uniform, elegant, a classic naval officer. He and I had virtually no common language. He ushered me into what I was to learn was his private house, and on through an interior door. With a broad smile on his face, he waved me into a lovely tiled bathroom, the shower fixtures gleaming, and a stack of fluffy white towels.

Once aboard the *Beagle II* at Wreck Bay, bound for the station on Santa Cruz, we learned that the ship was to go first to visit Genovesa (Tower) island for a team monitoring breeding seabirds. Genovesa was the island with the grand crater lake that William Beebe had found when he had talked the skipper of the yacht *Nova* into risking his ship to put him ashore in 1923, one of my prime targets.[12] The *Beagle II* was scheduled to be at the Genovesa anchorage for just two days, but the chance was too good to miss. Assuming the lake was ten meters deep, the sediments organic or soft, and the trek in only a couple of kilometers, bravado said we could do the job easily in two days with the five of us all pitching in.

The *Beagle II* anchored in Darwin Bay on the south side of the island in the early morning of July 16, and put us ashore with our rubber rafts and coring rig at 0800 hrs. Darwin Bay should have given us a hint as to what we should find at the lake, the bay being a great, circular collapse crater, larger than the lake but open to the sea on one side. Darwin Bay was slightly more than two kilometers across, with beautiful sheer lava cliffs rising sixty meters from the sea. And it was deep; the 1942 chart by the U.S.S. *Bowditch* found over a hundred fathoms in the middle (say 200 m). The *Beagle II* could anchor only close inshore over a shallow underwater ledge.

We packed our stores up a steep ascent, where the cliffs had crumbled at a broken part of the crater's edge, and were at once on the level summit. The island is the amazingly flat top of a single shield volcano, its surface nowhere more than seventy meters above sea level. Nearly circular; the longest diameter is about six kilometers. The lake is near the middle of this circular blob of lava but, since Darwin Bay represents a two-kilometer bite out of the island, the shortest distance from the edge of the lake to the edge of the bay was only about one kilometer. From where we had landed, we reckoned less than two kilometers of walking would be required.

We packed in. The island is covered with what can best be called scrub forest, all of drought-tolerant plants, many rooted in cracks in the lava. The tallest layer is of the tree *Bursera graveolens* that can grow to ten meters high elsewhere, but I doubt if many manage much more than six meters on Genovesa. They look like the wizened trees of an old apple orchard in winter — in winter because the Galapagos *Bursera* trees nearly always have no leaves when seen. The tree is "opportunistically deciduous," putting out leaves only when it rains, and dropping them again when the ground dries.

Low branches of *Bursera* trees are a pain to a man with a packload of drill rods; a pain made worse because the trees are well supplied with sticky resin, which richly transfers to the human hand and clothing. The lesser bushes under the *Bursera* trees are the standard cover of the drier regions of Galapagos islands, no more agreeable to porterage than Darwin's description of those he first saw on San Cristóbal. Without a trail through this chaos of bushes, we must navigate by dead reckoning and a compass.

But underfoot the going is not bad, mostly plate lava, the congealed surface of once molten rock that geologists call *pahoehoe* (a Hawaiian word pronounced something like pa-hoy-hoy). This is bad stuff for grazing knees, having grains of volcanic glass in its makeup, so don't fall when thin layers over old bubbles in the molten mess break through under your boots (don't try it in sneakers). By noon we had assembled our minimum equipment at the crater rim, where the descent seemed possible to the likes of us.

Curt McGinnis was the one mountaineer in our party. He lowered 50 m of goldline climbing rope down the cliff and made the end fast to a hefty rock. Then he tied off improvised harnesses for all of us and gave us a short lecture on the art of rappelling. He had found a place where the cliff was steeply inclined rather than vertical. We shouldered our 70 lb packs, clipped our "dee" rings to the rope, and down we went. Having absolutely no head for heights, I kept my eyes tightly shut most of the way. The rope was ten meters too short, but the bottom was just a scramble. And we were on the shore of the Genovesa crater lake.

Inflate a boat and get out there. I had one treasure, a neoprene yacht tender bought in 1963 from a firm in Northern Ireland that was in the process of going bankrupt for, as they put it, "making a Rolls Royce product in a Ford market." I had bought their last dinghy. It was a honey to row, superior to any I have found since from the plethora of rival manufacturers who now design their inflatables on the assumption that they will be driven by outboard motors; not real boats at all. We called my treasure the "black boat."

Flying the burgee of the Royal Victoria Yacht Club from the tiny mast, I rowed off to sound the lake. What a place to be afloat; just half a kilometer across, its banks the lava cliffs sixty meters high, a sense of echoing space, a feeling that the "black boat" might well be the first keel ever to disturb its waters. Overhead passed flights of red-footed boobies, which nested in the island bushes in their thousands.

In the middle I lowered my sounding line; no bottom; the line hung straight up and down. Slowly I paddled back to shore, sounding at intervals with the same result. So much for a quick coring operation in ten meters of water.

I tied one of the long anchor ropes to the sounding line, went out to try again, and this time found bottom at thirty meters. And that was it; the crater had a flat bottom under thirty meters of water, a depth more interesting in English measure: one hundred feet deep. Coring this lake would require 30 m of casing and 40 m of drill rod. I did have this much stored on the *Beagle II*, not because I was expecting so deep a lake, but because I had enough casing and drill rod to equip two drilling crews at once. So back to the *Beagle II*, post haste.

We should need more time. The *Beagle II*'s captain promised one more night at the anchorage, but then he must sail at first light. Being anchored close in on a lee shore surely had something to do with the skipper's reluctance to stay. The great crater that is Darwin Bay is wonderfully sheltered on all sides except the south, where the entrance to the open sea is the neck of a bottle. The island, although slightly north of the equator, is still within reach of the southeast trade winds as they push north of the geographic equator, and a good blow should make exit from the bottle uncomfortable. The skipper of *Beagle II* wanted us out.

Captain FitzRoy of the original *Beagle* surely had the same attitude, as his chart shows that he never closed with the south coast. Alone of all the islands, Genovesa is the wrong shape on his chart; the north, east, and west sides OK, the south side guessed. FitzRoy never saw the great harbor (or boat trap, depending on the wind) now called Darwin Bay, nor of course did Darwin, for all that Beebe named it for him in 1923 on the *Nova* expedition. Evidently FitzRoy, fine sailor that he was, chose not to risk his ship by sailing close to those frowning cliffs of the lee shore; he contented himself with recording what he could see from oblique views as he sailed past the small island to the east or west. The real *Beagle* did not have an engine to drag him off a lee shore as did *Beagle II*.

Had Darwin been able to land on Genovesa, he would surely have remarked on the red-footed boobies, variously estimated to more than a quarter million pairs strong. This is the colony where most, if not all, the world's red-footed boobies are bred. They fly out from the island to fish a huge area of the central Pacific Ocean, endless streams of the great birds coming and going into the distance of the west.

Resolving to meet our skipper's deadline, we packed in all the remaining equipment that afternoon: 33 m of casing and 50 m of rod. Early the following morning we set out our raft to three anchors made of bags of rocks, each attached to its cod line, more than 100 m long. Down went 30 m of casing. Not anticipating so long a string of casing, I had made no mechanical aids to hold it, so one man held this dangling weight while a second screwed the next section in place. But we made it without letting anything go. Then we cored, meter by meter, connecting and disconnecting more than 30 m of rod for every meter of mud gained, working through the day and by flashlight through much of the night.

We slept on the lake shore for a few of the darkest hours before our sleeping bags were bombed with guano by boobies of the dawn flight, then back to coring. We got two parallel cores, each of five meter-sections, both ending with the impact of rock. What little we saw of the mud as the tubes came up was lovely: organic, many colored, even with layers of hard white carbonate. Then we pulled out the casing, headed for shore, and began the process of hauling everything up that 60-meter cliff. We reached the *Beagle II* at nightfall, ready for the skipper to make his dawn departure.

Two raft timbers were left by the crater lake, hidden under some mangrove bushes. They were the nine-foot-long beams that crossed two small rubber boats to make a catamaran structure to support the drilling platform. I had bought the wood in Guayaquil, thinking it silly to ship American 2 × 4s to a country covered in forests. But the construction lumber in Guayaquil was eucalyptus wood, murderously heavy by the standards of American softwoods. And we balked at dragging those unwieldy, heavy things up that great cliff in our rush against the clock. Contrary to my oft-asserted principle of never leaving a trace of our passage in the wilderness, I said we should leave them. And so we did.

On our way out I was told that my student team had written on the two beams, in large letters, with permanent markers, "These boards left here by Herr Doctor Professor Paul Colinvaux."

Back at Santa Cruz, I chartered a fishing boat to take three of my party to the salt-mine lakes on Santiago (James) Island, the *Beagle II* being committed elsewhere. This was a precedent; Charles Darwin Station being then so new that the few visiting scientists had hitherto been accommodated on the *Beagle II*, or had sailed with one of the colorful European settlers who had made their homes on the island. We resolved to buy Ecuadorian on Ecuador's islands.

The charter was arranged in the office of the station director, Roger Perry, and was suitably solemn, the skipper, Señor Lopez, and I standing formally before the director's desk. Perry and I had agreed on a price that should reflect the full dollar value for a professional service; which we also intended to act as a precedent. I handed Señor Lopez a stack, more than an inch thick as I recall, of Ecuadorian currency, the sucre, and we shook on the deal. This was a significant infusion of cash into a local economy in those years before tourists.

David Greegor (Ohio State Graduate student), Curt McGinnis (Yale), and Will Horn (Duke) sailed in the morning, with a coring rig and supplies for a week, heading for the tricky waters of the hundred-kilometer passage between Santa Cruz and Isabela to Santiago. They found their skipper to be a cheerful companion, willing to do all in his power to accommodate their eccentric requests. And of course he knew those waters.

The salt lakes proved barren of records that might describe the long-ago ice age. The two most promising lakes were cored, but the sediments were little more than a meter thick, not without interest to the study of lake systems (limnology) but of no value to the ice-age historian. The salt-mine lake itself appeared to have nothing but salt for mining, as expected. Nor did my team find a skull in the bushes as Darwin had. But we had to check, and now we knew.

Meanwhile I set out with the remaining member of my party, Erno Bonebakker (Yale), a local guide, and two donkeys, bound for the tortoise ponds in the moist uplands of Santa Cruz, a day's hike each way. We went up the familiar trail to the moist region of forest, then followed another trail west-about round the island for several hours.

Our donkeys, of course, were laden with coring rig and camping gear, and we led them on rope halters. This is a slow form of transport for beginners in the art of donkey caravanning. I had not at that date read Robert Louis Stevenson's *Travels with a Donkey*, but I have read it since. Stevenson says it all. Donkeys do not care to exert themselves, particularly if they have loads on their backs, and particularly when they discover that they are in the hands of incompetent would-be drivers who know nothing of the trade. One thing I learned was that it is easier to pick up a packload by itself than when it is attached to a donkey that has decided to lie down; donkeys are heavy.

Stevenson's troubles were as bad as ours until a French innkeeper, unasked, made for him a goad. Stevenson wrote, *"Blessed be the man who invented goads! This plain wand, with an eighth of an inch of pin, was indeed a sceptre when he put it in my hands. Thenceforth [the donkey] was my slave. A prick, and she passed the most inviting stable-door. A prick and she broke forth into a gallant little trotlet that devoured the miles."*[13] But nobody told Erno and me about goads.

But Erno, the donkeys, and I did reach the tortoise pond, and spent the next day there. It was an irregularly shaped puddle, ten meters across at its greatest dimension; everywhere less than half a meter deep. The edges were sticky brown clay. Lava boulders reared out of water in many places, wet and slimy. An unlikely setting for a sedimentary record spanning to the ice age.

Three or four tortoises were in the vicinity, one of them big enough for me to sit on as a potential rider should. But it withdrew its head and feet, refusing to move until I got off. On attitudes to load-bearing, donkeys and tortoises were presenting a united front.

The tortoise hissed, as Darwin said they hissed: *"The instant I passed, it would draw in its head and legs, and uttering a deep hiss fall to the ground with a heavy sound, as if struck dead. I frequently got on their backs, and then giving a few raps on the hinder part of their shells, they would rise up and walk away; — but I found it very difficult to keep my balance."* Just as I had not known the importance of goads for the rump of a donkey, I had also not known about rapping a tortoise on the butt.

I pounded and augered my way into the mud of the pond, feeling my way between the rocks. Eventually I got about one meter of sediment, all of it the sticky, dark brown clay, more or less undisturbed (by me, that is). But surely the mud had been churned up by tortoises as they bulldozed their way through the pond! This was a poor outlook for any record of ancient climates.

There might nonetheless be some grains of ecological comfort to be gleaned from the sediments. Presumably the ponds had been used for drinking by tortoises for as long as ponds and tortoises had coexisted. The giant tortoise/pond ecosystem was a great rarity on earth, never, as far as I knew, studied before. A carpet of rich, bluish-green herbs, *Heliotropium indicum*, existed as an almost pure stand on the wet mud round the pond. Opinion at Charles Darwin Station was that tortoises did not eat this weed, which thus thrived in tortoise wallows. The plant is not peculiar to the Galapagos Islands; it is even common in damp, shady places in the southern U.S. But in the Galapagos you look for it in tortoise wallows. Could I use *Heliotropium indicum* pollen as a tortoise indicator? Doubtful. Anyway, I carefully packed my meter of mud for its journey back on a goadless donkey.

While awaiting the return of Señor Lopez with my team from Santiago, I prowled the trails of Santa Cruz. Beside one of the trails I found "the flower." It was on a tangle of vines at the edge of a clearing in the moist zone. I had no idea what it was, being still completely untrained as a tropical botanist. A purple center; a great green, branching style sticking out; white petals (or sepals or bracts, or whatever); a forest of stamens as a yellow fringe; the whole thing up to two inches across. A spectacular sight in vegetation that tended to unrelieved greens and browns. I added it to the growing collection in my plant presses to yield voucher pollen samples in due course. And of course I pressed samples of its curious boomerang-shaped leaves. For good measure I pickled a sample of its purple soft fruit in alcohol. And went on my way. Some six months later there came a letter from Ira Wiggins at Stanford, "Paul, it's new!" It was a new species of passion-flower that Wiggins named *Passiflora colinvauxii*. One down, but I wasn't going to catch up with Darwin.

Of the original lake target list only the Tagus Cove crater lake remained, to-gether with its neighbor, the lake where Darwin had sought to drink but found the water "salt as brine."

The Tagus Cove crater was a prime target, a lake that had been extolled by one of my early correspondents as the best chance for a long record in the whole archipelago. So I chartered the *Beagle II* to take us there. Because Tagus Cove was halfway up the far side of Isabela (the Albemarle of Darwin's account), going all the way round was no longer than retracing the outward route, it made sense to circumnavigate Isabela. We might get a chance to go ashore to look for other coastal ponds rumored at both ends of the island. So that is what we did.

Our course would take us through the narrow strait, Canal Bolívar, dividing Is-abela from the great and active volcano, Fernandina (Narborough), towering five thousand feet (1,700 m) out of the sea. In its central crater Fernandina was known to hold the largest lake in the Galapagos Archipelago. Fernandina had been seen to erupt in Darwin's century, foreclosing the idea that this great lake could hold ancient sediments. But we hungered to climb the mountain. If the Tagus Cove crater lake did not take long to core . . . ?

Tagus Cove itself is bizarrely beautiful, a channel of deep blue water between black lava cliffs. That blue water was so irresistible that I had dived from the side of the ship within minutes of the anchor going down. On surfacing came the an-noyed voice of the skipper shouting, "Sharks!" I swam back very quickly.

The narrow cove is blocked at the land end by a black slope of lava some 20 m high serving to hide the crater lake nestled behind it, a lake if anything even more beautiful than the cove itself, a startling, circular, deep blue-green pool in a black, sloping basin, both an aesthetic dream and a perfect pollen trap. We found it to be ten meters deep, an ideal depth, then set our raft in the very center of the lake.

Down went the aluminum casing until the tenth meter was screwed to the string and thrust down to feel for the bottom. A clanging sound echoed up the underwater pipe; we had hit solid rock. Beautiful Tagus Cove crater lake, once a prime hope of the project, had no mud in it; none at all.

The lake water was salt; very salt, so salt that I found it possible to float sitting up as in reading a book. The salt had to be marine salt; there was no other obvious source of salt in the lava rocks ringing the lake or forming the mountain above it. The sea was close, so very close in fact that an oblique view from the neighboring hillside let you see the lake, the sea, and the high ridge of hard rock that separated the two, all in one view or one frame of the camera's viewfinder. The problem was that the lake was visibly higher than the sea in Tagus Cove, at least ten meters higher by my estimate. So how in the world did seawater get up there?

I took flippers and face mask to have a direct look at the lake bottom; equipment kept handy for recovering things dropped overboard (no scuba). Ten meters (33 feet) is an easy skin-diving depth, except for the buoyancy of this very saline lake which required a certain effort to get underwater at the start of a dive. Visibility was good, even under ten meters of water. The bottom was like a crazy paving, each block one or two meters across; bare, hard rock with no more than a one-centimeter-thick gelatinous green fuzz on it. The cracks between the blocks were filled with hard, white crystalline deposits, taking the place of grout between slabs of a crazy pavement. I made a complete underwater traverse of the bottom, from edge to edge, to find it everywhere the same. I have no doubt that this bottom is the remains of a cooled lava lake, still retaining the cracks from the shrinking of the molten rock as it cooled and hardened. As to how the seawater got there, we had to leave it. A giant tsunami perhaps?

Months later, back in my Columbus laboratory, I got the results of a chemical analysis of our sample of Tagus Cove crater lake water from a government water-quality lab. As expected, the sample was basically concentrated seawater, essentially sodium chloride with additives, but with about two and a half times the normal salt content (80 parts per thousand as opposed to the usual range about 35 ppt). The salty mixture had one real oddity, however: there was too much calcium chloride included. So now we had to get seawater high out of the sea and find a way of enriching it with calcium at the same time. Evaporation alone will not explain the extra calcium because calcium salts are selectively lost from evaporating seawater, not gained.

The best answer we could find (can find, for that matter) to the double problem calls on the wild hypothesis of tsunamis swamping the lake and appeals to the grout of white crystalline deposits in the bottom cracks to test it. The first minerals to come out of evaporating seawater are calcium salts, notably gypsum. The "grout" I saw in my dive, therefore, is gypsum. At this stage a second tsunami comes, effectively to charge the lake with fresh seawater. Some part of the gypsum redissolves, temporarily enriching the lake water with calcium. But then evaporation and redeposition grow the grout deposits of gypsum once more. Imagine the process being repeated several times, through several tsunamis and you have a lake saltier than the parent sea, but with too much calcium and with gypsum concretions on the bottom.

An absurd hypothesis. Please give me a better one. On the other hand it does fit the facts. Tagus Cove is a long, narrow, fjord-shaped affair, pointing like the barrel of a gun directly at the sill behind which the crater lake lies. Less than ten kilometers across the Canal Bolívar lies Fernadina, one of the world's most active volcanoes, a fine source of explosive eruptions and hence tsunamis. The journal *Nature* published my tsunami hypothesis in 1968, and it still stands in the literature, unexposed as an absurdity.[14] I like it, but shall not be upset if someone finds a less flamboyant explanation.

That hint of tsunamis in the past was the sole contribution of the Tagus crater lake to the environmental history of the Galapagos Archipelago.

Only nearby Beagle Crater was left to core. Darwin's visit, you will recall, was described thus, "I hurried down the cindery slope, and choked with dust eagerly tasted the water — but to my sorrow, I found it salt as brine." Since we were packing three rubber boats and a coring rig, our hurry down the cindery slope, choked with dust, was even more labored. I remember particularly the lousy footing of that cindery slope. This was a lava field of the alternative kind to *pahoehoe* in modern geological classification: *aa* (*sic*), another Hawaiian word, pronounced something like ah-ah. It looks like, and probably is, cooled froth, with small cinders but also irregular blocks the size of footballs or even a man. It is great stuff for turning an ankle.

We did not taste the water, taking Darwin's word that it was salt as brine. The analyses of our water sample found the salinity to be 50 ppt, roughly halfway between that of Tagus Cove crater lake and seawater, confirming Darwin's estimate. We cored it at its deepest spot, 15 m of water, finding the sediment body to be two-thirds of a meter thick and that mostly sand: no go.

So we did have time to climb Fernandina. The mountain had been climbed a few times before but mostly by larger parties who established water stations at different elevations and took their time (several days). Roger Perry and the doyen of climbers in those days, Eric Shipton, had pioneered a better way: go up quickly traveling light, and rely on drinking volcanic water from the crater. Their technique had required that they do the ascent in one day or be dangerously thirsty: reach the 5,000-foot summit first, then descend to the 2,000-foot depth of the crater for their drink before nightfall.

Roger Perry had advised us on the climb before we sailed for Isabela, just in case we got a chance to try. He gave a vivid illustration of what traveling light means. Eric Shipton, who had got about as high on Everest as anyone except Mallory, long before it was finally conquered by Tensing and Hillary, was a compulsive pipe smoker. So his pipe had to go with him. But he calculated how many smokes he would take on the Fernandina climb and he counted out the number of matches this would need, removing the rest of the matches from the box before starting to save weight. We prepared for the climb in as close to that spirit as we could get.

Four quart water bottles each were the heaviest part of our loads but one that would quickly lighten. Four small plastic collecting bags (about 6 inches × 4 inches) full of a mixture of dry oatmeal, sugar, and powdered milk, and either a light down sleeping bag or a heavy sweater in which to sleep would do for three days. We should drink the volcanic water in the lake, knowing that it was laced with magnesium sulphate (epsom salts) and other volcanic "additives," but otherwise should be reasonably wholesome. We also had a small (15 cm) plankton net and some nylon cord (justify the jaunt as science). David, the party's photographer, was stuck with the weight of a 35 mm camera. I added a pair of leather palm gloves for clutching on to abrasive lava and used a tiny knapsack borrowed from Roger Perry. Thus equipped, we were put ashore at Point Espinosa at dawn with a promise that the ship would be back in three days.

Point Espinosa has the largest colony of marine iguanas, and abundant Galapagos penguins and flightless cormorants, all of them endemic to the Galapagos. The black iguanas congregate on the black rocks beside the water in their hundreds, now perhaps the best-known poster image for the Galapagos. The penguins are there in surprising numbers too, though less conspicuous; before 1966 so little seen that Roger Tory Petersen could suggest that only 500 existed.[15] In 1970–1972 P. Dee Boersma from my laboratory was to spend a total of 350 days camped on this coast and the facing shores of Isabela to study the penguins, finding the total to be between 11,000 and 23,000 breeding pairs.[16] But that was in the future; I cannot remember that I saw any in that first quick tramp across Point Espinosa on my way to the inner crater lake.

Keenest in my memory is sight of the flightless cormorants, particularly when in the water. Huge and heavy, they look like hunter-killer submarines, their long bodies nearly submerged, their necks erect like a nuclear submarine's sail. Tom Clancy should write about them.

Darwin mentions neither penguins nor cormorants, suggesting that Captain FitzRoy did not anchor at Point Espinosa, though he did sail past it. But Darwin was fascinated by the great black marine iguanas, which he found everywhere he visited; they are still everywhere, though the most spectacular concentration is at Point Espinosa. In the *Journal of Researches* Darwin describes his experiment of throwing one into the sea, and his highly plausible, but wrong, conclusion from the creature's refusal to go swimming for him:

"I threw one several times as far as I could, into a deep pool left by the retiring tide; but it invariably returned in a direct line to the spot where I stood. . . . As soon as it thought the danger was past, it crawled out on the dry rocks, and shuffled away as quickly as it could. I several times caught this same lizard, by driving it down to a point. And though possessed of such perfect powers of diving and swimming, nothing would induce it to enter the water; and as often as I threw it in, it returned in the manner described."

Poor iguana; it was cold and needed to warm up under the sun. But in Darwin's day nothing was known of thermoregulation and of how cold-blooded (what a misnomer!) animals must use the sun or other environmental heat to get their physiology up to operating temperature. Darwin's plausible alternative explanation was:

"Perhaps this singular piece of apparent stupidity may be accounted for by the circumstance that this reptile has no enemy whatever on shore, whereas at sea it must often fall prey to the numerous sharks."

Point Espinsosa is now a destination for tour boats, but in 1966 still a remote wonder. The fringe of life is only fifty meters wide, behind which the dark gray and black desert of lava begins. Darwin again, when the *Beagle* was becalmed in the channel between Fernandina and Isabela:

"Both [islands] are covered with immense deluges of black naked lava, which have flowed either over the rims of the great caldrons, like pitch over the rim of a pot in which it has been boiled, or have burst forth from smaller orifices on the flanks; in their descent they have spread over miles of the sea coast."

Our way to the summit, seven miles away and nearly one mile up, lay on these "deluges"; seeking those that were *pahoehoe*, trying our hardest to avoid those other stripes of cinders that are the *aa* flows. When walking over the cooled remnant of a "deluge of black naked lava" under an equatorial sun one imagines the state of an ant crossing a hot fry pan. Humans are not made for that environment. Incredibly, though, a few plants can exist there, inconspicuous and rare, but still living in an environment of sun-baked lava. Cactus one might expect, and there are a few small ones near the coast. But in what might be called the "middle passage" of the climb, the most deadly part, the rare plants are, of all things, sedges and ferns! Very, very few, hidden in nooks and crannies, brown, rustling, and dry; but grimly alive.

Higher on the mountain the incidence of moisture increases as fog trickles down from clouds on the summit. At this level our *pahoehoe* lava flow petered out, forcing us onto cinder cascades and winding our way through the rampart of *aa* lava blocks. Cinders served as virtual soil for cactus, supporting outliers of the organ-pipe-type *Jasminocereus*, lonely green giants in the black waste. Then more of them, some with egg-sized fruits that we cut open for their lemon-flavored juice.

Over the last hour of the ascent the way got steeper, sometimes needing hands as well as feet. A final scramble and we hauled ourselves over the cliff edge that fenced the summit. And there before us was Shangri-La.

A green forest covered the summit of Fernandina in 1966, utterly unsuspected until your head poked over the edge of the mountain's rim. To be sure, it was a low forest, certainly with few species; little variety. Yet a layered green canopy, undisturbed even by human thought, was there for an exploring soul to walk under and through.

The soil underfoot was copper-colored from the ash out of which it was formed. And in this soil were the gaping burrows of the land iguanas: big brown lizards, some three and four feet long, unperturbed by our presence, basking in the sunlight of open patches, evoking to the biological mind a vision of the Jurassic.

This lost world was quite enough to urge a naturalist to linger. But needs must when thirst drives and the nearest water was on the floor of the inner crater. We pushed through the quarter mile or so of forested tableland to where a sense of echoing nothing beckoned. And then we were on the lip of "the caldron," as Darwin would have put it; an impossibly vast hole, a void falling away for two thousand feet.

The crater was four miles across and a third of a mile deep. From where we stood, the far walls appeared as sheer cliffs, cut out with a cookie-cutter. The crater floor, where it was not under water, reflected the battleship gray of *pahoehoe* lava, but much of it showed the glint of distant water, the surface of a lake some two miles wide. Beyond the middle of the crater was the subordinate cone we had seen in air photographs, looking like a volcano within a volcano, complete with its own crater with its own lake in it. This inner cone looked like a pimple on the floor of the great pit, although actually a quarter of a mile across. We determined to sleep on it that night.

Curt planned a route down the series of terraces below us, finding diagonal traverses where even a two-footed climber like me could follow. So over the top, hands and feet down to the first ledge, and then the long scramble down. It took all afternoon to get to the bottom, by which time an ominous dark shadow was filling the west side of the crater.

They do not give you a decent twilight at the equator, they merely switch the light. We set off on the last mile and a half of *pahoehoe* at as close to a jog trot as tired limbs could muster. And then we were on the little cone, collecting dead sticks to boil volcanic water, in the hope that it would taste better flavored with tea, powdered milk, and sugar.

That was a night I shall remember to the end of my time. On my back before drifting off, looking up at a great black disk of sky, framed by the crater walls, studded with stars so many and bright that only an astronaut has seen better; and this was before there were any astronauts. Then wonderful sleep.

To make our third-day rendezvous with the *Beagle II* we had to climb out of the crater by nightfall of the second day, allowing only the morning to collect in the crater. We filled our water bottles from the two lakes; for drinking, and what we did not drink of the sulphur-tasting stuff would be our specimens for analysis. We counted fifty plus endemic Galapagos ducks on the big lake. And we took plankton tows from both lakes with a fine mesh net that should hold diatoms.

Nothing new was expected in the diatom flora because diatoms tend to be ubiquitous, able to disperse round the world only less readily than bacteria. What determines a diatom species list is not so much geography as the physical properties of the lake and its water. Even so, one Fernandina diatom was new; not because of Galapagos isolation like the other endemics, but because of the curious chemistry of the volcanic water. Ruth Patrick of the Philadelphia Academy of Natural Sciences, the principal diatom taxonomist of her generation, examined our plankton tows for diatoms, finding a new species of the beautiful genus *Amphora* in the Fernandina water.[17] She did me the honor of naming it for me but used my given name, bless her feminine heart. The new diatom is *Amphora paulii*.

I wanted mud samples from the lakes to measure pollen rain over the island. There was no mud in the large lake where I could reach, the water merely lapping on raw lava, nor on the sandy bank of the cone lake. So I tried to dive for it, or rather swim, deep into the lake in the cone. The water was pea green, utterly dark when I got beneath the surface. I swam down through the darkness as far as I thought I could, had some doubts about which way was up, was scared, got to the surface, and quit. With all our bits of line and string tied together, I sounded: about 15 m — beyond my naked swimming range without flippers; and perhaps it had no mud anyway; besides the dark down there was frightening.

We made a collection of all the plants growing on the little cone. The cone was cemented tufa: volcanic ash welded into rock by intense heat. The surface was still largely bare ground, with the texture of coarse sandpaper but layered from the original ash falls. Already it was weathered and cracked, in places with rudimentary soil. We collected twenty-five species of flowering plants, including bushes and small trees; quite a haul from this most barren, isolated, and recently sterilized spot in the middle of a most barren, isolated island.[18]

Not surprisingly, all twenty-five species belonged to groups known to be dispersed by wind (the hairy parachutes of Compositae) or birds (berries with seeds that pass through guts). All had previously been recorded from Isabela, the nearest neighbor just across the Canal Bolívar. But I find it curious that we found none of the *Scalesia* bushes and small trees that make up so much of the Shangri-La forest on the rim.

Thirteen families of plants are represented by those twenty-five species. Interesting to note that a very similar array of plant families provided the first colonists of Krakatau after the celebrated eruption on the other side of the world. The species were of course different (Krakatau is in southeast Asia), but no less than ten of our thirteen families provided first settlers to both volcanic islands after eruptions.

We walked out, after tanking up on the last of our tea-with-volcanic-water. As always on these ventures, the climb up that 2,000-foot cliff was easier than the climb down had been. Curt retraced the route he had found for the descent, and we hauled ourselves over the final rim in a cloud drizzle of the late afternoon, wended our way through the forest and iguanas, and chose our resting place for the night a small way down the outer slope to be as much as possible out of the rain.

Ever since that day I have wanted to live awhile in the extraordinary summit forest that I like to call "Shangri-La," after the fictional Tibetan utopia in James Hilton's 1933 novel, *Lost Horizon*. The Fernandina utopia is only for biologists, because physically not attractive; dusty and hot, or damp and cold, depending on the clouds. But like the original it is a secret place at the top of a mountain, a system of life cut off. To the naturalist, the forest of *Scalesia affinis* trees is wholly strange. Woody composites like *Scalesia* are straggly substitutes for trees, and the whole ecosystem has the oddness of being made only of pioneers. It is incredibly young, an ecosystem of pioneers that never grows up. Large lizards prowl here without rivals, as they did before the age of mammals: in another world with strange-looking plants.

Darwin did not penetrate to this particular Shangri-La, but the whole archipelago served him for an even greater wonder. He marveled that these volcanic islands were young yet had life building on them, entirely new life. And he recorded that wonder with the greatest sentence in biology:

> *"Hence, both in space and time, we seem to be brought somewhat near to that great fact — that mystery of mysteries — the first appearance of new beings on this earth."*

In the morning we walked down to where our *Beagle* was waiting for us and went home, our work done.

3

GALAPAGOS CLIMATE HISTORY: THE EASTERN PACIFIC AND THE ICE-AGE AMAZON

The Galapagos results were quite wonderful. El Junco, the lake that reviewers had said did not exist, not only existed, but had probably done so, on and off, since near the start of the last ice age. The whole ice-age story was there in outline in the 13–15 m columns of mud that were our longest cores. All we needed was the wit and technique to read it.

From Genovesa crater we had a shorter story, probably only of the last 6,000– 8,000 years, but potentially marvelously detailed. A perennial hope of palcoclimatologists is finding banded sediments, and in Genovesa were sensational bands, distinct in color or texture, easily visible to the naked eye. These bands offered the chance of a minutely detailed history, a surrogate weather station recording the last many thousand years of Pacific climate, compared to the broad-brush epochal history of climate in El Junco. This recording weather station was likely to be unique when deciphered.

Perhaps I had the wits to read these writings in the mud, I like to think so, but in the 1960s our technique had limits hard to grasp from forty years on. We had "bulk" radiocarbon dating, gross stratigraphy, pollen analysis, other microfossils (if we were lucky), spectrophotometric and wet chemistry. In addition, both X-radiography and remanent magnetism of cores were stirring. These six, but those most useful in the Galapagos were the first four. A property of these four was the consumption of large volumes of mud, ranging from cubic centimeters to handfuls. The cores were only 1½ inches wide, so a certain coarseness in resolution was inevitable when such large samples of mud were needed.

The old methods of radiocarbon dating that we dub "bulk" say it all. The method worked by burning a sample, collecting all the resulting carbon dioxide, inserting this in a chamber surrounded by Geiger counters, and counting the rate of beta particle emission.[1] In mud like El Junco's or Genovesa's we had to submit bits of core as much as 20 cms long — to supply enough organic matter, to hold enough carbon, to make enough gas, to boost the rate of emission of beta particles high enough to get a meaningful count. That it could be done at all was the miracle that made a dream of a climate history of the Amazon possible. But the crudeness of it was a sore trial. Each date took weeks. To conserve sediment we sent off samples one at a time to use the result as a guide for choosing the mud to be burned next, gradually building an age model.

Meanwhile, I had to learn the pollen of Galapagos plants, to become fluent in the pollen vocabulary, as it were. My goal was to know the pollen of all 600 species at least to genus. First build a reference collection. This must consist of all the Galapagos pollen types, mounted on microscope slides. NSF allowed me money to hire one assistant. I was lucky almost beyond belief in the assistant fate sent to me. Eileen Schofield was then assistant to the director of the Arnold Arboretum at Harvard, perfectly preadapted to the job by aptitude, work experience, and skill. She also was known in the great herbaria from which we would have to obtain pollen of plants not in my Galapagos field collections. Eileen did it all, assembled the pollen collection and made the slides. Even so, it took me a year, working with Eileen, to learn all the pollen by sight and to write a dichotomous key to identify the fossil pollen grains.

Two whole years, 1967–1968, were eaten up by the early work: equipping the laboratory, opening all the cores, radiocarbon dating, my experimental use of an industrial x-ray machine for making a permanent record of stratigraphy, learning pollen, and the rest, despite student help — longer than it took to plan, fund, and run the expedition.

El Junco was the big prize, and from it I set out to write the local climate history for a complete glacial cycle. The story must account for the most unusual stratigraphy revealed by the cores when taken together. The basin was an explosion crater at the very top of a mountain, looking like the crater of a very large World War II bomb. But it had been filled in for at least half its depth by mud and water. On the bottom a thin layer of red clay, in one core at least thicker (to some 3 m). This layer had apparently smoothed out the very irregular crater floor. On top of this more than half a meter of the organic lake mud known to the trade by its Swedish name of *gyttja* (I can't pronounce that either). This was rich in pollen and other microfossils, irrefutable evidence of standing water, a pond rather than a lake, because we found it in only one core, showing that the open water did not cover the whole crater floor. We burned most of this precious re-source trying to date it, finding it to be too old for the radiocarbon method, then limited to 30,000 years, although one dating laboratory dared to give us the result of >45,000 years. That the pond existed very early in ice-age time, or even before the last ice age began, was certain.

This layer of organic gyttja was buried by more red clay, some 7 meters of it in fact, before there was any more gyttja. This upper gyttja was the sediment of the modern lake, just three meters of it below water six meters deep at the time of our visit. Where the red clay joined this upper gyttja was easy to date even by 1960s technology, yielding an age of 10,300 BP. Near twenty more dates throughout through the surface gyttja on our whole suite of cores confirmed that an open-water lake, albeit with fluctuating water level, had occupied the crater for all the last ten thousand years of postglacial time.

A first-order conclusion seemed inescapable from this stratigraphic and chronological history. A gyttja deposit means interglacial times like the present, but red clay means glacial times. On this logic, the pond of more than 45,000 BP formed in a nonglacial period like the present, the best known being the last full interglacial, an interval now well dated round the world at about 110,000 BP, and the red clay was a deposit of glacial times. On this logic, I had a record of an en-tire glaciation: as Nero Wolfe[2] would say, "satisfactory."

Yet before the cheering could begin, I had to explain from whence the red clay had come. What does red clay in El Junco tell us about local climate? The knee-jerk response of many a geologist to a deposit like this is, "Slump." I still sometimes encounter that response. Were it a "slump," it would be useless for climate reconstruction, just an instantaneous event triggered by earthquake or storm, nothing but a jumble that came down in a landslide.

The red clay was not a slump. My first answer to those who think that way is, "Show me where the slump came from." El Junco is the top of the mountain; the only material higher than the crater floor is the gently sloping inner wall of the crater itself, now restricted to a narrow and irregular rim rising perhaps 20–30 m above the lake.

A second answer is based on the mineralogy of the clay. "Clay" is a name with many meanings. The simplest describes little more than the size of the particles, these being very small; we talk of the clay "size-fraction." The El Junco red clay certainly fits this definition. But many a clay so defined has a characteristic min-eral structure, being synthesized from oxides of iron and aluminum arranged in complex lattices; definite, identifiable minerals in their own right with names like montmorillonite, illite, and goethite. These have well-known physical prop-erties like the water-absorbing powers of montmorillonite that give many a "clay"

its notorious stickiness and boot-pulling powers. These clays with lattice structures are formed each in its specific environment, usually requiring plenty of water. Clay minerals are routinely identified by X-ray diffraction; El Junco's red deposit had few latticed mineral clays.

Instead X-ray diffraction showed the red matrix to be of shapeless small particles, in the clay size-fraction certainly, but not true clay minerals. In the trade
these are called amorphous clays. X-ray diffraction also identified concretions of
the iron minerals gibbsite and haematite. To find this mixture under an existing
lake suggests very different conditions when the clay was formed, the most
straightforward explanation being weathering of the volcanic debris of the crater
walls exposed to the air and to moisture both.

The red color itself is due to a coating of iron oxide that has stained the amorphous particles the color of rust. Thus the lake sediment sequence of brown
gyttja, red clay, brown gyttja can be understood as lake, empty crater, lake. This
translates into rain, drought, rain. More importantly, the reconstruction says that
glacial times were dry times, drier even than now, not only on the Galapagos, but
also for the whole eastern Pacific Ocean, whose climate the Galapagos Islands
share.

This hypothesis does not exclude moisture all together in glacial times, only
sufficient actual rain to keep the El Junco basin under water. Moisture continued to be necessary for the production of red clay from the volcanic material of
the crater rim. The enveloping cloud should be enough; equatorial sun, wind,
and fog should do the trick. When the rains failed at the beginning of glacial
times, wind, and perhaps rare showers of rain, would be enough to wash the resulting fine debris down the crater walls, eventually to accumulate seven meters
thick before the more regular rains of the postglacial period flooded the basin for
the second time. This history was confirmed by a study of thin sections of the
clay by W. L. Kubiena.

Because of a book he wrote, Kubiena was a hero of my undergraduate youth.
The book's title was *The Soils of Europe*, first published in Spanish in Madrid in
1950, with an English translation in 1953.[3] I carried my copy across France and
Spain in 1954 on my way to a summer stint surveying soils in the Minho of Portugal. Kubiena's account of the ground underfoot was like no other; a classification and understanding of what one could see in road cuts or holes in the
ground, all illustrated with splendid watercolor paintings far more evocative
than any color photograph. I had never met the author, nor corresponded with
him. His book was enough for hero worship.

Professor Kubiena walked into my office in Columbus one day. Some international arrangement in the soils world had brought him to visit Ohio State,

with its huge agriculture school, where rumors of the Galapagos work diverted him to my office. My memory of that one meeting is of the classic image of a venerable scientist, complete with shock of white hair, but perhaps memory plays tricks after more than thirty years. We looked at cores of the El Junco red deposit together. He took samples away with him.

Kubiena's report came in the form of a letter, handwritten in the elegant calligraphy of his generation. I keep it folded into my copy of *The Soils of Europe*. The deposits are described as, "a very dense clayey, originally yellow-coloured sediment stained by peptized amorphous iron hydroxide." Further, the clay-sized fragments were the products of weathering of basaltic rocks or tephra because magnetic grains had been well preserved in the fabric. Finally, flow lines were revealed in the thin sections, confirming that the weathered material had been worked onto the basin floor in surface water.

The origin of the red clay now seemed clear. It was the product of sub-areal weathering of the crater rim, supplemented, as Kubiena suggested, with similar weathering of wind-blown volcanic ash (tephra). There is no shortage of volcanic ash on the Galapagos, the place being built out of volcanoes. Over the many thousands of years during which the deposit accumulated, occasional rain showers had worked the finer parts of the weathered material down onto the crater floor, but the rain had never been sufficient to form a lake. Until, that is, climate change sent more rain at the end of the ice-age ten thousand years ago.

My conclusion was, I claim, inevitable. The environment of El Junco mountain, the high point of the island of San Cristóbal, was even more arid in the ice age than now, a desert essentially without rain most of the time. But did this history apply to the whole island, implying climate change, or could it be dismissed as just a freak result on a high mountain because of its relative change of elevation above the sea?

Sea level was down a hundred and twenty meters at the glacial maximum. This was true of every coast on earth, as ice sheets grew at the expense of oceans. Only in rare places, where the earth's crust sank for some reason, could the relative level of the sea not fall by this huge amount, and the Galapagos was not one of those places. From the viewpoint of El Junco the mountain and the lake were 120 m higher in the ice age. It could at least be argued that El Junco was raised so far above the sea that it projected above the clouds, losing much of its moisture and all of its foggy "lid." Then the drying would apply only to this one lake, and the drying of El Junco was explained away.

I thought then, and think now, that this argument had worms. But evidence that the drying was in fact widespread was at hand in the pollen and spore evidence from the two lake episodes, from before and after the ice age.

The pollen spectra, that is the relative abundance of different pollen types, was quite different in the gyttja of the two lake episodes. That in itself was not surprising (at least to me). To someone convinced that plants do not live in long-lasting communities or associations, the experimental colonization of the same place twice, from the same pool of species, should not produce the same result. Thus different pollen spectra before and after the ice age were entirely consistent with the idea that the drought had wreaked its damage far and wide, not just on El Junco mountain.

But the really striking pollen datum was from a single pollen type: the aquatic flowering plant *Myriophyllum* (water milfoil). This is one of the most familiar plants of ponds and the edges of lakes in cooler climates the world over. It grows rooted in the bottom mud, with its leaves also under water, not floating as do water lily leaves. Only its small flowers poke into the air to seek the service of bees. Water milfoil powers of dispersal are truly remarkable because it can live nowhere except underwater in a suitable pond and must travel through the air from one pond to another. *Myriophyllum* performs this feat by hitching rides on the feet or feathers of ducks. The results show that it is superb as a hitchhiker.

Myriophyllum is also a pollen analyst's friend, having a beautifully distinctive pollen grain that we all know well. When we start work on pollen from a new lake and a grain of *Myriophyllum* slides into the disk of light on our microscope stage, we think, "Aha!" with a satisfied familiarity. And there were *Myriophyllum* pollen grains in the mud of the ancient El Junco pond. Aha! Entirely reasonable to find them there. The pond was an ideal habitat; the plant is common in the Andes Mountains of Ecuador; waterbirds fly in from the mainland; and the endemic Galapagos ducks peregrinate round the islands for subsequent transport to every pond in the archipelago.

An interesting pollen discovery all the same because *Myriophyllum* is not known from the modern Galapagos Islands; never reported, not listed in the official flora, certainly not growing in modern El Junco Lake. Nor is there *Myriophyllum* pollen in the upper gyttja from the 10,000 years of postglacial time. The pollen was all in the lower gyttja from before the glaciation. That early population of water milfoil had been made extinct by the ice-age drought. If it had survived anywhere in the archipelago, surely the Galapagos ducks would have carried a sample to modern El Junco during the ten thousand years available. But they did not. None had survived the long drought anywhere on the islands.

Even more compelling is the story of the water fern *Azolla*, a plant found widely on suitable ponds in the wet Galapagos uplands wherever these exist; having been collected from Isabela, San Cristóbal, Santiago, Santa Cruz, and Floreana. All these collections are plants of a single species, *Azolla filiculoides*.

Being ferns, they of course do not produce pollen, but they compensate by leaving even more distinct fossils in lake mud. Their tiny trilete spores are abundant in lake mud under the pinkish floating canopy, and, what is more important, so are their huge (i.e., half a millimeter long) spore cases, called massulae. These are packed with the tiny spores and equipped with long hooks as an aid to transport on duck feathers. The massulae are absolutely distinctive, not only of *Azolla*, but even of the individual *Azolla* species. For this accuracy they are better than pollen. A quick first look at our pollen slides from both sets of El Junco mud revealed plenty of *Azolla* spores and massulae from both layers of gyttja, both the ancient and the postglacial. Fair enough. For a plant so easily dispersed from pond to pond; it had reached the archipelago early and had henceforth been ubiquitous in Galapagos ponds.

That might have been the end of the Galapagos *Azolla* story were it not for Eileen Schofield. She developed a personal interest in the Galapagos *Azolla*, an interest typical of the fascination with an organism or group that marks so many good naturalists. We agreed that she should add to her work an *Azolla* history by counting the massulae in our pollen preparations.

The day Eileen Schofield made the discovery was one of my big lecture days. I remember climbing the stairs to our laboratory and office complex rather ploddingly, in the tired state typical of the aftermath of a successful lecture to two or three hundred. Eileen was standing at our door, waiting for my return. She looked for an instant as if she were hovering a foot clear of the passage floor, suspended in a forcefield of excitement. "It is a different species," she said.

The *Azolla* of the ancient El Junco pond was not the *Azolla* now living on the islands or which made its home in El Junco for all of the last 10,000 years. The fossil massulae made this quite clear. Moreover, Eileen could name it from the structure of the massulae. All the fossils from the before-the-ice-age El Junco were from *Azolla microphylla*. There was no trace of the modern inhabitant, *Azolla filiculoides*, before the start of the postglacial.

Both these *Azolla* species are known from modern South America, though not from the same local region and never from the same lake. An ecologist should not in fact expect them to live in the same lake; their ways of life as a sort of tropical duckweed are just too similar. If two species reached a pond, we should expect a winner-take-all competition, with only one survivor. If one species already occupied a pond, the chance of a second species invading should be vestigial to nonexistent.[4]

Both A. *filiculoides* and A. *microphylla*, therefore, are available in South America as feedstock for the Galapagos Islands. Which one gets there first is decided by the chance of from which pond a migrating duck took off for its long flight. But when that duck arrived at a virgin Galapagos pond, the *Azolla* it carried should quickly have grown, expanded its numbers, and filled the pond. The plants are known to be able to do this very quickly, just one or two years from inoculum to complete coverage of a pond. And a few years later the endemic Galapagos ducks should have spread the *Azolla* to all the other ponds of the archipelago. The whole place would then be closed to fresh invasions of other *Azolla* species; no more species need apply.

Eileen Schofield's discovery is the strongest evidence that the great ice-age drought on El Junco was felt all over the archipelago, not just on this one high mountain. The drought had exterminated all the populations of *Azolla microphylla* that once lived throughout the archipelago, clearing the way for the invasion of a second species, *Azolla filiculoides*, when the rains returned at the end of the ice age.[5]

At the same time, the ice-age drought cannot have been so apocalyptic as to have wiped out a large proportion of Galapagos plants and animals. It was just the few species that needed year-round standing water that were hit so badly. The strongest argument for this might seem to be the sheer scale of Galapagos speciation: all those unique species, or even unique genera, of tortoises, finches, cacti, *Scalesia* trees and the rest cannot have been the product of the mere ten thousand years of postglacial wetness: they had to have been in the Galapagos far, far longer than that.

A powerful argument, certainly, but not by itself decisive. Most of these animals and plants live in the dry coastal regions. The cacti are preeminently adapted to drought; the tortoises, land iguanas, and even Darwin's finches can eat cactus or cactus seeds. Those that do not get their living from desert plants, the penguins, flightless cormorants, and marine iguanas, get their living from the sea.

These unique animals and plants of desert Galapagos lowlands are in fact good evidence for the prolonged existence of desert coastal climates. If the ice-age droughts had spread the reach of the deserts up the lower slopes of the larger islands at the same time that the glacial fall in sea level extended the area of dry land round the islands, then the opportunities for life and speciation in this desert natural history should have been greatest in an ice age. This is consistent with the observation that the greatest diversity of both plants and animals is in the desert coastal belts, not in the wet highlands. Not only has the coastal environment been stable for longer, but it has also occupied a larger area for most of Galapagos history.

To understand the climatic change that dried up El Junco lake and wiped out the ancient *Myriophyllum* and *Azolla* populations from Galapagos ponds we need to know what, if any, of the forest life of the interior survived. This is one area where a pollen analyst should be expected to help, the very possibility that drove me to the Galapagos in the first place. Yet I had failed. I did have ice-age mud; the red clay of El Junco; but there was no pollen in it.

All was not quite lost, however. I did have pollen and spores from the ancient wet period before the ice age, from the gyttja of that earlier lake in the El Junco basin. And in that mud were copious spores of the Galapagos tree fern *Cyathea weatherbyana*. This tree fern lives only in the clouded uplands of Santa Cruz and San Cristóbal. The characteristic view is of a crouching plant shrouded in mist, its huge fronds glistening with water. On Santa Cruz it lives beside the peat bogs; on San Cristóbal inside the El Junco crater itself. A rich deposit of *Cyathea* spores from before the ice-age drying is powerful evidence that these tree ferns survived the subsequent ice age in places foggy and damp, even as the rains failed to replenish the ponds.

The sum of the evidence now seemed to me to be clear. In ice-age time the rains failed on the Galapagos, leaving only cloud drip and drizzle to keep moist the higher parts. Nice to know for those studying Galapagos evolution and diversity, but even more critical, as it turned out, for those studying Pacific climate change. The climate of the Galapagos is also the climate of the eastern Pacific Ocean.

El Junco serves as a permanent rain gauge in the sea. Cores from the sea floor can yield detailed evidence of past water temperatures and have been a principal tool for the reconstruction of ice-age climate on a global scale. But deep-sea cores give precious little evidence for rainfall, as important a part of climate as temperature itself. And El Junco is a rain gauge providentially placed in an ocean where rainfall really matters, in the path of El Niño rains. My conclusion that El Junco held no water in glacial times can be taken to say that El Niño, as now understood, rarely or never came.

El Niño is now a term widely understood by the general public, almost a part of the vernacular: dark doings in the Pacific; changes in ocean currents; warm oceans off the coasts of Peru and Ecuador that kill cold-loving fish important to commerce; above all devastating rain and flooding on desert coasts. But forty years ago the term was much less known and much less used by climatologists. All those fearful events were called "crashes" or at best "El Niño crashes." And the connection of these events with events at the Australian side of the Pacific, now called the "southern oscillation," was little more than an embryonic hypothesis.

"El Niño" in those days still meant what it had meant ever since the Spanish conquest. It celebrated the light annual rains in coastal Ecuador and Peru that came every year from January to March, rains vital to agriculture on those dry coasts. This annual blessing followed closely after the 25th December festival of the birth of the Christ child, *El Niño.*

This El Niño was well understood by climatology in the 1960s. It was the result of the annual movement of the Inter-tropical Convergence Zone (ITCZ) southwards from its position north of the geographic equator for the rest of the year. The ITCZ is the wet place where the trade winds meet over the oceans, where warm air rises. When the ITCZ sweeps south to the Galapagos, it breaks the inversion and, by displacing the reach of the south-east trade winds, allows the ocean surface a modest warming. The blessed El Niño rains are the result.

To drain El Junco permanently, I had to stop it raining. Very well, abolish these annual rains. How? — prevent the movement of the ITCZ.

I finally sent these results off to *Nature* in 1972 as, in that journal's quaint nineteenth-century usage, a "Letter to the Editor," what we in the trade call a "note to *Nature.*" It had an elaborate title, taking up two lines of type. The editor changed the title to "Climate and the Galapagos Islands" and published it as a lead article.[6]

The "note to *Nature*" summarized the El Junco record, developed the argument for an arid ice-age climate, then it plunged into the proposed mechanism of displacing the ITCZ. Put bluntly, this model said that in the ice age the cold oceans conspired with southeast trade winds to maintain the drought-causing inversion year round, while the ITCZ remained north of the geographic equator in the winter months as well as the rest of the year. This model was parsimonious. It explained the El Junco history as I saw it. And it was naïvely wrong. The celebrated climatologist Reginald Newell pointed out the error a year later (1973) in another "note to *Nature.*"[7] In glacial times the ITCZ could not possibly have been held north of the geographical equator as I had speculated. Much more likely was the possibility that the huge ice sheets coming out of the north should have displaced climate belts southwards, driving the ITCZ far south of its present position year round.

Yes, of course! I was embarrassed. But the convergence would have had to be pressed really far to the south to be well clear of the Galapagos or the islands would have had rain, not drought. Apart from some quibbles, that is where the matter stood for the next thirty years: evidence that the Galapagos were dry in ice-age times accepted, mechanism in doubt.

Then came a thunder out of climatology based on the understanding of that other, very different, El Niño, the one known in the 1960s as a "crash event." Time to look again at the El Junco rain gauge in the sea. It was this that brought

Julian Sachs and his team, then at MIT,[8] and the Institute for the Study of Planet Earth at the University of Arizona under Jonathan Overpeck to suggest that we go back to the Galapagos, drill again, and look again. The stakes depending on another look were large.

"Crash events" were by now fully understood and explained. They are the result of an internal dynamic of the Pacific Ocean system called the "Southern Oscillation," the engine of the system being heating the equatorial ocean in the western Pacific off Indonesia and Australia. Every few years the accumulated heat is such that warm surface currents flow eastward, following the equatorial line all the way to the South American coast, everywhere displacing cold surface water with warm, breaking down climates that have cold air masses under inversions, and bringing rain. The climatic change is at its most virulent at the end of the ocean journey, as the flood of warm water disrupts even the upwellings that chill the southeast trades, thus breaking the inversion that keeps the Galapagos and coastal Peru arid.

No one in his right mind would call the catastrophic floods and the deaths of great fisheries that afflict the coastal people of Ecuador and Peru a blessing brought by the Christ child, El Niño. Nor do they. But both the annual descent of the ITCZ (cause of the real El Niño) and the intermittent arrival of warm currents from the west have in common that they bring rain, and they typically arrive at about the same time of year. The name came from a mild blessing given by the ITCZ. But good news is never as interesting as bad; climatologists effected a sleight of hand to transfer the pretty name to the disaster. For better or for worse, the disaster is now what is meant by "El Niño," at least to English-speaking climatologists.[9] (A less objectionable variant is El Niño Southern Oscillation [ENSO]. This at least separates the crash from the real thing.)

El Niño, as now understood, brings the Galapagos end of my long equatorial transect closer to the Amazon at the other end. It is now known that ENSO influences weather clear across the continent, increasing rain in some places, reducing it in others. East of the Andes the effects are not catastrophic as on the Pacific coast but they can be detected at long range. El Junco might well have something to say about the climate of the ice-age Amazon, but as yet I do not know what it is.

Recall that my reconstruction of ice-age El Junco was "no standing water." There was never enough rain to make a lake that would precipitate lake mud. Instead there was cloud moisture sufficient to permit the weathering of the crater rim to red clay which could be worked into the bottom of the basin. Kubiena's thin sections showed flow lines in the clay, suggesting occasional showers to help distribute the clay, but that was all the evidence for rain we could find. Hardly El Niño rains as now understood. If asked, I should have declared that catastrophic ENSOs were absent or rare in glacial times.

This conclusion of no glacial-age ENSOs is in sharp disagreement with conclusions of many of those reconstructing Pacific Ocean history from deep seacores whose data for ice-age sea surface temperatures are consistent with a model of Pacific climate stuck in an almost permanent ENSO condition. The Galapagos should have been wet! This is the model that led Julian Sachs and his team to find the El Junco data in a thirty-year-old issue of *Nature* and, as was only proper, to question my old conclusions.

If, as the sea surface temperature reconstructions across the ocean suggested, glacial times saw a nearly permanent ENSO, then El Junco should always have held a lake. How could I have missed finding evidence of this in the red clay? Indeed, how could the red clay have formed in such a lake with no trace of the gyttja deposits and no microfossils or organic carbon detectable by my primitive methods of long ago? Rienk Smittenberg of the Sachs team tried to find answers.

Suppose rain was so heavy and continuous that the crater was always full to the brim, in a constant state of overflow; quite possible because the crater rim is notched some meters above present water level and below the notch is the clearly defined channel of an old stream outside the crater and down the mountain. Imagine that the constant influx of water is cold (rain is cold) and sufficient to flush out the small lake continuously. This just might explain the absence of microfossils or gyttja in the sediments. Then find a mineralogical explanation in which the lava of the crater rim could be broken down to clay-sized particles under cold water saturated with oxygen (theoretically possible, so they tell me). Rather a heavy train of suppositions, but it gets to the desired end point of explaining the data as being possible in a rainy climate, an alternative hypothesis to my hypothesis of aridity.

Not a convincing hypothesis, on the grounds that it lacks parsimony or precedent in lake studies. Nor does it explain the microfossil evidence for the extinction of *Myriophyllum* or the replacement of one species of *Azolla* by another. However, the hypothesis has become testable in ways that were almost beyond imagination thirty years ago. The presence of organic carbon, algae, and other aquatic life forms can be detected by finding and extracting trace amounts of lipids and other biomarkers. Furthermore, these biomarkers can be used to measure such things as deuterium-hydrogen ratios, from which such useful knowledge as evaporation-precipitation ratios can be deduced. So what was for me and my team 35 years ago a purely mineral mass with no fossil traces has become an organic hunting ground for a new generation with equipment beyond our imagination.

After several sessions in the library of my Woods Hole house, Sachs, Smittenberg, and I were in complete agreement that fresh cores were needed to resolve the matter by modern analysis. We got them. In September 2004 we went together to El Junco, where my old notes guided us to the spot in the lake where that critical core B, with the ancient lake mud, was found buried under the red clay. Muscle as young as mine had been forty years earlier drove the Colinvaux-Vohnut version of the Livingstone corer, with its locking pistons, to secure two beautiful and complete sections of the ancient buried gyttja, as well as all that lay above it.

These are early days in analyses that will take years. But already we have an age model of the top 10,000 years of modern gyttja, based on accelerator mass spectrometry (AMS) dates of small discrete samples, an age model gratifyingly similar to that which I concocted with the crude methods of so long ago. Whatever the answer, rain or drought on ice-age El Junco, cannot fail to be exciting. Galapagos, Andes, and Amazon, all are involved, to say nothing of a big pond called the eastern Pacific Ocean, and models of global climate change.

One thing was certain from the old El Junco study and will remain certain if the new work revises my history to allow a wetter past (of which I am still skeptical): it confirmed that significant climate change at the equator in glacial times was real and could be measured. 'Twas well that I had taken Ed Deevey's advice of "Go south young man."

A bizarre consequence of El Junco, though, was that I had found drought at the American equator just when Jürgen Haffer was publishing his speculation that the Amazon was semi-arid with the great forest fragmented into refugia, of which, in its later guise of a paradigm, I was destined to be an agent of destruction. But because of El Junco, I was predisposed to believe in Haffer's claims for drought in the great forest. Thus I began my search for ancient sediments on the equatorial mainland with the possibility of finding ancient savannas real in my mind.

We also cored the Genovesa Crater Lake. This was the deep collapsed magma chamber on tiny Genovesa Island (Tower) where we had been obliged to rope down its near vertical walls and from which we had won complete cores to bedrock, in five one-meter sections, by coring night and day of a 36 hour operation. The sediments were wonderfully banded down their whole lengths, potentially offering the tremendous plum of a detailed weather record with decadal, or even annual, resolution. But both our primitive technology and circumstance conspired to delay the picking of that plum. Twenty-centimeter samples, including long rows of bands, were needed for radiocarbon. And when we did get dates they were not in sequence. We got 4,000 year dates under 6,000 year dates over more 6,000 year dates. The intact banding was as perfect a demonstration as you could wish that there had been no mixing of sediments. Thus there was something funny with the source of the carbon, almost certainly due to volcanic gasses, this was, after all, the entrance to the magma chamber of a recently active volcano.

We had two independent shots at the age of the record. The first was from knowing that the lake surface was at sea level. The water was seawater, with lake level shown to be responsive to the ocean tidal rhythm, though out of phase. This meant that the earliest that the basin could have filled with water was when the low sea level of ice-age times had come back to its modern level, possibly as early as 8,000 years ago. The second age estimate came from a pollen analysis done by Eileen Schofield. Pollen were sparse, both in number and kinds, but Eileen, with immense labor, managed to count just enough to make the result statistically respectable. The pollen diagram was in two parts, clearly differentiated, an upper and a lower. We were not able to say with any certainty what climatic event caused this change but there was one pollen event of comparable magnitude in the El Junco core, nicely dated at 3,000 BP. At Genovesa the pollen event was halfway down our record, so call the bottom 6,000 BP. We concluded we had between 6,000 and 8,000 years of finely banded sediment. No use at all for the ice-age history, and beyond the reach of our technology to reconstruct a detailed history of the last several millennia of eastern Pacific weather. For this the dating problem must be solved.

There was just one captivating thought for an ecologist, if not for a climatologist. Round this lake nested a giant colony of seabirds, mostly *Sula sula*, the red-footed booby, of which there were said to be a quarter of a million breeding pairs, nearly the entire world population. Their guano, dumped directly into the lake or onto the porous lava of the island, ought to be adding immense quantities of phosphorus to the lake system and hence to the sediment. Could we use sediment phosphorus to measure the size of the population of red-footed boobies for many thousands of years? If so, this would be by far the longest vertebrate history on record.

Dan Goodman took it on as a Ph.D. thesis project. He was a Woodrow Wilson fellow at Ohio State, with skills in both mathematics and chemistry far beyond those of his new major professor. Dan measured throughout the core all the chemical components that could be measured in small samples by the technology then available to us, including X-ray fluorescence spectroscopy for calcium, iron, potassium, silicon, phosphorus, aluminum, and magnesium, adding to these nitrogen by the classic kjeldahl method; and measures of three organic pigments. From this massive data set it was possible to demonstrate that phosphorus was present in far greater excess than could be accounted for if its origin was, like that of the rest of the list, from the parent seawater or solution from volcanic rock. Calculation of phosphorus production on the island by a quarter of a million breeding pairs of boobies showed the source to be entirely adequate to account for the excess phosphorus input to the lake. The final token that we had a record of the booby population came from comparing Lake Genovesa chemistry with that of Beagle Crater lake.

Beagle was the crater on Isabela Island to which Darwin had struggled in hope of a drink, only to find the water "salt as brine," and to which we had struggled in hope of a long core, only to find the sediment less than a meter thick. The beauty of comparing the two lakes was that both had the same salinity of evaporated seawater, but Beagle had no breeding seabirds in its watershed. It was a natural model of what the Genovesa lake chemistry should have been without the boobies. It both gave us confidence in assigning the excess phosphorus in Genovesa to guano input and allowed a crude translation of phosphorus content into bird numbers. Dan had an estimate for percent of phosphorus with zero birds breeding (Beagle) and for 250,000 pairs breeding (Genovesa). Using simple proportions, he could plot bird numbers throughout the long history of the Genovesa lake. This calculation described the population as remarkably constant, a finding of great theoretical interest. Unfortunately, though, there was, and is, a catch.

The data were all percentages, which is to say the amounts were all relative. To convert them into actual numbers (grams present or bird numbers) it is necessary to know the rate of accumulation, ending up with grams of phosphorus per unit area per year or, our real object, birds per year. But we had no reliable radiocarbon dates, our only ages being a rough consensus on the date of sea-level rise and a possible pollen correlation with a dated event on another island. Not good enough. Dan Goodman has refrained from publishing this lovely, but provocative, record until the dating problem can be solved.[10]

New cores from El Junco were only part of the 2004 Galapagos agenda. New cores from Genovesa were the other part, in fact the part of the project which was of longer standing, with a genesis some fifteen years earlier in my long discussions with Jonathan Overpeck, when I was at Ohio State and he was at NOAA, about the dating problem. Do that, and we could resolve the cause of banding to test the hypothesis that these recorded the heavy rains of ENSO events.

We remain convinced that Genovesa holds the dream of a detailed history of El Niño several thousand years long, as well as my lesser dream of resolving the population history of red-footed boobies. If those bands are the rains, and if we could date them accurately, this long record just might approach the accuracy of the instrument record of modern times. That successful dating should also free the bird population record would be icing on the cake. Overpeck is now director of a new institute in Arizona whose mission is the history of climate change. His team was negotiating that forbidding cliff to core the Genovesa crater lake while the Sachs team was hammering at El Junco.

Overpeck had with him at Genovesa Michael Miller, who had been my essential companion on most of the Amazon and Andean work, to say nothing of side trips coring in Siberia and the Russian far north. With the Sachs party at El Junco had been Miriam Steinitz-Kannan, who had led every one of our expeditions on the mainland of her native Ecuador and had organized the whole of these latest Galapagos ventures for both the Overpeck and Sachs teams.

I should like to be able to conclude this chapter by reporting the results of Overpeck and Sachs for the climate history of the Galapagos, the eastern Pacific Ocean, and not a little for climate theory and climate change. But I cannot for it is far too soon. A reality of our trade is that a week in the field can mean years in the laboratory before your results are certain.

4

PLEISTOCENE REFUGE AND THE
ARID AMAZON HYPOTHESIS

The Pleistocene, literally "the last of the recent," is the top bit of the geological section, the bit spanning the million years or so before the short warm period in which we live. Like all geological periods, the Pleistocene is formally described by its included fossils, and begins with certain extinctions seen in an exposure in Italy. That is all the term "Pleistocene" really means. But we northerners think of the Pleistocene in terms of the terror that comes by ice. However correct our geological minds, "Pleistocene" also tacitly means "the epoch of ice ages."

Talking about the "Ice-age Amazon" is clearly a misnomer because the continental ice sheets reached only to near the 40th parallel in North America. But I do it all the time, "ice age" is more expressive than "Pleistocene." Whatever caused the climatic upheaval of the continental glaciers must have had some impact on those warmer climates farther south, or the obliteration of northern lands under ice must have itself caused far-reaching climate change, or both.

My mentor, Daniel Livingstone, had brought the impact of the ice ages south to the plateaus of East Africa, where it was manifest as drought. East African lakes had lost much or all of their water even as the north was afflicted by ice. The vector of Pleistocene climate change in Africa, therefore, was aridity synchronous with the advance of ice sheets in the north. Thus the idea of aridity as a key property of the Pleistocene tropics became plausible and popular.[1]

Meanwhile, our reconstruction of the climate history of El Junco was suggesting parallel drought on the Galapagos Islands. Moreover, I was toying with movements of the intertropical convergence zone (ITCZ) as driven by glacial events to account for the presumed failures of the Christmas rains over the Galapagos (chapter 3). This fitted the African model nicely; the islands too, in my model, suffered from Pleistocene tropical aridity.

This interpretation of the Galapagos record brought a history of aridity right to the American equator, only an-Andes-and-a-strip-of-sea away from the Amazon basin itself. Dan Livingstone had shown that East Africa was arid. Could the Amazon have been arid too? From an intellectual point of view that would be shockingly nice. In the Columbus laboratory I mused about it as the Galapagos history unfolded.

I remember imagining an Amazon basin that was only forested in the foothills of the Andes, while the bulk of the basin, like East Africa, held nothing but savannas stretching eastward to the sea. My mind's eye even had savanna on the drained continental shelf to form a coastal plainsland fit for humans to live on at the edge of the Atlantic Ocean, a sort of tropical Bering land bridge. How we are driven by our own experience! My imagined Atlantic coastal region was something like the forest-savanna borderlands of ancient Africa postulated by anthropologists as the homeland of the first humans. Mere musings of course, and embarrassingly far-fetched. The Galapagos and the facing continental coast were in a quite different part of the climate system to the Amazon basin across the mountains. The Amazon got its weather from the Atlantic. A shocker kept for dinner party conversation, O.K., otherwise forget it until we have pollen records from the Amazon basin to say "yea" or "nay."

Then all such musings were made redundant. In 1969 there appeared in the journal *Science* a paper by one Jürgen Haffer proposing a far more radical form of an arid Amazon hypothesis.[2] Haffer had no knowledge of the Galapagos results, which were not published until 1972, arguing instead from the African data that tropical aridity was plausible for the whole tropical belt, including the Amazon. Perhaps so; perhaps not. What made this suggestion in Haffer's paper respectable was that he was able to offer evidence from the Amazon basin itself for his arid ice-age Amazon, biogeographic evidence.

Bird distributions first prompted Haffer's hypothesis. He collected birds by shooting them until, as a German colleague tells me, his collection numbers thousands of skins. His key observation was that there existed, within the great lowland forests of the Amazon, local areas with their own, unique species or subspecies of birds.

Local bird species confined to a limited area in the forest is curious. Birds can fly; of all Amazon creatures, we intuitively expect to find the same birds everywhere. The forest is one forest, right? Even though more than two thousand miles across: we know this because it says so in the family atlas, where the whole thing is colored green. The climate everywhere is wet and warm, with gradients of rainfall to be sure, but gradual gradients. The more common trees can be found from one side of the forest to the other. That a bird species should occupy just one local patch of this great forest, never to be found in the rest, seems decidedly odd.

Odder still was that these forest patches usually had not just one but many unique bird species living there. Haffer could draw maps with smallish areas (by Amazon standards smallish; some were as big as Rhode Island or even Ireland) where a distinct and unique suite of bird species could be found. Not all kinds of birds; the Amazon has very many bird species that live over huge areas of forest. The larger and more conspicuous birds are as widespread as common sense says they should be. But in some select groups of small forest birds the patterns of local "endemism" as argued by Haffer seemed to be real, as if these animals had carved little nation states out of the great forest.[3]

Haffer looked at his discovery with a clear evolutionary perspective, asking how this peculiar pattern had evolved by natural selection of individual species of birds. His problem had two main components. The first and greater of these was *how had these unique species evolved, each in just one part of the forest?* The second and easier question was *how, once evolved, had each been confined to its original region?*

Ecologists have no difficulty answering the second question. We are familiar elsewhere with species distributions that abut with minimal overlap. Such lines of separation are particularly familiar where environmental gradients are steep, on mountainsides, for instance, or the banks of great rivers, but they can also be set within long gentle gradients in geography or climate. Species with similar needs (that is, which occupy similar niches) meet along the environmental gradient where the subtle environmental change tips success in the delicate balance of competition from one species to the other.

In the Amazon, with its near incredible number of species, competition for just about any possible resource must be well nigh certain. If, therefore, some peculiar local circumstance equipped a place with unique species and if there were subtle gradients in the supply of critical resources between that place and neighboring regions, the local species might well be confined to their original locality.

It turns out that Haffer's places of bird endemism are mostly associated with what little high relief the Amazon basin offers, many on the flanks of the hills that form the rim of the giant basin. To the extent that local endemism reflects the shape or slope of the land in this way, the second question is answered. The endemic birds persist in their isolation because their ranges are set by competition along subtle environmental gradients within the forest.

The really brilliant part of Haffer's conception was to provide an answer to the first question, *how had the species evolved in the first place?* Haffer found his answer in the ice age, a most reasonable conclusion for a researcher trained and employed as a geologist. The blurb accompanying the 1969 *Science* paper described Haffer in these words: "The author is research geologist in the Field Research Laboratory of *Mobil Research and Development Corporation*, Dallas, Texas. He worked for several years in South America prior to his assignment in the United States." That was all. *Science* did not say that Haffer had been a field researcher in the Amazon and Andes for the Mobil Corporation for the best part of ten years while also being a first-class field ornithologist.[4] He had already published some dozen papers on South American birds, was well read in the fine points of evolutionary theory, and knew what we ecologists were arguing about even as we were ignorant of his geology. He was a master of the "modern synthesis" of evolutionary theory, with the discovery, or at least insistence, by Ernst Mayr that the physical separation of breeding populations was necessary before natural selection could effect the origin of new species, a process that came to be known as "vicariance."[5] "Research geologist for the Mobil oil company" seems a rather inadequate guide to this author.

The German oil man from Texas gave a geologist's answer to his bird problem. The Amazon basin had not always been covered by the great forest at which we marvel. Once upon a time the rains had failed so that the Amazon lowlands became the home of savanna instead of forest. The last remaining patches of forest in those times had covered only the flanks of the hills and selected bits of the lowlands that had remained moist during the great drought; perhaps where rising air on the flanks of hills released its rain in what we call *orographic rainfall.* In these isolated bits of forest, natural selection should inevitably work its magic as it does in any isolated community, on the islands of the Galapagos Archipelago for instance.

Haffer called his imagined islands of forest in a great sea of savanna "forest refuges." The regions of bird endemism that he circled on his maps were the heritage of these ancient refugia, in each of which his endemic bird species had evolved in isolation.

But when had this happened? And what was it that had cut up the forest into isolated patches? To a geologist the "when?" was easily answered: "Why, in the last great geological convulsion of course, in the Pleistocene, in the ice age!" And the "what did the isolating?" was equally obvious: Pleistocene tropical aridity did it. The rains failed over the Amazon basin just as they had failed in East Africa. Tropical rain forest could not stand failure of the rains and should have been replaced by something like savanna. Only where orographic rainfall watered hills and hillsides could forest survive in this scenario. And in each of these "forest refuges," natural selection worked to produce new and different species of birds, as the "modern synthesis" said it should. When the ice age ended and the rains returned, forest spread back over all the Amazon basin, to leave local collections of unique species as testimony to the days when forest survived only in a few local patches of moisture.

This was the Haffer hypothesis as set out in *Science* in 1969. The excitement to biologists was not that it explained those odd collections of endemic birds; we had never heard of these anyway in the years before Haffer. The excitement was that the hypothesis provided the missing ingredient for theories of Amazon diversity. If birds speciated in forest refuges, so should everything else. The process should have been repeated in synchrony with every northern ice age throughout the last million years or so of the Pleistocene.

What Haffer had invented was a species pump working with the rhythms of ice ages. With every glacial advance in the distant north, climate change confined the Amazon forest to refuge "islands" in a sea of impassable savanna; with every interglacial the forest patches merged into a giant forest. With the rains of each interglacial a few new species from the refuges would remain isolated, like Haffer's birds, but most would disperse, going to increase the total species number. The speciation engine needed to explain the spectacular diversity of the Amazon had been found.

Add to this engine of speciation our growing understanding of the Amazon capacity to hold huge numbers of species through intermediate disturbance and predatory cropping (chapter 1) and the diversity problem was solved. Haffer's engine pumped out species, and the capacious warehouse of the Amazon forest preserved them. It was good. Before long the Haffer refuge hypothesis received the highest accolade scientists can give to a hypothesis: it was said to be beautiful.

The prime appeal of the refuge hypothesis was that it provided geographic isolation of populations and what is more did so repeatedly. Prolonged isolation that prevents crossbreeding is the vital prerequisite for the making of a new species. Refuge theory gives us repeated isolations for stretches of the 80,000 year duration of a glaciation each time. There have been some twenty glacial periods in the Pleistocene.

Adding to the plausibility of Haffer's scenario was Haffer's use of the term "refuge." The idea of "refuges" (*refugia* to the classical minded) is hoary in biology. Find some tundra plants growing in central Ohio? This is a refugium of plants that lived there during the ice age and somehow kept going when all about them were overwhelmed by later invasions. Wonder where the Alaska spruce forests came from when we know that no real forests survived north of the ice sheets in glacial times? Must be from a forest refugium that existed some-where on the Bering Land Bridge, now inaccessible to our question because flooded by the Bering Sea. The simple anthropomorphic concept of refuge from terrible adversity has done much good (or, depending on your point of view, ma-lign) service for biogeographers. The profession was quite ready for Haffer's claim that all the wonderful complexity of the Amazon rain forests survived his postulated catastrophic ice-age droughts, again and again, in forest refuges.

Thus did the "refuge hypothesis" become a talking point whenever biogeog-raphers met, wrote, or taught. Disciples of refuge theory were to appear who would sift distribution maps of other animals, plants, or even ancient peoples in the Amazon, to see if they could be fitted to Haffer's maps of endemic Amazon birds.

Yet that unerring judge called hindsight can see legitimate reasons for disquiet in even the earliest statements of the refuge hypothesis. Did we really know the details of bird distributions in a partially explored forest larger than the whole of Europe? Haffer had shot a lot of birds, and those who came before him had shot their share, but was this really enough to preclude finding the local birds else-where? Much must be taken on trust. What about the size of his refuges? Could life in a block of land the size of Ireland, as some refuges were, really be treated as "isolated"?

And had Haffer really explained how his refuges had been rained on in the ice age when all the surrounding land was dry? Orographic rainfall? Perhaps. But he drew some refuges near the central lowland where it strains credibility to find orographic (i.e., mountainside) effects. And what of the evidence for aridity it-self? Was it reasonable to extrapolate from the East African plateau, a very differ-ent place today from the low-lying, much wetter Amazon basin? But so Haffer proposed. And he got away with it. His vision was to dominate Amazon studies for thirty years.

Haffer, the geologist, was obviously aware that what he was proposing should inevitably have left its print on the surface of the ground. In South American sci-entific literature he found reports consistent with a more arid continent at some time in the past. Brazilian soil scientists working in agricultural southeastern Brazil, for instance, had identified structures that they described as fossil *pedi-ments* in coastal districts from Rio Grande do Sul northwards to Minas Gerais.[6] In geology, a pediment is a block of old deposits left standing when erosion has cut down the surrounding landscape. They are familiar features of parts of the southwestern United States, where they form under modern arid climates of Ari-zona and neighboring states when rare heavy rainstorms cut gullies to leave blocks of more resistant soil or surface rock standing. The Brazilian claim is that no such pediments form in modern climates, suggesting they are fossils of a dry time, when rain was rare but heavy when it came. Even if climate change was indeed the cause of the Brazilian structures, their age was still not known and they were a continent away from the Amazon lowlands.

South America has known fields of fossil sand dunes, like the huge dunes of the Llanos in Venezuela and Colombia, said to have been seen by the great explorer Alexander von Humboldt long ago. But when these formed has never been determined. Haffer cited these dunes and the Brazilian pediments to suggest that they corroborated his contention that the land surrounding Amazonia bore evidence of dry times in the past. Offered as evidence in a court of law this "evidence" that the soaking wet Amazon was once arid land would not be enough to "hang a dawg," as the American legal colloquialism has it.

The power of Haffer's refuge hypothesis as originally stated, however, rested on evolutionary arguments, not on the hard evidence of geology. The evolutionary arguments were what enchanted a biological community needing to explain Amazon diversity. Crudely put, we were willing to accept the lack of hard geological data in the cause of a beautiful hypothesis that explained so much biology. Or else we assumed that more geological data were there in the Brazilian scientific literature that we did not read anyway.

For someone already committed to reconstructing the ice-age Amazon as I was, Haffer's refuge theory had bittersweet advantages. The bitter part was that he had got in first, and his ideas would thus set the terms of debate. The sweet was the stronger part, however, because the refuge hypothesis was also the arid Amazon hypothesis; and the prediction of aridity in the Amazon lowlands at the defined dates of the last northern glaciation was inherently testable by the time-machine methods of core drilling and pollen analysis.

Haffer's refuge hypothesis stated unequivocally that what is now wet tropical lowland subject to intense humid weathering was once a dry plainsland covered with savanna. Moreover, this great savanna was the dominant vegetation between 30,000 and 20,000 years ago, a time interval during which the northern glaciations had been shown by radiocarbon dating to have reached their maximum extent. This prediction of Haffer's hypothesis was eminently testable by the very methods of core drilling of lake sediments, pollen analysis, and radiocarbon dating that we were already set to employ along the whole American equator.

A license to go ahunting ancient lakes in the Amazon was inherent in Haffer's claims. That license would be cited in my grant proposals. What I had intended to do anyway could now be done in the guise of testing a great and popular hypothesis.

5

The Republic of the Equator

As the warship *Jambeli* brought us back from the Galapagos in 1966, we planned our first go at the mainland. We had a coring rig already in Ecuador, free of the costs of transport and of that peril by customs that haunts expeditions with heavy equipment. David Greegor, who had led the Santiago (James) Island detachment, needed a thesis topic; the history of an Andean lake should do nicely. I counted my pennies, staked him with the balance, and left him with a coring rig in Guayaquil.

Dave's mission was to core a lake, any lake, in the Andes; and somehow to get home with the results. I have a vague memory of pointing to the interior as the *Jamboli* docked and saying something like, "Over there are the Andes, go to it." I hope this is memory playing tricks and that I never said anything so corny, but such was our mood.

David Greegor did it. He found a truck for rent in some corner of Guayaquil, recruited an assistant, quizzed the locals, drove into the mountains, and found Laguna de Colta. The lake was inland of the first cordillera (Cordillera Occidental) on the inter-Andean plateau, 3,288 m (10,784 feet) above sea level. It was high above the tree line yet in the rain shadows of the cordilleras on both sides. The climate of the surrounding countryside is dry as a result, with most of the less than 400 mm of rain falling in just five months, the other seven months having almost no rain at all. A rural population grows the tasty, mealy potatoes of the high Andes and other hardy subsistence crops.[1] The natural vegetation is *paramo*, tussocky grassland that serves in lieu of tundra for the arctic-alpine of the high Andes.

Laguna de Colta is a lake of great beauty, two and a half kilometers long by one wide, occupying a naturally dammed valley between highish hills. The lake is shallow, so that Dave was able to core it in two places on his brief visit. But the cores were short, just over two meters long; moreover, only one meter was good lake mud, with the rest bottom peat and sand. Underneath that was the basement gravel.

The oldest radiocarbon date we could get was just sixteen hundred years. Dave's photographic record explained this distressing youth by showing the dam as a fresh-looking lava flow. Apparently the lava had poured out of a volcanic fissure to dam the valley as recently as two thousand years ago. People probably saw this happen.[2] We had had our first lesson that Andean landscapes, though grand in scale, can be distressingly young.

Laguna de Colta lies at 1° 45′ S latitude, making this lake core the first raised close to the equator on the American continent. David Greegor keeps that honor from the exploit. His pollen history showed that from the start the lake was surrounded, as now, by *paramo* and agriculture.[3] Dave went on to other biological topics in far away places, including Antarctica. He is now a professor in Nebraska.

Clearly my technique of "find a lake, any lake" had worms; or put with formal stuffiness, it "invoked a grave risk of expensive failure." I should have to do homework on the ground before launching coring teams at the American continent.

Meanwhile, I fell victim to a version of Parkinson's Law as the Galapagos research expanded to take on the ambitions of students and colleagues eager to work in the Galapagos.[4] I prefer to explain my conduct without reference to Parkinson by citing instead the military maxim, "Exploit success," for that is what I was doing, exploiting success. Call it what you will, the Galapagos would not let me go.

P. Dee Boersma came, having written for an interview. She opened the interview with, "I have decided to do a Ph.D. on Galapagos penguins and that you should be my advisor." Or words to that effect; my memory has surely massaged that sentence in thirty-five years, but that was the message; and so it was to be. My protestations that I knew nothing about penguins were to no avail. I was now in the penguin business. Boersma's research and adventures with the Galapagos penguins led to a seminal study and deserve a book of their own. She was to make herself an authority on penguins and now holds a named chair at the University of Washington.

Dwayne Maxwell was working for his doctorate under an assistant professor of marine biology, Llewellya Hillis, who also happened to be my wife. Marine thesis subjects were not exactly easy to come by in Ohio; unless you played pretend with Lake Erie (as many did). The Galapagos Islands had ocean galore, and the productivity of that ocean was vital to penguins and other Galapagos folk. As co-advisor to Maxwell, I was in the ocean productivity business.

Dan Goodman was all this time developing his history of the breeding population of red-footed boobies from phosphorus in the Genovesa cores (chapter 3). Thus three Ph.D. students doing pioneering work in the Galapagos but none working on my primary goal of the ice-age environments of the equator. Faculty colleagues wanted in too. Jerry Downhower, having finished a study on the mating habits of marmots in the Rockies, needed a new animal. The Galapagos? Hmm — Darwin's finches. I found myself in the Darwin's finch business. Charles Racine, a botanist, also needed a new organism. Soon I was in the giant cactus business.

All this Galapagos research led to theses or publications or both.[5] Hugely interesting, even exciting, it all was. The daily lunch meetings round my laboratory table in Columbus are among my more satisfactory memories. I was learning ecology at my lunch table even as I taught ecology to ever-larger undergraduate classes (400 students in my majors class one quarter). I wrote my textbook in the evenings after the children were put to bed.[6]

My grant proposals were now titled "Ecological Studies in the Galapagos Islands." Working with Eileen Schofield, I finished publishing the main Galapagos results,[7] which contributed nothing more to the grand design of exploring the ice-age equator and Amazon, the design that had brought me to the Galapagos in the first place. Haffer published his speculation in 1969 that the ice-age Amazon was arid (chapter 4). This event should have revitalized our attempt to drill for ancient sediments on the continental equator, for Haffer had done the enterprise a great service. He had laid down a hypothesis for us to test. I should have dropped everything else and gone for the jugular of an ice-age history of the continental equator right away. The Galapagos were to have been my gateway to the continent; instead I had let them seduce me from my purpose.

The 1960s ran their course, merging into the year of the riots and into the 1970s. The smell of tear gas was known in Columbus. Students were shot down at nearby Kent State University. My noon classes on environmental affairs were attended by more than a thousand students, packed into our largest auditorium. It amused me to think that I might be drawing away the angry crowd besieging the administration building so that the staff had a little peace at lunchtime. The administration did not recognize this service, but the students voted me a teaching prize. Then I took a sabbatical year on a Guggenheim fellowship, writing my book on an ecological model of human behavior, *The Fates of Nations*.[8]

The day of reckoning came when my third big Galapagos "ecological studies" proposal was rejected by NSF. The results for the laboratory were dire as I simply ran out of funds. By 1974 I could no longer "meet my payroll." Times like that are bad. But hindsight can find a silver lining in anything. The silver in this crisis was that it freed me to go back to my original purpose of hunting the ice-age Amazon; after, of course, the necessary months of trauma.

The first glint of silver came in 1975 in the form of another letter from a student asking for an interview for Galapagos research, but this student was from Ecuador and her interest was in lakes. The "Republic of the Equator," as Darwin had called Ecuador,[9] was coming to me.

Miriam Steinitz, a native of Quito, was completely bilingual in English and Spanish. The fascination that brought her to biology as a life's purpose was with diatoms; not the birds, butterflies, or beetles that catch the collector's instincts of so many of us: diatoms. Not many people have a passion for diatoms, but Miriam was, and is, smitten. She was to take the known species list for Ecuador from seven, when she started, to twelve hundred and counting in her present status as a Regents' Professor in Kentucky. It was a young version of Ruth Patrick, the doyenne of American diatomologists, who came to my office that day.[10]

Diatoms live in water, tiny photosynthetic plants floating in the plankton or lying on the bottom mud, mostly smaller than pollen grains. The highest magnification of a light microscope with an oil immersion lens is needed to tell one diatom from another, a good enough reason for the scarcity of diatom fans.

But diatoms are stunningly beautiful, in their way as beautiful as orchids and even more diverse. And, like pollen, they keep their shape and beauty when dead. Whereas pollen grains are preserved by an outer coat of near-indestructible carbon-chain compound, diatoms go one better. They live inside a fitted box of silica, the transparent, rock-hard chemical relative of glass. Their silica skeletons lie in lake mud alongside the pollen, an independent record of the past.

Think of a diatom as an elegant crystal box made in two halves, called in the trade *frustules*: one half slightly the smaller so that it can slide inside the other to close the box. A dead diatom can fall apart into its two halves, hence the name "two atoms," but one frustule is enough to name a diatom because it is the markings on the silica frustules themselves by which we classify them. Almost unique among the living things of the earth, diatoms flaunt their identity with a near-indestructible mineral dog tag that they carry in full view. We can measure the diversity of diatom populations by counting the kinds of frustules, dead or alive. And the dead frustules in a sediment core can yield a complete history of the diatom community and thus of the lake environment in which they lived.

Miriam and I put together a start-up budget from the university's development fund and a doctoral improvement award from NSF and together explored Ecuador for ancient lakes and diatoms.

As Julius Caesar said of Gaul, all Ecuador is divided into three parts; to plagiarize outrageously the opening sentence of the "Gallic War" I might try "*Omnia Ecuadorae in tres partes divisa est*" or perhaps "*Omnia res publica Ecuatori in tres partes divisa est.*" The three parts are at least as distinct as those of ancient Gaul.

First are the plains and low hills of the Pacific coast, about a third of the country; except for the coast range and the foothills most of this region is no more than 300 m above sea level. Next is the center slice of the Andes Mountains, another third of the country, a rectangle of some of the highest mountains on Earth, with rows of peaks approaching 6,000 m (20,000 feet). And then the remaining third, inland, is the westernmost portion of the Amazon lowlands.

Ecuador looks to be a tiny country on the map of South America, only rivaled as a midget by Uruguay and the three bits of Guyana. Don't believe it. As a block of land roughly four hundred statute miles across in both directions, it is about the size of much of England, Ireland, and the Irish Sea combined. The coastal third is about the size of Ireland, the Andean third roughly equivalent to the Irish Sea and Wales together, and the smaller Amazon bit about the size of the top half of England as far south as the Wash. A significant bit of real estate to search for ancient lakes.

This simple three-part structure is complicated by a minor mountain range within the coastal plain. These hills rise to about 600 m (2,000 feet), roughly equivalent to the mountains of Wales but almost negligible in an Ecuadorian context. And then there is the great fold within the Andes mountains themselves, as the central mountain block of the Andes is divided down the middle by the inter-Andean plateau at 3,000 meters (9,800 feet), a crease between cordilleras to west and east. The interandean plateau would itself be a spectacular mountain range over most of the earth, say the equivalent of the daunting Brooks Range of Alaska. But it is discounted by the flanking cordilleras that rise as high again above it (to 6,000 m). The capital city of Ecuador, still going by the old Inca name of Quito, lies on this preposterously high plateau, and many a visitor has gasped for breath when leaving a pressurized aircraft at Quito's Marshall Sucre Airport. The city is nearly twice as high as mile-high Denver.

All these features run roughly south to north, in parallel. From the sea inland along the equator an imaginary journey would be across the low coast, over a hump of coastal mountains, across the low (300 m) coastal plain, steeply up the flank of the Cordillera Occidental, through a pass between the 5,000–6,000 m peaks, down to the inter-Andean plateau at 3,000 m, through another high pass across the *cordillera oriental*, and down through the foothills to the Amazon lowlands at 300 m elevation again.

If the traveler crossed Ecuador's eastern frontier to continue across the continent, the remaining journey would be through Amazon lowlands all the way: a few hundred kilometers of Amazonian Peru and Colombia, depending on where the frontier was crossed, then the vastness of Brazil for three thousand kilometers before reaching the Atlantic Ocean. By far the larger part of this journey would be through lowland tropical forest.

Ecuador thus sits on the more outrageous relief of the equatorial slice of South America. This relief is the product of a titanic crash between drifting plates of oceanic crust and the South American continent, a real-life example of an almost irresistible force meeting a near immovable object. The land is still crumpling under this continuing impact to be concertinaed into a series of mighty folds, with the Republic of the Equator taking the brunt of the blows.

Mass is added to the mountain folds from once molten rock that leaked up through the fault line of the crash. The mountain peaks themselves are all volcanoes, many of them still active; the highest of them, *Chimborazo*, being 6,310 meters (20,697 feet) high. These volcanoes, like the folding, are a consequence of plate collisions. Crust ruptures in the collision, molten rock is forced up under pressure, and mountains of cooled magma and fused ash pile up.

This spectacular geography has climates as varied as its relief. Much of the coastal plain is dry, verging on cactus desert near the sea. This is the same dry climate that inflicts the Galapagos Islands.

North of the equatorial line, as the coastal region approaches Colombia, the climate is moister and the vegetation is mostly woodland, some separated from the sea by palm-fringed beaches. The whole coast is breached at intervals by rivers draining the Pacific flank of the Andes, their estuaries clothed in mangrove swamps.

The largest of these rivers, Rio Guayas, reaches the sea near the south of the country in a splendid great estuary debouching into the Gulf of Guayaquil. It is the terminus of many merging streams whose headwaters are high on the western slope of the Andes, flow both south and west, and drain through half of the coastal plain. Imagine something like the Shannon River and estuary, draining an area comparable to the Shannon's half of Ireland but with more input streams and the muddy water of tropical lowlands. Unlike the Shannon, the Guayas brings water from afar, flowing through lowlands with little rain of their own, a great river system in a fairly dry land.

The Gulf of Guayaquil is large enough to be a familiar shape on world maps;

it is that little nick where South America bulges into the Pacific Ocean. The port city of Guayaquil, already a commercial center in colonial days, was where Simon Bolívar met San Martín, both with armies in train. In the Ecuadorian school system they speak of the time when the liberator of the north met the liberator of the south in Guayaquil.

Away from the coast, the high face of the Cordillera rises up through the inversion into dripping clouds and falling rain. The result is a rich forest, in many ways made unique by its long isolation from forests on the Amazon side of the Andes. Among the botanical marvels is one of the highest incidence of novel orchids on Earth. This treasure of a forest has largely been cut down, even as the first thorough botanical survey was being made.[11]

The steepness of the ascent into the Cordillera Occidental is astonishing to those of us raised in flatter lands. From the coastal plain altitude of 300 m to the summit of Chimborazo at 6,310 m, for instance, is only 50 km (30 miles). The world behind the mountains is thus nicely insulated from coastal doings by this rampart.

Beyond this first mountain range, the Cordillera Occidental, the inter-Andean plateau lies in rain shadow from the Pacific Ocean. Perhaps more important still, the plateau is also in the rain shadow of the mountains to the east, the Cordillera Oriental, which bars the plateau from the Amazon basin. Parts of the plateau can be very dry, even to bare ground with thorn scrub in some places. But this plateau was the seat of one of the great civilizations of South America, culminating in the Inca empire. For the last Inca, Atahuallpa, Quito was said to be his favorite city, though, as is noted in Ecuador, his murder by the Spaniards was in Peru.

Civilizations are not founded in deserts, nor are emperors likely to live there from choice, powerful evidence that the inter-Andean plateau is not all dry. Nor is it. Gentle rains spill onto a large part of the plateau through passes in the cordilleras, both to west and east. Quito in fact has months of cold fog as clouds overflow the passes from the Pacific to settle on the city, making legitimate the local version of grumbles about the weather.[12]

The countryside of Quito smiles; at least to an Englishman it does. It is cool by altitude and green under its enveloping cloud. Driving out of the city in sundry rented trucks, the sights of green fields and hedgerows again and again would invoke memories of English countryside, the County of Dorset perhaps. Not only the verdure of the place, but also the sense that the land had been laid out by the hand of man, by farmers at work for centuries on end. It has that feel to it.

Until you look up when the clouds lift. Then one or more of the mountain peaks appears, high above, mountains that were gods to the Incas: Cotopaxi, Cayambe, Antisana, Chimborazo, Tungarua, Sangay. I have seen people stop in the street when the clouds lift, to point and shout: "Cotopaxi!" "Antisana!" The County of Dorset myth flies away.

The eastern cordillera (Cordillera Oriental) comes next, another fold from the great land crash and with more than its fair share of the great volcanoes. On the rare occasions when Sangay can be seen from civilization on the plateau, a reddish halo caps it from the fire below. The inside (western) flank of this easternmost cordillera is in the rain shadow from the east, its forests upslope, where clouds spill over from the Amazon. This is the climatic divide of the continent. On the Amazon side of the cordillera, rains come mostly from the east. Falling water hails from the distant Atlantic Ocean, not the far closer Pacific. The water is fated to return to the Atlantic down long, winding rivers across the Amazon plain.

Rain is very heavy on this eastern flank of the Andes so that the feeding rivers of the Amazon rush headlong down the steep slopes. Feeder rivers the size of the Thames, mere minor tributaries of the colossus downstream, tumble into the lowlands with the exuberance of trout streams, plunging into the tropical rain forest.

This remaining third of Ecuador, Oriente Province of the Amazon region, belongs to the trees and the rivers. Travel is by motorized canoe. This is a country far off from the high and ancient civilization of the plateau, just a few hours' drive away through a pass in the cordillera.

Miriam and I took counsel together about where in Ecuador to begin. A long core from the coastal plain had immense appeal because it should test the accuracy of our Galapagos climate history. We had found the Galapagos to be dry for a whole glacial cycle, with the sea cold and the air inverted. Therefore the dry coast facing the same cold sea should also be dry. Unfortunately the necessary lake was not obvious in the coastal desert!

Even more appealing were the Amazon lowlands; a long Amazon record was my holy grail and would remain so. But here, too, target lakes were decidedly not obvious. The whole region was soggy, true, but Ecuadorian cartography of the 1970s gave no promise of the sort of real lake we needed; ponds and shallow floodings, yes; convincing ancient lakes, no.

Thus both these parts of Ecuador were poor prospects, but the third possibility was the lake-rich region of the Andes. If you want lake sediments, go where the lakes are. This policy might even be more than a mere chickening out of the two difficult thirds; because Andean lakes should get their water from one ocean or the other so that their histories might serve as proxies for a Galapagos test or for an Amazon history.

There were excellent reasons for expecting a truly ancient record from a well-chosen Andean lake. The Andes offered volcanic lakes, glacial lakes, and even the delirious prospect of a small rift-valley lake in a graben. A "graben" is part of a geological faulting system. It is the depression made when a section of crust falls down between the sides of an active fault. The grandest, deepest, and most ancient lakes on earth, Lake Tanganyika in east Africa and Lake Baikal in Russia, occupy systems of grabens. Long, narrow and deep, Tanganyika and Baikal have lengths measured in hundreds of kilometers, depths of more than a thousand meters, and ages more easily measured in geological epochs than in years or millennia.

Lake Yambo is not made on this scale; it is a graben lake writ small. The shape is the same as that of Tanganyika and Baikal, long and narrow, but it is little more than a kilometer long. Air photographs showed the steep sides of the fault leading down to the water, and the fault itself could be traced on the images, running north and south, parallel to the cordilleras on each side, neatly bisecting the inter-Andean plateau close to the zero latitude line. Despite Yambo's symptoms of grabenhood, Miriam found only seven meters of water at the end she sounded from her rubber boat. This was enough to make one believe in fairies.

As an accessible mini-graben lake on the equator, Yambo appeared perfect but for one oddity: it was in a desert. It lay in a section of the inter-Andean plateau with near complete rain shadows of the cordilleras from both east and west. This was real desert, not as bad as the Sahara, with its blowing sand dunes; more like parts of Arizona, with scattered shrub and cactus and flowering bromeliads; but definitely desert. Lakes in deserts sustained by groundwater from distant catchments have their own quixotic appeal. Who could resist something so odd as water in a desert? A desert lake could even yield powerful evidence of past water levels, and hence of climate, from the chemistry of its sediment. I was sold on coring Yambo.

Volcanic lakes were perhaps more promising still, for they were plentiful in all parts of the plateau, the wet as well as the dry. There were chances of finding them on the outer slopes too, overlooking the coastal plain on one side and the Amazon on the other. Holes made by volcanoes come in several flavors, including valleys dammed by lava flows, like Laguna de Colta. Most attractive are the craters.

Volcanoes make craters of every size, from huge magma chambers (calderas), emptied out by flowing lava and big enough to hold a city (Bogota in Colombia lies in one such), to small explosion pits looking like the bomb craters I saw round London in my boyhood. Explosion craters of middling size were what I hoped to find. El Junco was one such. I asked Miriam to find an El Junco in the Andes.

The peculiar advantage of explosion craters (maare) is that they are unlikely to have inlet streams. In the vernacular of our trade they tend to be "closed lakes." They collect water only from their own rim or as rain on the water surface and lose it mainly by evaporation, possibly by seepage, and only occasionally by overflow. Among the many benefits of closed lakes is that sediment buildup can be very slow, making for short cores to the ice age.

Miriam found a lake that fitted this pattern precisely. It looked to be a dead ringer for El Junco, a round explosion crater lake just one hundred meters across, the rim of its crater higher than the surrounding land so that the only source of water was rain falling into the crater itself. A notch on the low point of the rim suggested overflow in some distant wet times. Miriam brought back both her own photographs and air photographs supplied her by the Ecuadorian military. The crater had the weathered look of an ancient volcanic surface, familiar from El Junco. Better still, the lake lay in a wet, pastoral part of the plateau at 2,800 m, just 14 minutes north of the equator. The cone in which it lay was tucked in against the inner face of the cordillera oriental, where it should get moisture from the Amazon side. The local name, culled from the few locals who knew it, was Lake Cunro.

Lakes Cunro and Yambo between them ought to give us what we needed, so off we flew by Pan Am to Quito, with rubber boats and the paraphernalia of coring in our airline baggage. Those were still the days when bundles of half a hundred metal tubes looking like gun barrels of various calibers were no handicap to air travel, provided you paid the modest fees for excess baggage. If your host who met you at the airport was Miriam's father, Hans Steinitz, who had long run a local import business, your passage through the airport to a waiting truck was swift. Gone were our days of seeking consular help in Guayaquil. We secured our Quito base in the Steinitz family home, built our raft from local lumber, and were driven to Cunro by a family friend with a truck.

Lake Cunro was all I had dreamed it might be. The approach was a narrow paved road that wound its way through green hills, moist with hovering clouds, sparsely scattered with the simple homesteads of an agricultural community. We drew off the road by a grassy bank and walked up it. Lake Cunro was below us, a perfect teacup of a lake, the circle of water just one hundred meters wide, the sloping rim of comparable width, the whole thing a rounded pimple on the side of a green mountain.

Launching rubber boats, anchoring the raft, and lowering the casing went like pleasure boating on a summer's day in England. The coring, too, was almost as easy, with water about 5 m deep, winds negligible, the water surface flat. We punched two cores through to the gravel basement of volcanic debris under about 5 m of lake mud. But we encountered, not obstacles exactly, but frequent obstructions to be overcome.

Several times we had to stop the steady push-drive of the piston sampler to clip on our anvil and smash our way through something with our ten-pound sliding hammer. Each time the corer broke free after a few hammer blows, and we could push our way again. The steel cutting shoe we had used as piston gripper was proving its worth as a penetrator too. The steel shoe came up polished rather than nicked after these hammering sessions, and the mud behind its shoulder held traces of fine sand. I guessed that the obstacles were thin layers of tephra. And was foolish enough to greet these "ash" layers with joy. Perhaps we could add to our other pending triumphs a history of Andean volcanism, dated by radiocarbon of the surrounding mud of each ash fall. Or even better take advantage of the new technique of tephra chronology to develop a regional timescale.

That the sediments were only five meters thick was disappointing but not a real worry because encouraging precedents for great time spans for thin sediment sections in explosion crater lakes were many.

Running through my head was the Monterosi story. Lago di Monterosi is a maare lake near Rome, a tempting site for an environmental history and thus one of the first lakes to be attractive for coring. The field party returned from Rome chastened because their cores were less than three meters long. These were the days when the coring fraternity had learned that to get a complete Holocene (postglacial) record in the typical small lakes of the temperate zone needed anything between seven and twelve meters of sediment. The field party must have wondered if their cores even spanned the history of the Roman Empire, let alone postglacial times. In fact radiocarbon showed they had a continuous mud record, spanning more than 25,000 years and thus the first ice-age history of Italy.[13]

If Monterosi, why not Cunro? Thus reassured, we set out for the even better bet of Lake Yambo. The desert round Yambo was astonishing to people fresh from the smiling countryside of Cunro. Though at a comparable altitude on the inter-Andean plateau and 80 km from Cunro, the Yambo landscape is semi-desert, with the harsh glare of tropical sun reflected off rocks and sandy soil. This is the heart of the arid portion of the inter-Andean plateau, in rain shadow from both sides so complete that rain is rare. Plants must make do with the chance of mist for much of the year, rather as does the coastal belt of a Galapagos island.

Yambo is a narrow ellipse of water about a kilometer long, on the floor of a deep rift through this desert place. The sides of the rift soar high above the water to channel a rush of wind the length of the lake, always from the same direction like a steadily aimed peashooter.

A spiraling rocky apology for a road gave access to a narrow shoreline bench on the east side. We launched our rubber flotilla, rowed out against the wind like a pea in a peashooter going the wrong way, and found water over 20 m deep and still shelving down. This was far more than the 7 m reported by the early reconnaissance, which had been deceived by a broad shelf at this depth. All the equipment we could muster would let us core in no more than 17 m of water. By brute force against the wind, we dragged the raft to where we found 17 m, anchored, and cored. We got less than a meter, sandy mud, with slanting bedding planes, showing we were still on the slope down.

And so back to Ohio to write a real grant proposal, after storing the equipment for the duration in the hospitable Steinitz home. Coring the lakes of Ecuador was going to need more resources, more coring equipment, more round trips to Ecuador, renting vehicles rather than begging from friends, per diems, and student stipends, to say nothing of the costs of laboratory work. Off to NSF we go with a proposal called "The Environmental History of the American Equator since 18 K."[14] The NSF gave us the money.

X-rays of the Lake Cunro cores showed fourteen stark white mineral bands (white like the bones on a medical X-ray because opaque to X-rays). Thirteen bands were, as we had guessed, tephra, each the ash of a separate volcanic eruption that had settled on the lake bottom. The fourteenth was an ancient piece of pottery, just big enough to fit in the core tube, as it had been buried lying flat on the lake bottom. Mud around the potsherd in the tube looked to be undisturbed, showing that chance had allowed perfect alignment of the drill. The chance of a one-and-a-half-inch core barrel snaring such a trophy in a lake 100 m wide must be near vanishingly small, but it happened.

Radiocarbon soon took the pleasure from these discoveries, showing the core to span something less than 6,000 years. This was our maare lake core that ought to have reached the ice age. For our grand design it was as useless as Laguna de Colta had been.

The piece of pottery was of some solace. The depth-age curve generated from our series of radiocarbon dates assigned it an age of 2,850 BP. Its presence seemed to indicate people living round the lake long before Inca times, not surprising, as Cunro adds good water to a site green with pasture. But there is no visible trace of habitation in the crater now.

Miriam called at an establishment that became a mecca of the Ecuador field parties, the *Instituto Geografico Militar* in Quito and bought stereoscopic pairs of air photographs of the Cunro region. On all the bare fields on the slopes beyond the crater rim quasi-rectangular marks are visible from the air; footprints of an ancient settlement? No chance for us to dig and find out; NSF funds real archaeologists for that; so does Ecuador. But I still speculate on how that potsherd got there, 50 m from the shore. Perhaps a small boy from that ancient village, so long before the Incas' time, skipped bits of broken pot bouncing across the lake. I should have done so in his place.

Armed with the new NSF funding, Miriam drew up a list of possible lakes to try in the Andean third of Ecuador, many of them with delicious names still in the Quechua language of the Incas in which the suffix "-cocha" literally means "lake." Yaguarcocha (lake of blood), Cuicocha, Chicopan (a Quechua name now sadly being usurped by "Lago San Pablo"), Limpiopungo, Caricocha, Huarmicocha, Papallacta, San Marcos, Quilotoa.

A second shot at Yambo had the benefit of a lightweight outboard motor with which to drag the raft against that channeled wind, a sonar bottom sounder, sundry new limnological equipment, and enough core rods and casing to reach through 40 m of water, all flown down as excess baggage and kept in the accommodating Steinitz home. A tented camp, driven to the lake in a rented truck, let us set up on the bank until the job was done. I had an excellent team for the job: Miriam and another young woman, Kris Olson, from the Columbus laboratory, for whom a spot of fieldwork broke the monotony of making my pollen slides. The sonar revealed the deepest water close to the side with the high cliff, where an oblong strip of flat bottom lay under 23 m of water. We were a little upset not to see the image of an old locomotive said to have plunged off the cliff from the railroad above long ago, not a trace of it; don't listen to tales that people tell. But that flat bottom to the deep hole was the place to core; that we did, hammering our way through a mix of mud and sand until we struck rock or gravel at about three meters. Nothing more; three meters for a supposedly "ancient" graben.

The measly three meters of Yambo sediment was almost without organic matter, a serious business for the old beta-decay and Geiger counter method (chapter 3) of radiocarbon dating. Dating laboratories demanded at least two grams of carbon in those days, preferring ten grams. But the "richer" (i.e., least poor) lengths of core had just enough carbon for an approximate date. The whole core, thus Lake Yambo itself, and presumably the graben also, was no more than 2,500 years old.[15]

Thus perished our dream of an ancient graben lake in the equatorial Andes. Actually there is one such, though not on the equator where we wanted it: legendary Lake Titicaca in southern Peru, at four thousand meters elevation, more than 200 m deep, the highest of the world's great lakes. I used to dream of Titicaca. But the technical effort with the resources of those days, even if the very large funding were possible, was not justified by the likely results. The elevation, high up in the highest paramo at 4,000 m hinted that pollen in Titicaca mud would be sparse and uninformative; a cold high place would have been a cold high place then as now, another version of Ed Deevey's, "Paul, you showed that it was cold in an ice age." So adventure had to bow to reality, and Titicaca was left alone.

A quarter of a century later Titicaca has been cored at last by others, with huge drilling equipment hauled by tractor trailers and based on experience of the ocean drilling program. The histories of water volume, isotope chemistry, and lake levels, as well as pollen, are yielding moisture profiles of the ice-age climate of pressing utility to the Amazon reconstruction.[16] But that is now. Thirty years ago we were not ready.

Meanwhile, the much better-placed graben Lake Yambo was not old. The implication of this had to be that the seismic upheaval required to split the mountain plateau into a great fault was itself recent, a few thousand years old at most. There could scarcely be more provocative evidence that we were seeking an ancient record in a landscape subject to explosions and violence.

Our first two tries had failed for the same reason: recent geological mayhem. Laguna de Colta lay behind a lava dam no older than Yambo, and the explosion that had blasted out the basin of Cunro was only twice as old at 6,000 years. Worse still, the Cunro mud recorded fourteen subsequent blasts with its layers of tephra fallout. It took some time to realize it, but the writing was on the wall.

My reaction, I still think the proper reaction, was to core every lake I could. Surely the entire surface of the Ecuadorian Andes could not all be only a few thousand years old, as if it were part of the earth imagined in 1658 by James Usher, Bishop of Armagh. Usher's scrutiny of the Christian Bible led him to declare that the six days of creation began at nine o'clock in the morning on Monday, 23 October 4004 BC,[17] which I take to be as near as dammit just 6,000 years ago, as I write on April 14, AD 2003. James Usher would have known that I was wasting time and money searching for ancient lakes in Ecuador. I am glad he was not on the NSF review panel.

NSF let us plunge on. Yaguarcocha, the Inca lake of blood, whose waters do have a reddish glow at times, had gloriously banded, very sticky, red mud, lots of it. We hammered through 9.5 m of it from a raft washed over by white-cap waves; felt we had been in a fight and had won it. The oldest radiocarbon dates were between eight and nine thousand years; better but not good enough.[18] Lake Chicopan (San Pablo) forced us to core in white-cap conditions too but in 30 m of water. We got three meters of sediment, some of it sand, and then were beaten by a tephra layer of great thickness. After hours of pounding, we had driven the best part of a meter-long sample tube into this, managing with difficulty to pull it out again. The core barrel came up empty, the ash having poured out on the ascent through the water. Tephra had also flowed into the hole we had beaten through it, making further progress impossible. Radiocarbon dating of the overlying mud gave an age for the major eruption spewing this thick ash of 3,000 years ago. Smaller lakes like Limpiopungo and Papallacta had no sediment worth bothering with.

We had a look at the high mountain lakes, many of them occupying magma chambers (calderas) that no interpretation of the geology could convince us were ancient. We did launch a rubber boat on the great caldera lake, Quilotoa, largely to compare its water and planktonic life with that other great caldera lake on the Galapagos, Fernandina. The descent from the Quilotoa rim to the lakeside was not on the scale of Fernandina, though a significant plunge carrying gear. But the rim itself could be reached by a mountain road, giving the whole operation the feel of a holiday outing. Putting together all the rope we carried into the crater for sounding, anchoring, etc. gave us a sounding line 200 m long. We had, as the leadsman's chant from the days of sail would put it, "No bottom with this line." Quilotoa is a deep crater lake. Its water has the same taste of epsom salts (magnesium sulphate) as Fernandina had and is the same primary volcanic water of a very fresh eruption. Day's outing over. But in the plankton tow Miriam found *Amphora paulii*, up to then known only from our type specimen from Fernandina.

High-altitude but shallow crater lakes Caricocha and Huarmicocha had us struggling, both for breath and to reach them over difficult trails with the impedimenta of rubber rafts and coring rig. Cold, wet, but triumphant, we cored them, getting no more than 2,000 years of mud.

In desperation I looked at the other main suite of lakes likely to be found on high mountains: the kind built by glaciers. The mountains are so high that several are capped with ice even now. Chimborazo still has a fine ice cap, or rather it did have before recent greenhouse warming sent much of the ice tumbling down as liquid water in the last few years. But in the ice age these mountain glaciers grew and spread, thrusting down in places almost to the inter-Andean plateau (which is not all that far toward the tropical lowlands, of course; the plateau is itself at 3,000 meters). The marks of this ice mayhem are heavy on the highest land so that the descent of mountain glaciers was one of the two ice-age facts about the equator of which we were certain back in the 1960s (the other was that sea level fell). Some splendid lakes now occupy the highland hollows that were scooped and dammed by glaciers, lakes that are beautiful analogs of say, the Lake District of England.

It was quite clear that we could not get a long record of ice-age times from a glacial lake which by definition has existed only since the ice melted. Yet these lakes were not quite hopeless. One critical datum is inherent in sediments from a glacial lake: the environment immediately after the retreat, a useful parameter for assessing what went before. More than this is the chance of finding a lake left by an earlier, more extensive glaciation. There were plenty of hints of this sort of thing in writings of glacial geologists of the Andes, of marks of old glaciers well below moraines left from the last glacial maximum.

One glacial lake was obvious, though not particularly low. Lago San Marcos is one of the grandest lakes in Ecuador, a beautiful fjord of a lake formed in a classic U-shaped valley left by a mountain glacier bulldozing its way down from the heights. These glaciers push heaps of spoil before them, either tumbling ahead of the ice or mixed in with the leading edge of the ice mass. When the glacier eventually melts, all this debris is dumped as a great mound of jumbled rock and soil that ought to be called just a "spoil heap" but, this being science, is given an insider's special name of "moraine." This was the name long given to piles of rock and dirt on existing glaciers by peasants of the French Alps. Alexander Agassiz lifted the word at the foundation of glacial geology early in the nineteenth century. Agassiz's successors have since extended, refined, and defined the meaning, as is the habit of science.

The spoil heap (sorry, end moraine) that dams the water of San Marcos lies at 3,414 m (11,198 feet) on the southeastern flank of snow-capped Mount Cayambe. This elevation is in the mid-range for Andean end moraines, suggesting that it dates only to the last glacial maximum. But at 10,000–11,000 years this would be a huge improvement on what we had got, with so much labor, from the volcanic lakes.

We got to San Marcos via a bulldozed track through mud and rocks, a "road" that turned into a river (more or less) when it rained and that wound up the side of Cayambe. Two vehicles and a tow rope made this drive possible, one with 4-wheel drive to get through mud holes first and then tow the load-bearing truck through. Rain was nearly continuous for the three days our little sleeping tents were snuggled into the bushes by the morainic dam.

We cored in 37.5 m of water twice, the deepest water I had cored under by hand, each time hitting gravel in the fourth meter. Another disappointing result; conventional wisdom of the trade says there ought to have been at least ten meters of mud in such a lake. But there wasn't. Not to worry; the moraine dam stipulated the age at in the 10,000 year range.

Packing up to go, I yielded to the childish custom I had developed, that of swimming in all "my" lakes. Certainly it would be cold, probably 7° or 8°C, but it might well be the first Andean lake to have yielded a respectable record. I dived off the dam into ten meters of water. Folly! A naked body, in cold drizzle, at high altitude, when plunged into icy water, gasps for breath. But I had not been up high long enough to be properly acclimated. Little air to be got by gasping; this almost felt like trouble. A desperate dog-paddle back to the dam and my lowered feet felt mud; such bliss!

Radiocarbon dating of the San Marcos core gave me a gentle reminder that even the age of end moraines could not be taken on trust at zero latitude, and this one was almost exactly at zero latitude. The bottom of the core turned out to have the depressingly familiar radiocarbon age of about 3,000 years.

I don't know why the San Marcos sediments are so young; a Holocene readvance of the mountain glacier? The pervading set of explosions around three thousand years ago to throw gravel as well as finer tephra into the lake? The dam is not a moraine but a recent lava flow? I do not believe any of them and I do not know the answer. Instead of worrying, we had to get on.

Miriam talked to Ecuadorian geologists, lamenting our woes in the explosive landscape. It was said that in the southern Ecuadorian Andes, going toward the Peruvian frontier, the volcanic landscapes were older. She did her own survey out of the city of Cuenca at three degrees south, found a list of prospects, chose two lakes, and took us there.

Both were glacial lakes, one high above the other. Unlike San Marcos, they were easily accessible by a good road, the upper lake being a beauty spot on the tourist route, complete with a small cabin for the use of an alpine club. Wonderful amenities, but I did not like the lake; it was just too high. It was a glacial cirque, a place where streaming ice met an obstacle to be diverted into a slow-moving vortex analogous to the little whirlpools that develop in mountain streams of water, but this time moving literally with glacial speed. But if the speed of motion seems minuscule, the force exerted is not. The ice clutches boulders on its bottom and grinds out a hollow. Give it a few thousand years to work, melt the ice, and a deep circular lake results.

I gave this one a miss. It would certainly have been easy to core; we could even live in the hut. But the bleak surroundings of the high paramo made the possibility of a useful pollen record dubious. But the second lake, many hundred meters further down the mountain, seemed a better bet, even for a Holocene record, let alone for our goal of reaching ice-age sediments.

Someone else was to core a similar high lake in this southern lake region nearly ten years later, finding beautifully banded sediments going back to a radiocarbon age of 12,000 BP and undisturbed by the volcanic mayhem that had dogged us. They found it at what was apparently their first try. They counted and dated bands, from which they interpreted the relative strengths of El Niño rains. The high altitude that had let me reject our high cirque was no handicap for this study of postglacial rainfall and may have been an advantage because the lake was sensitive to Pacific rainfall "slopping" over the top of the mountains just above it. Their rainfall record was strikingly similar to histories of major El Niño records elsewhere, letting them claim the first detailed and precisely dated history of El Niño (climatologists' version) for the equatorial Andes. The work is now cited informally as "the Ecuadorian climate record" or "the data from Ecuador." It was a brave and hard-earned triumph.[19]

We were right about Llaviuciu, or Surucucho, as the lower lake is variously called. It lies behind an end moraine which, at 3,180 m, is the lowest glacial trace in Ecuador, thus dating the basin to an early glacial advance. The sediments confirmed this; 12 m of mud in our cores spanned somewhere in the range of 20,000 years.[20] Our field party, led by Mark Bush (of whom more below), cored Surucucho only in 1989, marking the end of our 14-year struggle to force the Ecuadorian Andes to yield a long climate history. I had learned the hard way that ancient sediments do not lie undisturbed in an intensely active volcanic landscape.

I had also failed to crack the coastal third of the three parts of Ecuador. There were no obvious lakes known to geographers of this large region, not surprisingly for a coastal plain, much of which is dry, if not actual desert. Conventional wisdom clearly said, "Drop it." But it would be nice to have a coastal record to set beside that of the Galapagos; I would not let it go.

Miriam's hours at the *Instituto Geografico Militar* yielded a lake in the desert region of the southern Ecuador coast plain; just one, but an undoubted lake of a size good enough for coring, 100 m or so across. There it was, a bright blue disk on the largest scale map sheet (1/25,000), near a tiny settlement called Chongón. Two or three hours drive from Guayaquil, going north and a bit west, ought to do it.

A lake in the most desert part of Ecuador; no idea how it got there; just thank the fates and go to have a look. Miriam and I stole a day from the Andes, flew down to Guayaquil on the early morning flight, rented a car at the airport, and drove through the desert for three or four hours. Temperatures in the shade were in the 90s or worse; the car did not have air-conditioning. Mostly I remember the wonderful thirst relief from a papaya as large as a watermelon. Those who know papaya only from the fruits of commerce have no idea of the wondrous giants of the species grown in Ecuador.

We got to the lake. It had been bulldozed. All the sediments had been dug out and piled on the banks as a circumferential rampart, now baked into sun-dried brick. The prints of a large-tracked vehicle were heavy in the hollow where only months before a miracle of water had lain in the Ecuadorian desert. I do not know why it was done.

The other lake was geographic myth and legend, existing mainly by word-of-mouth tales that spoke of a big, swampy jungle lake in the forested northern part of the coastal plain. It was unknown to the *Instituto Geografico Militar*, an organization having at its disposal the reconnaissance arm of an air force. It could not exist. But we found an old map in a book (so help me, we really did) that showed a largish lake southeast of the northern port of Esmeraldas, between the coastal mountain range and the Andes.

The old map gave the lake a name to tempt a fidgety exploring soul: Laguna los Ciudad; a name sounding to this Anglophone individual hauntingly like "lost city" (shades of Colonel Fawcett!).[21] But if military geographers possessed of an air force did not know about it, it did not exist. So forget it. And yet suppose the old timers blew up reports of a swamp! — a swamp up there in the jungle was not completely absurd. And swamps can yield pollen records of the past too.

Fanciful lakes on old maps apart, the northern section of the coastal plain did offer the most realistic hope for a lake west of the Andes. The 100 km stretch of shoreline between the mouths of the Rio Esmeraldas in northern Ecuador and Rio Mataje at the frontier with Colombia, for instance, is known for mangrove swamps, with jungle in the narrowed strip between mountains and the sea. A likely place for standing water, and so few roads or people that little was known of it.

Miriam and I took another jaunt away from the Andean work to fly to the town of Esmeraldas, at the mouth of the eponymous river; a wonderfully cheap flight I might add, on TAME, the subsidized airline operated by the Ecuadorian air force. In Esmeraldas we were able to rent a vehicle more suited to our needs than the family sedan we had been stuck with on that desert jaunt, one of Toyota's early copies of the original jeep. And off we went to hunt for lakes.

Our method was simple: drive wherever the jeep could be made to go, and when we met people, ask. Miriam is brilliant at this. From peasant to police chief as targets for our nonsensical questions, she can come back in ten or twenty minutes, bursting with news. I was that background figure, the driver, a specimen of the breed cowering in the rear: Anglophone, Anglo-Saxon, and embarrassed.

We searched almost to the Colombian frontier in this way, mostly close to the coast, actually on the beach in those many parts of the coast without even an apology for a road. On one significant stretch, where small rivers ran down to the ocean, we actually took the jeep out to sea (up to its hub caps, that is), finding the sandbars parallel with the shore the best way of crossing small rivers. Habitation attracted us as chances to ask questions, likewise bits of road heading inland. We did not find much; there was nothing much to find.

A drive inland led to the hacienda whose landholdings were likely to include the site of the mythical Laguna los Ciudad. The house itself did exist, and Miriam's inquiry there sent us down a trail to find the owner. And on that trail we met a *pistolero* on horseback; at least he was on a horse, he did have a loaded revolver in a holster at his waist, and on his head a hat with a brim suited to one who spent tropical days on a horse. But he was not a *pistolero*; he was the owner of the hacienda.

Yes, there was a large stretch of water "over there," and, yes, he could take us to it, but only in the dry season when horses could cross the marshy land. Learning what our coring equipment was like and the logistic needs of a coring team for a week, he saw no problem; ten or twenty horses, he had them. Aha! vision: a real mounted posse off to drill a lost lake in the jungle; an adventurer's dream come true!

At the same time the landowner's description of the place could well fit my hypothesis of swampland better than an actual lake. He spoke of islands of grass as well as open water, and he could only reach it in the dry season; the heart of a great swamp, where the last shallow patches of water were still present in the dry parts of the year? Back in Quito, I chartered a small aircraft to fly me over the region. Nothing, not even a glint of water under the trees. I still suspect swampy bottomlands, but they had nothing for me in my increasing desperation to drive my drill into ice-age mud.

The hunt for ancient sediments had failed on the coastal third of Ecuador even more thoroughly than had the hunt in the Andean third. This was reality in the 1980–1981 field season, five years after the Republic of the Equator had come to me in the person of Miriam Steinitz and fifteen years since David Greegor had cored Laguna de Colta.

If a moral lesson is to be found in this double failure, it can only be that negative results are as common in science as in other human endeavors. He who wants ancient sediments had better go where ancient sediments are, but he who seeks to answer a problem had better go where the problem is; and be prepared for the possibility of failure. Resulting bruises to the soul can be soothed by the Band-Aid of thought that says, "We were pioneers."

6

REFUGE THEORY EXPANDS IN BRAZIL

Brazilian scientists had been collecting in the Amazon long before Haffer. They were already well aware that many species could be found in some parts of the great forest but not in others, particularly animals likely to be sensitive to climatic change, like amphibians and reptiles. They, too, had strong suspicions that ice-age climates had something to do with the species distribution, suspecting arid intervals. P. E. Vanzolini's thesis on the subject, published in 1970, reflected previous years developing the theme, which he continued to do in ever more scholarly detail during his career as a professor at the University of São Paulo.[1]

These distributions, whether, like Haffer's birds, endemic (restricted to a single area thought to be their place of origin), or, like Vanzolini's reptiles, disjunct (with widely separated populations living in special habitats), pose real problems. How did they get there? Vanzolini and his Brazilian colleagues were surely the first to spot the simplest and most direct answer: "In drier times." Already some of their geological colleagues, notably J. J. Bigarella, were suggesting that much of Brazil was drier in glacial times, and Bigarella had even given a paper to that effect to the International Quaternary Association in 1965.[2] Thus the most complete answer became, "In the dry times of the last ice age." A few thousand years of relative dry times in an ice age should do nicely for the animals to spread even from the dry *caatinga* forest of one dry periphery to another. This reasonable thinking could well have prepared the way for Haffer's brilliant conception that any wet spots surviving the drought became forest refuges where his birds evolved in isolation.

The question of, "How did these animals get across hundreds of kilometers of hostile wet forest?" was nicely solved by a dry ice age. But the evidence for a dry ice age was still extremely slender. Fitting facts of biogeography to a supposed arid Amazon basin of the past was little more than saying, "We could explain this if only . . . ," and Bigarella's suggestion of general ice-age drying in Brazil was based on his coastal studies of pediments and stream channels (chapter 4). These data were not, of themselves, hard evidence of past aridity in the Amazon basin itself, saying, "It might have been," not "It was." The need that Brazilian biogeographers must have felt for corroborating evidence from the Amazon of prolonged past aridity on at least an ice-age timescale is palpable. But help was at hand, and it came to them in two ways.

The first source was the air force of the military dictatorship then governing Brazil, with its program of mapping the whole of the Amazon from the air with side-scan radar, a technique that yields images of the land surface under the trees. This *Radambrasil* project revealed a deeply dissected land surface under some of the richest forest of the central Amazon. Deep dissection suggests erosion of the kind seen on bare ground. Good enough. That was when generations of lizards crossed. For good measure the Radambrasil maps showed land forms under some stretches of forest that looked uncommonly like sand dunes.

But the observation that did more than any other to make ice-age aridity believable was "stone lines." You see them in road cuts, on riverbanks, in excavations in the Amazon basin. Some particularly fine specimens were visible from bus or taxi along the highway from Manaus to its airport. The shear face of a cutting exposes the red clay of tropical soil, but buried deep within that soil are pebbles of variable sizes, seldom large but arranged in long lines or layers, not just scattered through the red clay face. These are the celebrated stone lines. And a hypothesis that used them as evidence for past aridity was ready to hand: they were old desert pavements.

On the bare ground of a semi-desert, like parts of Arizona, rain, when there is any, or wind sweep fine particles away, leaving pebbles and larger stones behind. Eventually the land is paved with them: a desert pavement. If a desert pavement was buried for some reason, a slump perhaps, the result would be a line of stones running through the soil. By this argument, the Amazon stone lines were made in the same way in a dry time lasting the tens of thousands of years of an ice age, and were buried by mud movements driven by the rains of the wet times in which the forest now grows. If this sounds a little far-fetched, it is. Many competent tropical soil scientists had quite other understandings of stone lines in lateritic soils. None of them, however, were in the Amazon.

The desert origin of stone lines was trenchantly argued by a school of Brazilian soil scientists, particularly by the eloquent professor Aziz Ab'Sáber.[3] Now everything fitted together: biogeography, deeply dissected soils, sand dunes under the forest, stone lines, and the refuge theory of Amazonian speciation.[4] It was good.

The great beauty of the refuge theory in explaining the extraordinary diversity of the forest was boosted by this barrage of opinions from the realms of geology. None of the phenomena were shown to be of glacial age. You cannot date a stone line by radiocarbon, nor what might well be a dune-like structure glimpsed on a radar image. But what other timing was likely for the synchronicity of events now accepted? The ice age it was.

Perhaps refuge theory and ice-age aridity applied to all tropical forests. Cycles of ice ages, with their supposed rhythms of wet and dry, were global. Haffer's hypothesis became one of the most fascinating ideas in a generation. Biogeographers in other tropics began to look to "refuges." In the Amazon, the hunt was on to find other animals or plants whose centers of dispersal, and hence origins, fitted Haffer's maps of refuges.

The Amazonian hunt was helped when the Brazilian government established a research institute and field station at Manaus, the classic Amazon city a thousand miles up the Amazon where nineteenth-century rubber exploiters had built up wealth sufficient to support a world-class opera house in this far-off jungle. With no more obstacle than bureaucracy, it was now possible for students and teachers to go to work in the central Amazon.

A team from the New York Botanical Garden led by the Garden's then director of research, Ghillean Prance, made particularly energetic use of the new station. Distributions of a few forest trees on which the group was expert were plotted from the garden's collections and shown, more or less, to meet the footprints of Haffer's refuges for birds.

But the really big push to refuge theory was given by butterflies.

Butterfly hunters have usually been the first into new territories if the bird watchers do not get there first. Both groups have prowled the earth just a few steps behind the treasure hunters and adventurers, but of the two it is the butterfly hunters who get most specimens. They can show off their results with the hardest data of all: dead insects impaled on a pin.

Between 1972 and 1977, K. S. Brown published at least seven long and data-rich papers showing the distribution of heliconid butterflies across and around the Amazon basin. Heliconid butterflies are lovely things, clothed in glorious bright colors. What is more, the majority are weak fliers, flittering slowly across the way, and thus easy to catch. Brown's data set is breathtaking in its detail. And it reveals the same sort of oddity in distributions that Haffer's jungle birds did: local patches of endemic butterflies in an apparently continuous forest homeland.

No question the butterflies set out the same problem for evolutionary theorists that Haffer's forest birds did. They seemed to have formed and maintained distinct local species without physical isolation and within a seemingly endless and continuous habitat. Like Haffer's birds, this seemed to violate the new evolutionary synthesis of Ernst Mayr and others (*op. cit.*) that speciation requires physical separation. Call in Haffer's forest refuges and assign the butterflies to speciation long ago in this same archipelago of forest patches in the ice age, and the problem is solved. The butterfly maps are more detailed than the bird maps, with more collecting points. Bird and butterfly maps do match, however, more or less. Room for argument here, in fact quite a lot of room; but the story does hold together.

None of these biogeographic games were of interest to us climate historians. We wanted hard data, not the musings of naturalists about what might have been. We needed time machine data: accurately dated proxies for temperature, precipitation, and plant cover. We were not to get it; at least not yet. The 1970s gave us only the insistence on stone lines, or dissected earth and sand dunes sensed remotely by side-scan radar, but their collective force in what might be described as a climate of universal satisfaction with refuge theory was powerful.

I actually had accurately dated proxies for precipitation in ice-age times from the America equator, even if from the wrong side of the mountains in the Galapagos. And my special relationship with Dan Livingstone gave me direct knowledge of the accurately dated records of glacial aridity from Africa.

Twenty years later I can gauge the effect of these reports on our thinking by a news article that appeared in the scientific journal *Nature* in 1979. The headline was "The Ice-age Amazon." I quote extracts from it as follows:

> *A map has been published by Ab'Sáber which shows the Amazon basin to have been largely without closed forest during the last ice age. . . . The importance of this to both climate dynamics and ecological theory is very great.*
>
> *The most persuasive direct evidence for environmental change in the Amazon basin is the existence of dissected and gullied land surfaces now covered by closed forest. These surfaces were revealed by side-scan radar from aircraft in surveys made by the Brazilian government. . . . I was assured in Brasilia that the coverage is extensive. . . . Bigarella and Becker . . . have given accounts of the fossil land forms, showing that at times in the Pleistocene wide stretches of the Amazon cannot have supported closed forest. Ab'Sáber's map is a further compilation of these data.[5]*

The news report went on to talk of the problems of butterfly and bird collectors, whose evolutionary appetites would be slaked by these "hard" data for Pleistocene aridity. In short, the refuge theory lived. The report even mentioned the Galapagos climate record as still, at that date, being the only properly dated record of ice-age aridity at the American equator; not surprising, really, as the author of that *News and Views* piece was given as Paul Colinvaux, Professor of Zoology at Ohio State University.

Most of the reports out of Brazil in these years were in Brazilian journals, which even major research libraries did not possess, and many were written in Portuguese and thus effectively hidden from monoglot Anglo-Saxons. I learned of the work from a sneak preview of the Ab'Sáber map, which I saw only because my old work on the Bering Land Bridge made me a consultant to a National Geographic Society map of the Americas, both north and south, in an ice age. This lead was all I needed to find the rest.

The news report in *Nature* is evidence enough that I was persuaded by the Brazilian studies into accepting the general thesis of dry times in the ice-age Amazon, quite possibly with fragmentation of the forest by savannas as the refugialists wanted. But my reporting concluded by showing doubts of the evolutionary claims. Bird and butterfly distributions were indeed hauntingly the same, but they also reflected underlying geography. A bird map or a butterfly map was also a map of hills. Refugialists used this fact to say that hills were places of increased moisture during the ice-age drought, thus preserving their patches of forest. But what, I wondered, if it was instead some property of the modern climates of the hills that favored local speciation? Then we need not go to the ice age for our evolutionary answers at all.

Even starker was the fact that no one had yet produced a single radiocarbon-dated proxy of ice-age climate. Still no "time machine" record. I had been profoundly influenced by the conclusions of the Brazilian soil science community, but even they had nothing to which they could point and say, "This is an ice-age object." The arid times of which they wrote could, if correctly inferred (which, at that time, I accepted), have been at any unknown epoch of the past, so why the ice age?

The only data set that would really settle both the biogeographic yearnings of refugialists and the climatic changes of an ice age would be properly dated pollen histories of the vegetation of the Amazon lowlands. After wasting a significant part of a scientific lifetime vainly hunting old sediments in mountains and coastal plain I should henceforth throw my laboratory at the vast Amazon lowlands until I had the answer.

———————◆·◆———————

7

THE PARADIGM AND THE PROPHET

In 1980 refuge theory was about to become a *paradigm*, and thereby hangs a tale.

The word "paradigm" has taken on a special meaning in science, rather loosely based on its roots as special meanings often are, but clear and powerful to the practitioners of science. Crudely translated it means, "You had better believe it because everybody else does."

Thomas Kuhn had given this new meaning to an old word, explaining by it the very human property of scientists to think alike until devastating new data force them all to change their minds together. Then "everybody else" believes the new thing. This revolution in scientific thought leads to a large increase in knowledge.[1] Many a scientific tale illustrates what Thomas Kuhn saw, from the advent of atomic theory to the discovery of evolution by natural selection. But the tale best worth telling for the Amazon adventure is a geological tale of drifting continents.

Humans, geologists among them, walk ancient lands of rigid rock. As early as the mid-1800s, the geological trade knew that these rocks had heaved and buckled and moved, had been flooded by ancient seas, then dried again. All this was fine, because by then we had realized that far under our feet was a fiery mass of molten rock. Our world was the cold rind of a hot molten sphere. When the crust cracked we felt the earth quake. A crack was also a good enough explanation for a volcano; it let out molten rock under pressure. It was reasonable, also, to expect that crust could deflect into the softer substance beneath, like a dimple on a ripe lemon. We called this "subsidence."

Thus the geological mayhem around us was simply explained as an inevitable consequence of physics when a cold and rigid crust has its outer surface exposed to the cold black body of space while its underside is bathed in impossibly hot molten rock. This was rational, even parsimonious, thinking. "Choose the simplest explanation; abandon the simplest only when shown to be untenable!"; thus says the *principle of parsimony* that guides so much success in science. The essential geography of the earth could well be thought of as fixed; a continent was a continent, an ocean was an ocean, and both (roughly) always had been. A powerful worldview this, comfortable and according to common sense.

But eccentric thinkers muttered that this could not be right. The great Alfred Wegener made issue of a fact that many a schoolchild before him must have noticed: in an atlas the shape of eastern South America would fit nicely into the shape of west Africa. If only the sea were not there! Just squeeze out the Atlantic Ocean and Brazil's Nordeste nicely matches the Bight of Benin. Other bits of continent could also fit together like the pieces of a giant jigsaw puzzle. Doubtless an importunate schoolchild bold enough to want to fit continents together would have been (and was) told by teacher not to be silly.

But Wegener, explorer and traveler (he was to die a hero's death on the Greenland ice-cap), took a hard look at the coastlines and wondered.[2] He found similar rocks in similar order on the opposite shores of the oceans. Not only did the shapes fit, but the rocks had haunting similarities to opposite pieces of a break. It was as if an original pot had broken and its shards had been thrust apart, drifting away across the sea. Thus the conclusion of outspoken Wegener, plus a few sympathetic souls in the early twentieth century. The idea was called *continental drift.*[3]

But drifting continents? Did the things float like rafts, ploughing their way across the fluid mass of molten rock beneath them? The notion, shall we say, lacked obvious credibility. One is tempted to ask, "What about the bow wave and the wake?"

So leave the question of how continents might drift to others, and look for more evidence that the continents actually had moved apart. Wegener and those who thought like him appealed to biogeography. If indeed the continents had drifted apart, floated as it were across the oceans, they should have carried, raft-like, the living things of the old homeland. Just as the fracture lines of the continental breakups matched across the seas, so should the animals and plants "match." Granted the breakups were an awfully long time ago, but at least the ancestors of modern life should match so that close relatives would live on distant continents. Biogeographers intrigued by this theory scanned their data; dare I say, "In hope"? Other biogeographers, probably most of them, reviewed their data in a more skeptical spirit.

Biogeographers actually held their own worldview of more or less fixed geography firmly based on that same principle of parsimony. When an animal or plant was found on the "wrong" continent, as sometimes happened, it was decidedly more parsimonious to float the living things across an ocean than to float the whole continent as a ship. Darwin himself had scolded those "who built land bridges as easily as a cook makes pancakes." Biogeographers could play it safe by assuming fixed continents as the gameboard on which life moved about in evolutionary time.

Moreover, biogeographers had had their greatest triumph by dwelling on the distinctness of continents, on the unique properties of life in different parts of the earth. This was the prime signal that evolution must have occurred. Darwin, Wallace, and their early imitators had sailed the oceans and collected from the world. What they found was unrelatedness, with unique species on distant real estate. Origins were local; species had evolved in isolation; sometimes the isolation was clearly of whole continents (think kangaroos in Australia).

In fact there was no inherent incompatibility between this worldview of biologically unique continents and the idea of continental drift, because all conceded ultimate common ancestry. Australian kangaroos and the African primate *Homo sapiens*, for instance, are both mammals, descendants of some ancient common stock. This common stock had lived in a place or places unknown, long long ago. Early migrants to both Africa and Australia from this common stock are all that the data require. We didn't need continental drift for this so ancient connection.

And yet! Grand patterns of distribution would sometimes be the more easily understood if a continent or two had bumped together in the past, letting whole faunas cross rather than single migrants. Australia's oddities would surely be less of a challenge if the founding population had been one with the fauna of South America through the long gone animals of Antarctica. "All" that was needed was to join these three continents together, at the same time shipping them off to a warmer latitude. Wegener's jigsaw puzzle of the continental shapes allowed just this. If only it were true!

Orthodoxy struck back. In 1957 Philip Darlington of Harvard published his treatise on zoogeography, widely acclaimed, that demonstrated, once and for all, that the distributions of animal life on earth could be comfortably explained without any call for continental drift.[4] The text expounded brilliantly and in detail on how every known distribution, living and fossil, could result from migrations without moving continents. Some migrations might, it was true, be easier to explain if the continents had moved about, but given a hundred million years and much luck, an animal was more likely to get from point a to point b than have its homeland continent moved over.

A graduate course at Duke University in 1959 used Darlington's treatise as a text. I took that course, my first in graduate school. I spent a whole semester studying Darlington. I was utterly convinced. The learning was impeccable. The arguments were Popperian, models of parsimony.[5] The conclusion was as iron-clad as you can get: no unimpeachable biogeographical evidence for continental drift exists; and, what is more, there never could be any such evidence because moving an animal would always be easier than moving the land on which it lived.

Meanwhile, geologists were, almost literally, proceeding to move the ground under our feet by showing not only that the land did move but also how it moved. New techniques let ancient rocks be examined in ways hitherto impossible. And, as is so often the way in science, the new techniques speedily led to the destruction of old ideas. Critical were studies of remanent magnetism of rocks, particularly when coupled with radiometric dating, and then of studies of the floor of the deep sea.

In flowing lava and other molten rock, iron minerals in the liquid mix line up their own tiny magnetic fields with the earth's magnetic field. Once cooled, with its iron minerals immobilized, an ancient slab of lava preserves the ancient magnetic field, faint but still readable with a magnetometer. This remanent magnetic field acts as a fossil compass needle, telling us where "north" was. In ancient lavas, as it turned out, the direction of "north" was often not where it ought to have been.

The only plausible hypothesis for the whole earth to have a magnetic field with a north-to-south axis is that the earth works like a dynamo. The spin of this huge mass of viscously fluid interior under its hard outer crust generates a magnetic field lined up with the axis of spin. That this interpretation is broadly correct we now know by measuring magnetic fields on other planets with spacecraft, then comparing the results with the planets' rates of spin. Venus, for instance, has a very weak magnetic field; which is reasonable because Venus rotates very slowly.

The earth's magnetic field must always have been lined up parallel to the north-south axis of spin. In theory north and south could be reversed without violating the logic (the names are only a geographic convention), but the lines of force must be aligned with the axis of spin. The first measurements of remanent magnetism quickly showed fields in ancient rocks pointing in impossible directions. Obviously the rocks must have been moved since the lava cooled. But whole formations of rocks, countrywide, recorded the same impossible directions. Therefore the whole countryside must have been moved as one rock to "point" in a different direction. If vast chunks of landscape could move, why not continents too? In this way did the hypothesis of continental drift come alive again.

Soon it was possible to make time-lapse jigsaw puzzles of drifting continents. Measure remanent magnetism on rocks of the same age on a global scale, rotate and move each continental slab to align the fossil magnetic fields with true north, and the result is a map of where the continents might have been at a point in time. Repeat for other geological periods identified by fossil stratigraphy or radiometric dating, and *voila* a moving picture of drifting continents. Slowly, very slowly, the movie was produced, showing, among other mysteries, exactly the early fusing of Australia with South America through Antarctica needed to account for Australian oddities.

But to capture universal belief the drifters still had to show us "how." Geophysicists themselves had long since discounted, with their own data, the idea of continents rafting their way over a sea of magma. Seismic evidence demonstrated that continents were far too thick for any conceivable mechanism of ploughing their way across the mantle. Do your best at inventing heat engines or convection cells in a sea of molten rock, the answer was still, "No, continents are not rocky ships ploughing a magma sea." Thick continents could not drift by rafting, even though remanent magnetism had shown that they had.

It was oceanographers, mappers of the sea floor, who broke this dilemma. Sonar scanning of the sea floor found the great ocean ridges, underwater mountain chains, as high from their base as the mighty Andes but larger, far larger. The mid-Atlantic ridge is 2,000 kilometers wide in places and so long it stretches much of the distance from pole to pole. It neatly divides the north and south Atlantic oceans into eastern and western halves. These great underwater mountains, and their vast foothills, are almost bare of the sediment that covers the rest of the sea floor, at once suggesting extreme youth (geologically speaking). Sediments get ever thicker (older?) far away from the ridge.

The mid-ocean ridges are spewed lava mountains, as hot magma pours into the sea to be quenched by the great heat-sink of the ocean. So much is evident, and it explains the paucity of sediment on the mountain slopes; the cooled magma is too young to have collected much sediment. That the eruption, once started, should be virtually perpetual is then entirely reasonable because cooling at the seabed should set up a density gradient in the underlying rock. The result of this continued, endless production of new sea floor parallel to the ridge should be a sideways push, what we now call sea-floor spreading. Spreading of the sea floor then becomes an attractive hypothesis to explain how continents might be moved about after all. They are pushed.

The clincher came with another discovery in paleomagnetism, that the earth's magnetic field reverses itself at intervals. The field remains parallel to the axis of spin, as it must, but north can switch with south and vice versa. This is from an obscure property of fluid dynamics within the earthly dynamo; the direction of spin is not changed, of course, but turbulent flow within the fluid interior upsets the dynamo effect. The field collapses, quickly to be reinstated with polarity reversed.

Finding this impossible to understand, I once appealed to a mathematical colleague to tell me why this could be and learned that complex systems in hydromechanics were "predictably unpredictable." And with that I had to be content. But magnetic reversals happen. Stacks of old lava flows have been found in which the remanent magnetic direction is normal over reversed over normal over reversed, etc. Sometimes a single direction has prevailed continuously for millions of years, at others only for thousands. The sequence is global and has been reconstructed for a large part of geologic time.

Soon magnetometers were taken to sea, dragged behind ships just above the sea bottom. On a traverse away from the mid-ocean ridge of the Atlantic Ocean toward the distant continent to the east the magnetometers recorded the magnetic sandwich seen edgewise: normal — reversed — normal — reversed, and on and on and on as the mother ship sailed towards the land. Only one explanation was possible: the sea floor was built from molten rock poured continuously out from the volcanoes and faults of the mid-ocean ridge; it took on the contemporary magnetic direction, then was pushed aside by more molten rock. If, meanwhile, the field had changed direction, then the remanent magnetism in the new rock should record it. And so it had.

Thanks to the magnetic reversals, *sea-floor spreading* was now a discovery more than just a hypothesis. To complete our understanding, we needed only to know what the end might be. Sea floors could not spread forever! So what happened? Oceanography had the answer ready to hand. Far away from the mid-ocean ridges, alongside the continents, are the abyssal trenches, the deepest places in the sea. The trenches mark the line where spreading sea floor meets continental mass, now thought of as a "plate." The continent is nudged, but the spreading floor is driven down and under in a process called "subduction." The deep trenches mark the line of the subduction nose dive. Sea floor that was once hot magma becomes hot magma once again. And a nudged continent "drifts."

Geologists had found a new lease of intellectual life: patterns of geography and volcanism were explained in ways not possible with the old simplistic ideas of continuous cold crust over a molten interior. Earthquakes were understood as descending sea floor grinding against resisting continent; geological strata were reinterpreted with many fewer puzzlements; a template for the inner workings of a planet was found in time for the first explorations of the rest of the solar system. Needing words to explain our euphoria, we could do little but talk of a "revolution" in geological thinking.

For those of us who would understand the distribution and abundance of life on earth — biogeographers, ecologists, evolutionary biologists — the geological revolution was shock therapy. Continental drift lived! No more should we argue that it is easier to move an animal than a continent. All had to be rethought. I, who had so recently been a skeptic, utterly convinced by Darlington's book, remember clearly how I got the news. I was standing with a copy of the latest issue of a journal in my hands, I think *Nature*, and there before me was a map of magnetic stripes on the floor of the Atlantic Ocean; parallel stripes, normal, reversed, normal, reversed, in long progression from the mid-ocean ridge. Realization was instant and total. I gasped out loud, "So it is true!" This was forty years ago now, but I remember the incident as clearly as I remember what I was doing when I learned that President Kennedy had been shot, also forty years ago.

At the time we had to call the intellectual upset of plate tectonics a "revolution" because we had no better word. Now we call it a "paradigm shift." Thomas Kuhn's book, giving the new name, was being written even as the twin concepts of "seafloor spreading" and "plate tectonics" were being established, and he used this geological experience as a prime illustration in his *The Structure of Scientific Revolutions* of 1962. Despite that title, the philosopher Kuhn disliked calling upheavals in science "revolutions," a word more suited to violent changes in government, even blood in the streets. "Revolution" was hyperbole. What actually happened was the replacement of an old belief with a new one. So Kuhn reached for the word "paradigm," and talked of a "paradigm shift" instead.

But "paradigm" meant something different then, and always had done since it first entered the English language in the work of William Caxton in 1483. It comes from Greek and Latin words meaning "pattern" or "example," and could eventually be used to convey the idea of "typical example." It was in this sense that those of us, even scientists, who became literate before the 1960s thought of paradigms. Carl Sagan once wrote, "There is a generation of men and women for whom . . . the Moon was the paradigm of the unattainable."[6]

To those with impossible ambitions we said, "Don't reach for the moon." Good advice, even now that astronauts have walked on it. "Reaching for the moon" is still a good paradigm, or exemplar, of the unattainable in personal ambition. But Thomas Kuhn meant much more. His "paradigm" described the hypothesis selected by social consensus among specialist scholars as the one most likely to be correct. This paradigm was not merely an example of a pattern of be-lief; it was a social contract, inevitably entailing social enforcement. This is what I mean by my own definition of Kuhn's "paradigm": "You had better believe it because everybody else does."

A paradigm shift from one socially condoned hypothesis to another does have aspects of true revolution in that social pressures to believe are paramount. Pres-sure can even loose the temptation to fit data to hypothesis in defense of a fa-vored paradigm rather than to use data in an attempt to falsify, as a scientist should. The danger of this social process is what leads to pseudoscience, best il-lustrated by the pathological paradigm of special creationism.

American special creationists are like-minded people whose prevailing para-digm is taken from scriptures of long ago. Their community produces a long se-quence of writings trying to fit scientific data to their paradigm, even in face of a reality that nothing fits. This self-called "scientific creationism" is an intellectual pathology, but an amplified echo of what can happen in real science. Paradigms in real science, however, can be overturned by truly devastating data, like mag-netic stripes on a sea floor.

Kuhn's great achievement was to point out that these social processes are in-separable from scientific advance. Science is a human thing.

When first put forward in 1969, the refuge hypothesis of Amazonian specia-tion was nothing more than a highly speculative and tenuous hypothesis, relying on events of long ago that were not in fact known to have happened. But it cap-tured the imaginations of so many that soon, "You had better believe it because most people do" was directing laboratory research. Almost without discussion or argument a new "paradigm" had been born. Social behavior had overlain the cold dispassion of science.

Refuge theory met powerful needs of the human psyche: it swept away a mys-tery, it gave a story line to what you did, it called upon events of long ago, and it used a word "refuge" beloved of human sentiment. Start with the evocative word "refuge."

Biogeographers had talked of refuges since the earliest when they started to map life on earth. These biogeographers came out of the north, from universities or museums of Europe and America, in countrysides known to have been over-ridden by the glaciers of "the Great Ice Age." Land that had once lain under hun-dreds of meters of ice was now green with plants. Where had the plants come from, and the animals too?

Somewhere the ancestral animals and plants must have had "refuge" from the killer ice. So where were they? Obvious answers: "somewhere down south where it was warmer." This answer was safe enough; the animals and plants of temperate Europe and North America did have ancestors who lived further south. Beyond these simple truths of ice-age survival, though, we went wrong, and remained wrong for a long time.

We wanted refuge for forest, not just for trees, and were eventually to learn that there wasn't any. Trees survived the ice age down south, and many other ice ages before the last one too, but familiar forest did not. The textbook example of this discovery comes from a triumph of pollen analysts in eastern North America who had set out, quite explicitly, to find the forest refuge and plot the forest's journey north. They found a very different story: instead of the march of forest out of refuge they found showers of immigrant trees. The job took fifty years.

Partly it took so long because the scholars involved were inventing new techniques and a whole new discipline (pollen analysis of American lake sediments) as they went along. Partly because we started at the wrong, or northern, end of the presumed march of the forest: because the research universities of the early twentieth century were in the north, Yale prominent among them. Partly because the work can be time consuming, tedious, and poorly rewarded. More importantly, the work was beset at all stages with controversy.

These were the years when ecological theory was in its most holistic phase. Most ecologists talked not of single plants but of communities of plants. Whole communities were to be the units of study and were given special names: *associations, sociations, formations*, and the name most prominent still, *biomes*. These communities were societies of species with interlocking needs, living always together.[7] The ice-age refuge we sought was for the whole forest; for trees "living together" not just for outcast trees.

So the early pollen analysts in the New England states looked at their simple pollen diagrams showing the percentage change in a dozen or so pollen types, and extrapolated what they saw into the nearest equivalent pollen mix from a modern forest. They found, as expected, a land without trees immediately after the glaciers melted, then a time of coniferous forest reminiscent of modern Canada, and finally the spread of whatever forest is there now. Details led to argument, but the story held together: the forests had come from the south after the ice melted. And the fifty years of effort droned on.

An end to this easy belief came when Bill Watts, an Irishman from Dublin, crossed the sea to core lakes in montane Georgia and coastal Florida where the glaciers had never been. These were the lands of the expected refuge; but it wasn't there. Watts found only the pollen traces of scattered trees of the New England forests.[8] There never had been a forest refuge "down south."

The only alternative was that Yankee forests must have been built out of immigrants from hither and yon; a good American idea that! And it was so. The proof finally came only in the 1980s when Margaret Davis and Tom Webb, working independently, used the whole expanding data set of pollen and radiocarbon dates from all the lake cores taken in eastern North America to plot arrival times for each individual kind of tree since the ice age, doing this for stations all the way up the eastern seaboard. In their maps you can see the forest growing, species by species, as the trees come in from whatever unlikely place they survived the ice.[9]

Thus, although the hypothesis of a "forest refuge" from the ice was already set to fail in temperate lands, the concept of a refuge for forests against hostile climates was still venerable and popular in 1969 when Haffer published his version for the Amazon. Only Haffer's proposed peril was different: instead of the terror that comes by ice, he conceived the tragedy of a great drought. But it was still a whole forest community that found refuge, the trees and all the animals living together in safe havens provided by the vagaries of geography. Sanctuary! A satisfying idea, that.

The mystery that was "solved" by refuge theory was the origin of Amazon diversity. Instead of trying to find ecological ways for more species to coexist in the Amazon than anywhere else, the great puzzlement of our generation, Haffer gave us a simple species pump unique to the Amazon. It was a wonderful story line. I used to explain the idea in university lecture classes by saying it was Haffer's claim that Amazonia breathed out Darwinian species with the tempo of ice ages; refuge — merge, refuge — merge; in and out of forest refuges for all the million years of Pleistocene time.

Others used the story line more seriously, especially in Brazil where the great Amazon forest was more than an amusing abstraction. Unknown to those many of us who do not read Portuguese, a school of Brazilian science had been musing on the possible effects of ice-age climate on Amazon rainfall well before Haffer burst into *Science* with his refuges. I described some of the Brazilian work in chapter 6, with its title *Refuge Theory Expands in Brazil.* That title is a bit of a misnomer because Brazilian scientists were thinking in terms of a drier ice-age Amazon, and of how it might explain modern biogeography well before Haffer spread the word in *Science*. Haffer's was the concept of grafting the idea of refuges onto the dry ice age and thus tickling our imaginations with his evolutionary scheme to explain high Amazon diversity. But Brazilian biogeographers, Paulo E. Vanzolini to the fore, were already using glacial aridity as a tool to explain otherwise puzzling distributions of Amazonian animals that they knew so well. And it was Brazilian initiatives that looked for evidence of the dry times on the ground surface, leading to the discovery of dissected land surfaces revealed in the side-screen radar survey, and to what might be called "Ab'Sáber's poem of stone lines as evidence of past aridity." Within Brazil it might well be said that an Amazon forest fragmented by dry times in an ice age as a driving force for the distribution and abundance of modern Amazon animals was already achieving paradigm status.

For refuge theory itself to reach the status of a paradigm required much more. It required repeated success, the adherence of noted scientists, and above all publicity.

Even before publication, Jürgen Haffer had found encouragement for the evolutionary implications of his model, shown by his acknowledgment in the *Science* paper of discussions and encouragement from evolutionists like Ernst Mayr himself. For biologists long puzzled by the teeming diversity of the wet tropics and its evolution, the refuge hypothesis had that grand property called "generality." Ice ages affected the whole earth. Whatever local quirk of climate managed to fragment populations did not matter; it was enough that climate change and fragmentation were likely in all the tropics. For the most puzzling, most diverse system of all, the Amazon, Haffer offered a truly beautiful hypothesis.

But the stride from "beautiful" to "paradigm" requires spreading the new thing, particularly to the global communities of biogeographers and ecologists. "Word of mouth" advertising goes a long way in science, but the refuge theory did not have to wait this slow process. An international conference in the news, and the resulting symposium volume, made "word of mouth" unnecessary. Ghillean Prance, then director of research for the New York Botanical Garden, both organized the conference and edited the symposium volume that followed with none of the delay usual for conference symposia to be published.

Prance was ideal for this task. His own skills as an Amazon botanist had taken him to INPA (*Instituto Nacional de Pesquisas da Amazônia*) in Manaus, where he not only matched the distribution of plant families that were his specialties with Haffer's refuges, but also was well informed of the Brazilian work on biogeography of the Amazon, radar mapping of soils, and aridity hypotheses of Brazilian soil scientists and geologists (described in chapter 6).

The conference was probably the first high-profile international scientific gathering on the Amazon ever organized. It was given under the auspices of the Association for Tropical Biology, and funded by the Smithsonian Institution and the U.S. National Science Foundation. And it was to be held in the heart of the Amazon at Manaus, a circumstance that turned out to raise the profile even more when the military government of Brazil forbade this foreign invasion of the

Amazon. All the major scientific journals, together with some leading newspapers, had leading articles about this turn of events. The conference now had the publicity of the "banned in Boston" variety, as it went ahead in Caracas at the invitation of the Venezuelan government.

The high profile certainly had much to do with the conference title, "Biological Diversification in the Tropics." Nary a mention of Pleistocene climate, or the refuge hypothesis round which every invitation to attend had been based; just the dominant issue of "biodiversity." It was indeed high time that Amazon diversity was addressed in a very public forum in the presence of biologists from Amazonian countries who actually knew what the place was like.

I was not invited. When I saw from a list of the invited contributions that the conference was really a celebration of the refuge hypothesis, not a dispassionate inquiry into biodiversity at all, it did occur to me that I was the only researcher who had actually published radiocarbon-dated evidence for aridity in equatorial America with the El Junco record from the Galapagos. True, the Galapagos are offshore islands, but they are a great deal closer to the Amazon than the East African records with which Haffer justified his claims for aridity. I was a bit narked, but hindsight tells me I was well out of it. As people talked refuges and ice-age aridity all round me, I should have been bound to state my unique hard data showing that a neotropical and equatorial land had indeed been arid in ice-age time, giving credence to manifold speculations. I should soon have regretted it.

For a "celebration of the refuge hypothesis" was indeed what the conference turned out to be. The proceedings volume, still with the same misleading title, is a written record of this celebration.[10] Prance, in the opening sentence of the book, frankly belies the title by saying that they had decided to use Haffer's refuge theory as the symposium topic *"because of its influence on biogeographers."* The conference and the book were not about the *causes of biodiversity* at all; they were about an untested theory of ice-age climate and evolution.

The book has thirty-seven chapters by a total of forty-one different authors. Prance and Haffer state the refuge theory forcefully in the opening chapters; then come the Brazilian soil scientists with their stone lines, gullies under the rain forest, and soil pediments all over Brazil. All as previously published in the Brazilian literature, but with one intellectually important addition, the availability of the first radiocarbon dated Amazonian section with data from glacial times. What is more, this record was published by Rhodes Fairbridge of Columbia University, one of the most revered geologists of his generation.

The Fairbridge laboratory had samples of sediments from the Amazon fan, taken with a Kullenberg piston sampler from the deep sea off the mouth of the Amazon where the great river dumps its mud. Thorough radiocarbon dating demonstrated without question that they had ice-age mud. And in that mud were feldspar minerals. Feldspars decay rapidly under tropical humid weathering and thus should not be present on the Amazon lowlands where, it was assumed, the Amazon River got its sediment. As the hypothesis required, mud deposited by the river over the last few thousand years of known wet climate was found to have no feldspars. Eureka! the ice-age feldspars were from the great arid time when there was no humid weathering.[11]

This discovery, especially as it had the Fairbridge name attached, forced its way into our consciousness. It flattened me. It was more than anything responsible for my writing that *Ice Age Amazon* news piece for *Nature* (chapter 6). Rhodes Fairbridge is an inspiring man. In my young scientist days I was an ardent member of an informal organization called "The Friends of the Pleistocene" whose purpose was to tramp together round ice-age relics on the lands of New England, afterwards enjoying a vinous banquet in honor of the Pleistocene. The best memories are of Rhodes Fairbridge and Richard Foster Flint, the Yale guru and author of the contemporary leading textbook of Pleistocene geology, standing before cliffs of ice-age debris arguing about what it all meant. When you are young, arguments between the greats, the more heated the better, are the most privileged education there is. So Rhodes Fairbridge's "arkosic sands," as the feldspars are known, were to worry me long after I had reason to think something must be wrong. I did eventually see where the error lay as the Pleistocene history of a long succession of ice ages began to unfold. Whether Amazonian climate was arid in the last ice age or not, it was impossible that the lowlands could ever have held "arkosic" sands because they had been wet enough in earlier interglacials to leave them feldspar-free. So the feldspars in the fan sediments had to have come from somewhere else; from deep in the earth perhaps or else from far away. Few doubt now that that they came from the Andes, whose lavas are rich in feldspars. They were carried all the way to the distant sea in a river system running a little faster as the hundred-meter fall in sea level provided a steeper slope across the lowlands. But for refuge theorists the arkosic sands in the fan were something like proof of their thesis: authoritative, radiocarbon-dated evidence of ice-age aridity.

And that was it. For the whole Amazon, a region as large as the continental United States, Prance's symposium could muster only one dated record of dubious interpretation from the sea and a few observations of soil scientists of landforms they found puzzling unless there had been a drier climate at some unknown time in the past.

Next call on pollen analysts for evidence that the postulated aridity was reflected in vegetation. They had one response, not from a properly dated section but from discrete drill samples of a geological survey in what appears to have been the infill of old valleys or swales. The pollen analysts, working in Europe, had never seen the site, which was near the edge of the forest, within 100 km of open savanna, nor were the samples radiocarbon-dated.[12] The pollen they found was nearly all grass, except for the swamp tree *Mauritia*. The most they could show was that the forest-savanna boundary, always in a state of flux, had at some time in the past moved 100 km to the north at that place. No matter: Eureka! once again. This was the savanna required by the hypothesis to have replaced the whole forest.

I have done a careful rereading of the Prance book, all 680 double-columned pages of it. After the introductory chapters, one-third of the papers cite this thin list of reports as having *demonstrated* fragmentation of the forest by savanna and aridity in ice-age times. The same thin list always; I took to using the notation "the usual suspects" in my notes. Having thus established an ice-age forest fragmented by savanna as given, the biogeographers among them then looked to see if their own distribution data matched. This group of authors tended to find their matches. They were helped by the Ab'Sáber map, the one that triggered my essay in *Nature*, described in chapter 6. It showed the whole of South America imagined with sufficient drought to produced the arid land condition in the Amazon lowlands required by Ab'Sáber and Bigarella's interpretation of soil profiles. As such, it too depended on "the usual suspects." It reappears at many places in the Prance volume, each time with the inferred ice-age distribution of another group of animals or plants. The repetitive suggestion from these papers is powerful.

Conservationists aided the push. What should they tell politicians to conserve? The bits with the most, surely. But where were these places of super diversity? We did not know. But those spaces already mapped by biologists as sites of ancient refugia, where theory put the very origins of many species, seemed as good a bet as any. Well before Prance called his conference, Brazilian conservation planners were mapping out their preferred "refuges for the future" round Prance and Haffer's "refuges of the past."

Anthropologists were more problematical. Recruiting them to the refugial team faced the hazard that human populations were (and are) thought not to have reached South America until after the ice age was over. So why should human cultures or language patterns in the Amazon basin reflect ice-age geography? A virtually impossible correlation; but two scholars gave it a go. Start with the "usual suspects" to show that the land was once arid, be convinced that hunter-gatherer peoples needed open land rather than forest, expect that the forest was slow coming back so that the open land period lingered on until after human arrival, then look for the match. One archaeologist even set possible courses of colonization onto Ab'Sáber's map, drawing arrows of advance like the symbols for offensives on military charts; a brave attempt to link modern cultures to what happened before the people came. In effect, the chosen scholars signed off for the discipline of anthropology as endorsing the theory.

Three of the chapters had little patience with refuge theory because their data plainly do not fit; papers on Amazon fish, Amazon frogs, and Asian forests. OK, exceptions. And a few chapters fitted the title of the symposium without contributing to the Amazon refuge debate; the biodiversity of Mexico, for instance.

But two chapters were by eminent scholars whose presence should be overwhelmingly necessary for a conference whose declared intent was to examine Pleistocene refuge theory: Alwyn Gentry and Daniel Livingstone. Both men have entered my story in earlier chapters, Gentry as the finest neotropical botanist of his generation, Livingstone as the pioneer paleoecologist of Africa, responsible for the key evidence for ice-age aridity there.

Gentry's primary work had been on the Pacific flank of the Andes, particularly on the northern part of that flank called Chocó. Now the Chocó was included in the Prance and Haffer maps as a Pleistocene forest refuge, but it was a refuge with a difference in that it was separated from the lowland forests of the Amazon not by postulated savannas but by real and concrete mountains. Gentry's analysis of the highly diverse Chocó forests let him conclude that the forest biome there was ancient, certainly not fragmented in Pleistocene times. Gentry left it at that. From the refugialists' point of view, OK, as far as it went, they had always claimed the place as a refuge for persistent forest. But there was a sting in the tail for those who wanted to find it; Gentry's argument was that the rich diversity of interacting life in the Chocó was evidence for forest antiquity, so why not similar antiquity for the lowland Amazon forest also? This unspoken question was not answered in the symposium.

Livingstone went about his business of showing that East Africa was arid at the last glacial maximum with implacable data, much of it gathered by himself; hard, radiocarbon-dated evidence of drying lakes, coupled with radiocarbon-dated pollen sections demonstrating the replacement of forests by dry woodland or savanna. This was a far cry from the "usual suspect" sort of data being offered for past Amazonian aridity, making it a vital crutch for refuge theory. "If you want to criticize the evidence from the Amazon as inadequate, okay, give us time; but just look at this hard stuff from Africa!"

But Livingstone took pains to point out that demonstrating the ice-age spread of African savannas did nothing to show that forest was confined to refuges in the Pleistocene even of Africa, let alone the Amazon. He told the tale of the forest refuge that never was in southeastern North America, how he had accepted the hypothesis when a graduate student at Yale, later, like his peers, having to abandon the idea in face of Bill Watts' pollen evidence (see above). He then went on to summarize his experience in Africa in these words: "*We have some information in range changes of forest trees. And yet, I would be unwilling to undertake the specification and location of a single Pleistocene forest refuge in Africa. In tropical America, where the greater biotic richness might reasonably be construed to cast doubt on the whole idea of forest reduction by Pleistocene climatic change, I would be more unwilling still.*"[13]

Livingstone goes on to offer his hope that his unwillingness should be contagious, a hope fated to be denied. Refuge theory was the new thing, rapidly becoming the only thing. Even before the conference it had become something like a national cause among scientists in Brazil, who had done most of the Amazon work and had a valid claim that they were first with the hypothesis of an arid ice-age Amazon influencing modern biogeography. Prance's prestigious conference proceedings gave prominence to all this work, acting as publicist and prophet, aided by the delicious contrariness of denying that the huge, timeless tropical forests of the Amazon were anything but, being mere products of the last few thousand years like the temperate forests of Europe and North America. All this gave refuge theory a strong following, yet it was the solution offered to an evolutionary question that was decisive. Refuge theory seemed finally to settle an argument that had simmered ever since Darwin. It was that which turned guess into paradigm.

The concept of Amazon forest refugia had not been invented to solve the problem of high Amazon diversity but to explain endemicity in forest birds without violating the principles of the modern synthesis. It was necessary for those forest birds to have lived in *allopatry* (physically separated) when their specific differences were fixed by natural selection. Haffer invented refuges as the only plausible way to isolate populations of such motile organisms as birds from the *sympatry* (in the same place) of life in a vast forest. In doing so, he removed a potentially nasty thorn from the side of the modern synthesis.

For the astonishing biodiversity in the Amazon forest was as big a difficulty to the modern synthesis as it was to ecologists. Ecologists were struggling to explain how that vast concourse could coexist in tropical forests when other environments had so much less, but contemporary evolutionary theory met the problem of how that same vast concourse living in sympatry was fashioned by natural selection that required allopatry for each individual species to have formed. In offering a possible solution to the evolutionary problem, Haffer had inadvertently rescued ecologists from their diversity problem as well.

Thus two of the most testing challenges of contemporary biological thinking were met by this one brilliant hypothesis. No wonder it was grasped so rapidly by so many. Even the title chosen by Prance for his conference and volume *Biological Diversification in the Tropics* becomes a clear statement of what in fact was happening. The title was a bold way of stating that "These great problems are solved."

Ernst Mayr, in his 1963 treatise, *Animal Species and Evolution*, had summarized the thoughts of a generation and resolved a hundred-year debate on the mechanism of speciation: physical separation of populations was the essential prerequisite.[14] Mayr supplies a record of how strong, even virulent, had been the debate over more than a century before the final consensus was achieved. Isolation of species by physical barriers had been noted in Amazonia as early as 1863 by the legendary Amazon naturalist Henry Bates, particularly when species limits were set by rivers the animals could not, or would not, cross. Soon afterwards, in 1869, the German naturalist Moritz Wagner proposed a formal theory of speciation by geographic isolation based on his own direct observations in the tropics. Penetrating observers these two must have been, because the first stunning view of the tropics gives most of us northern outsiders a confusing impression of an immensely diverse array of species living together; you must know them well to discover separations.

Mayr notes how Darwin's writing reflected the crucial influence of the physical isolation he had seen on islands at first, but also how he had wavered later when coming to grips with how natural selection actually worked. But Mayr's most devastating insight, at least to this reader, is how great population geneticists were later to waiver on almost the same issue. Even Thomas Hunt Morgan, with his fruit fly *Drosophila*, and R. A. Fisher, architect of the mathematical theory of natural selection, missed the fatal flaw in the idea of new species being formed in sympatry. Their pioneering studies of experimental animals in laboratory containers (Morgan) or models of how individuals should exchange genes in sexual species (Fisher) inevitably concentrated their minds on how selection works *in situ*. They were studying evolution in the sense of how the individuals of a population can be changed by natural selection in response to changing environments, rather as had Darwin himself in his later work. They succeeded brilliantly in showing how it was done. But the scope of their experiments and models obscured how traits, even apparently favorable traits, would be diluted to impotence in huge, widespread, cross-breeding populations. It is this reality that made speciation in sympatry so improbable to the modern synthesis.

When Haffer postulated his Pleistocene forest refuges, he was armed by Mayr and his codevelopers of the modern synthesis. Speciation was possible only in allopatry; his refuges met this requirement in spectacular style. This is what lay behind the efforts of so many biogeographers to see if their plants or animals "fitted." And it was the reason that the Prance volume served to elevate a speculative idea into a paradigm necessarily included in textbooks, despite the dubious evidence for the actual geography of the ice-age Amazon and despite the poor fit of much, if not all, biogeographical data.

In the euphoria of the new theory, none of these worries had an impact; they could essentially be brushed aside, both in Prance's summary chapter and by the general acceptance of the refugial paradigm that followed.[15] Only one alternative model had real sting: L. A. Endler's model of speciation along clines.[16] The model took advantage of the immensity of the forest where animals could live so far away from each other that it was unrealistic to treat the whole population as one cross-breeding unit. But this was soon knocked out of the intellectual race by Ernst Mayr himself in the journal *Evolution* in 1986.[17] In effect, Mayr's paper probably served many as a seal of approval of the new paradigm.

Ernst Mayr walked into my office one day *circa* 1993. I was then working as a staffer at the Smithsonian Tropical Research Institute (STRI) in Panama and was in the process of publishing my laboratory's first pollen histories from lake sediments in central Amazon lowland forests. Quickly he came to the point. He had heard that I was disputing the refuge hypothesis and wanted to know why. We spent the morning talking of my group's efforts to test the Haffer hypothesis directly with the record of ancient lake sediments. I remember so well Mayr's parting remarks, standing by the door to leave: "Paul, if you get the cores from the Amazon that you are trying to get, and if they show, as you expect, that the Amazon was never arid, nobody will believe you because the refuge theory is so beautiful." And then he was gone.

Such is the power of a paradigm.

8

AMAZON'S BITTER LAKES

The decision to throw the resources of my laboratory at the Amazon lowlands until I had the answer did not stipulate that I leave immediately for Brazil. Ecuador had Amazon lowlands too, and in Ecuador I had a base. Remember Clausewitz: "First secure your base." Miriam once again led the charge.

The Amazon third of Ecuador is a tongue-shaped chunk east of the twin cordilleras. Most of it is just 300–200 meters (900–600 feet) above sea level. Several of the major tributary rivers of the Amazon system flow through these lowland forests of Ecuador, significant rivers to northern-bred eyes, the Thames at London, say, or the Charles between Cambridge and Boston, but trivial compared to the giant they will join in the next thousand miles (1,650 km) before their combined waters' final thousand-mile run to the sea. On this colossal journey from Ecuador to the Atlantic the water will fall only 300 meters, an almost imperceptible gradient but enough to do the job.

We chose the rivers Napo and Aguarico as our access highways. Fed by cascades down the eastern flank of the Andes, they are tamed as they wind through a forested landscape rich with lakes. We had little hope of these lakes, predicting correctly that most would be young remnants of old river channels and meanders. But the Napo and Aguarico dance their way along the very equator itself, the ideal place to be. Just possibly an older lake might be hidden in the youthful array. The bitter truth was that we had no idea where an old lake might lie in Ecuador's lowland forest, nor Brazil's, for that matter. Back in the last millennium, in the 1980s, we had no satellite surveys that should let us find lakes in the 3,000 km expanse of lowland forest, let alone data that would show us which were old. We must explore the old-fashioned way: go, try, talk to people, then go try again.

Rivers like the Napo and Aguarico are shallow, winding, and as afflicted with mud shoals as the Mississippi of Mark Twain's day. But they pack a heavy punch of raw energy from their plunge down the mountains. They cut way into the weathered clays of the lowlands, chopping down ten, twenty, thirty meters or so of steeply sloping bank to the rivers' own floodplains. Imagine a giant hosepipe thrashing about loose from the pressure of water pouring through it. With this sort of energy it was not inconceivable that deep basins could be scoured out, afterwards to become ancient lakes. If any old basins existed, the problem was to find them among the many more recent puddles. If we found only puddles at the first try, at least we should have sediments to let us begin the dread business of finding climate signals in Amazon pollen. (I still had a frightening awareness of those 80,000 species of plants and the trees pollinated by animals who would keep the pollen for themselves.)

My first lake was Lago Agrio. In February 1980, Miriam, then a few months past her Ph.D., and I went alone for a first look, as was our custom, using seed money and Miriam's formidable skills at coaxing help from her native land. The oil company Texaco had a base at the jungle town of Lago Agrio, a place named after the eponymous lake. Miriam called on the Quito office of the company, finding, as she nearly always did, the occupants to be friendly. They found room for us and all our equipment in the company aircraft in the morning, with lodging at their Lago Agrio base to follow. Bless them and thank you.

"Lago Agrio," literally "bitter lake," is a rectangular stretch of water more than two kilometers long but narrow, like a giant lap pool in the rain forest. The water fits its name: *agrio*; black from the air; it is brown from close to and so opaque that a bright white "secchi" disk lowered into the water vanishes at 25 cm.[1] Plunge your arm over the side of a boat, and you cannot see your hand. It is color, not sediment, that makes the water so opaque, the rich brown color of bog water, *agrio* indeed. Brown and opaque and tending to acidity: these properties of Amazonian lakes were to become familiar over the next eighteen years; it is not only Lago Agrio that has them.

Having started soft with the hospitality of an oil company, we continued soft, taking a taxi to the lake. Overloaded with all our gear, the ancient vehicle pounded over corrugated road like a wooden wagon without springs for a few miles, and there before us was the dirt track to the lake. But someone had dug a deep ditch and rampart, totally blocking the dirt track, presumably part of a pipe-laying project. Get out and walk, only some 500 meters, but several round trips needed to move all the coring stuff. I remember sicking up beside the trail, possibly dehydration. But we got in and built our raft on the lake shore at a place obviously used by local fishermen on days when the approach track was not cut. A hand-painted signboard on a post said, "No dynamite."

We rowed across and back, and a few hundred yards each way down the length of the lake, me at the oars, Miriam sounding and collecting as we went. No doubt at all that this was an old river channel: shallows with emergent tree stumps at the sides; a flat bottom under four meters of water in the middle; a long, straight, and narrow lake. The emergent tree stumps were a worry; was the lake as young as that? But an answer was on the shoreline in what was clearly a waterline nearly two meters above the water on which we floated. Locals had told Miriam that the weather had been the driest anyone could remember, with no rain at all for a month. Falling water level had exposed old tree stumps that can endure submerged for centuries.

This could not be our kind of old lake, but we cored it just the same, anchoring the raft, lowering casing, and coring through to the sandy basin by five o'clock that first afternoon. The mud, lovely, richly organic gyttja, was just 1.75 m thick. It lay over sand, clearly river-bottom sand with water-rounded grains, proof that the lake was a stretch of abandoned river channel. Despite this, we now had a core with Amazonian forest pollen in it, and from our very first day in the Amazon lowlands. Good for a start.

The twelve hours of daylight were fading when we had hauled out to the bank, so we left the gear, trusting to the blocked trail to guard it for one night, and walked out the few miles to the comfort of the Texaco camp.

Nothing had been touched when we went back to work in the early morning. First task take the vital second core to confirm the stratigraphy of the first, then do some Amazonian limnology.[2] We must have a temperature profile. The proper way to measure temperature is with a probe on a long cable from the instrument in the boat. Not only did we not have the probe with us in this low-cost effort, I had not so much as a thermometer. Not to worry: human skin is thick with temperature sensors, so I simply swam to the bottom, making records in my central nervous system as I went. Sophisticated science at work.

The top water layer, a couple of feet thick (about half a meter), was warm, unpleasantly so, like submerging your head in the bathtub. But go a little deeper and find the delights of refreshingly cool water. I get this information from my field notes, dictated at the time to a machine so that must have been what my body thermometer told me. The actual temperature range measured on a later visit with a submersible thermocouple was only four degrees Celsius. Four degrees' difference apparently seemed pretty good to the overheated human when the surface water was nearly 29° C (84.2° F).

Whatever the thermometer used, Lago Agrio was thermally stratified; steeply so, with a sharp discontinuity between the warm upper water and the colder depths. I was not ready for this. In the temperate northern lands we are used to stratified lakes with a fairly thin layer of warm water floating over a mass of much colder water. Those who fish for lake trout in a Canadian summer can cool their six-packs by dropping them over the side attached to a long line.

Northern lakes cannot help being like that in summer. In spring their water, emerging from the winter ice-time, is cold through and through, ice-water temperature all the way to the bottom. Then the summer sun warms the surface, the warm water expands and, being lighter than the cold water below, floats like a rubber duck in the bath tub. Sinking a rubber duck is hard; so is sinking a parcel of warm floating water. Nothing but a real gale of wind can do it. But the gales seldom come until the fall of the year when the surface water is already cooling under gray skies. Then the lake water is mixed again, one temperature all the way to the bottom, waiting for the ice to form.

None of this should apply to an Amazon lake, or almost none. The "almost" is because all lakes are heated from the top by the sun. But the Amazon has a hotter sun, and it doesn't go away except at night. You might expect that Amazon lakes would be heated right down to the bottom, a single temperature all the way from the surface to the mud. A lake like that would have no rubber duck effect and could easily be mixed by wind and wave, carrying oxygen all the way to the bottom. But no, that is not what happens.

The rubber duck is still alive and well in an Amazon lake because it needs less of a temperature contrast to work in warmer water. The relationship of density to temperature is not linear but exponential, which is to say that for every degree of warming, the expansion or "making lighter" of a water mass is greater. Surface water, being lighter because slightly warmer, floats over slightly less warm water in an Amazon lake just as surely as warm water floats over cold in Canada.

All you need for an Amazon lake to be stratified is an occasional influx of cooler water, perhaps from evaporation at night or from groundwater, though the big source is likely to be rain. Heavy rain showers come down bitterly cold, so cold that we were to get into the habit of jumping off our coring raft into the lake to keep warm until a shower passed. Cold rain dumped into the shallows should flow to the bottom under the warmer surface water. But Lago Agrio is so close to the Andes that I suspect groundwater from the hills contributes to that 4° C spread. Two degrees turned out to be more usual in the lowland forest further from the mountains.

This habit of Amazon lakes to be thermally stratified despite the absence of winter was crucial to using the mud for environmental history. The bottom water is cut off from the air by the floating surface layer, starving it of oxygen. Oxygen supplies of the bottom water are quickly used up by the respiration of the quick and the decomposition of the lake's dead, making the bottom anoxic. The resulting water without oxygen over mud without oxygen is the perfect system for preserving pollen shells that would be destroyed in the presence of free oxygen.

With the survey of Lago Agrio finished, it was time to go, particularly for Miriam, who wanted to chase down a rumor of a far more interesting lake at the nearby village of Santa Cecilia. She left, but I gave myself a quiet hour in the forest, on the far bank of the lake where there was no sign of human presence.

Although thinking "Amazon diversity" for so many years this was the first time I had been in the lowland Amazon forest, having slowly made my way there via Galapagos and Andes. The local forest was a shock, not for its grandeur, but from the lack of it. I had in mind soaring giants of trees, with buttress roots tall enough to hide a man, with stems more than a meter across above the buttresses, a canopy spreading thirty or forty meters overhead with open ground beneath, as I had seen in Nigeria in my student days. There are such trees in the Amazon, plenty of them, but not in the riverain forest at Lago Agrio. This was jungle, dense, dark, and resisting entry.

There was no place to land the little rubber boat. So I tied up to a straggling mass of branches drooping over the water, climbed a tree, and played Tarzan for ten meters or so into the forest before dropping to the ground. I could see very little, certainly not the lake from which I had come, getting a curious feeling that there might well be many miles of unbroken jungle in front of me in which to get lost. I took a compass bearing from where I thought the lake was and pushed on.

Such a place is wonderfully silent. Keen ears can detect rustlings, but the main sound is the crunching of one's own boots. Stand still, and silence again. Perhaps there should be shouting birds or noisy troops of monkeys; we were to enjoy many such in the years to come. But not everywhere. And Lago Agrio was close to its eponymous frontier town where shotguns were plentiful. Whatever the reason, the forest was deadly quiet. And it was amazingly dark. The human eye, with its magic adaptability, can see well enough, but a camera cannot. At noon in the forest I was later to use a Nikon FM camera with an f2.8 lens and 800 ASA film to find that I still needed an exposure of one second at noontime; the equatorial sun directly overhead shaded almost to nothingness by the five layers of canopy.

I collected samples that I hoped might yield a measure of pollen fallout from the forest, some moss from tree crotches, which can be a useful trap for falling pollen, and a bag of pinch samples from the soil surface. The compass directed me back to the boat; a second, briefer game of Tarzan, and I was rowing "home."

I saw my first caiman, a small one, about a meter long. It floated belly-up, a great gash across the white scales of its underside, its guts spilling out. A boom and splash down the lake before I reached the home shore told of the fate it had met in the last week. A man appeared, poling along on a simple raft. As he went he threw short sticks of dynamite, looking keenly at the resulting turbulence. I landed by the "No dynamite" notice and packed the coring rig out to the rampart and ditch.

Miriam arrived at the rampart in a truck provided by the oil company. On with the coring rig, while she bubbled over with the good news of the lake rumor turned into a real lake, this one on a "military base." The oil discoveries had set down in the jungle all the characteristics of a booming frontier township, reminiscent of images of the Old West of U.S. fables. The Texaco compound was close by, but so was the "fort," in the form of an outpost of the Ecuadorian army. And near the army base at the village of Santa Cecilia was a lake, a fine elliptical basin with steep banks, where soldiers relaxed by fishing.

Miriam's informant had driven her out to see the lake. Her eyes gleamed as she reported that "it looked a bit like El Junco or Cunro." She had arranged transport, for that very afternoon, of just our second day in the Amazon. The road to the village of Santa Cecilia passed within 300 meters of the lake, which was in plain sight across cleared land. We man-and-woman-hauled our gear and had a raft on station by 7 PM. Coring was easy, a textbook operation. We had got through ten meters of sediment under seven meters of water and were hammering at something harder when the sun went down at 10 PM. The hard base of the sediments felt like the old weathered land surface, not the river sand of Lago Agrio. A good sign. We hauled out in the brief twilight, to be driven back to camp to get ready for our promised rendezvous with the Texaco aircraft in the morning.

Santa Cecilia lake was a splendid improvement over Lago Agrio. It was an ellipse of water seven meters deep, some 50 m across, and with steeply sloping sides ten to twenty meters high. No sign of old stream channels entering or leaving that I could see from the raft. From that brief look-around, it could even be one of those highly desirable "closed-basin lakes," the property that had let Miriam compare it to Cunro, or even El Junco. An El Junco in the rain forest, what a prize that would be! And we had a core ten meters long, not like that old river channel, Lago Agrio, with its meter and a half. These ten meters were good for the spirit.

Miriam named her find Lago Santa Cecilia after the village, and it was in high good humor that we flew back to Quito in the morning.

Euphoria at having got a decent core from Santa Cecilia lasted long enough for the writing of the next grant proposal, and the longish core gave hope of even better from larger lakes further out in Ecuadorian Amazonia. We had limnology data from two Amazon lakes (water chemistry, stratification of the water column, plankton collections, basin morphometry) when very few such data were available from the entire Amazon basin. Our proposal promised to hunt for still longer cores from lakes in the real rain forest well away from the mountains. At the same time we had assembled a team for full limnological studies: water chemistry, plankton diversity, the physical properties of bathymetry, temperature and oxygen profiles, productivity measurements, and shoreline or aquatic vegetation. Our worries that none of the lakes could be truly old in that water-swept plain remained, but anything we got would be new. In 1982 NSF funded us to go ahead.

Before we left for the new venture, the bitter properties of our first two lakes were expanded. Lago Agrio had a radiocarbon age of just 675 +/− 190 years BP. Worse yet, the bottom of that lovely 10 m core from Lago Santa Cecilia was without sufficient carbon for a date.

Our method of measuring the carbon content was crude but standard for those days: we took it to be half the "loss on ignition." We weighed a sample, burned it, and weighed the ashes. Half the weight lost from the burning was then assumed to be carbon. We refined the process a bit, but that was the essence of it. For the lower meters of the Santa Cecilia core the loss on ignition was little more than 1%, showing that the carbon content was far too low for the methods of radiocarbon dating then available. We had to move up to the fifth meter to find enough (barely) carbon, where the radiocarbon age was 750 +/− 100 BP. Allowing for the stated measurement errors, this was the same age as the bottom of Lago Agrio.

Thus ended the euphoria for the long core; worse still, it was an augury of other results to come.

Two years later Miriam and I were to fly low over Santa Cecilia and Agrio in a light aircraft and saw with our own eyes how wild were our early hopes for an ancient basin. We could easily trace the course of a single river channel of which both lake basins had once been a part. The deep water and steep banks of Santa Cecilia probably formed in a bend where the current had gouged out a pool, whereas Agrio was a long straight stretch of rapid water. Tropical erosion, slumping, and forest growth had all but obliterated the lengths of channel running in and out of Santa Cecilia, leaving only traces, easy to see from the air but nearly invisible on the ground. Erosion of the steep local landscape had put more sediment into Santa Cecilia in the same limited time. That was the only difference.

What we did have was evidence that a river, probably the Aguarico itself, had made a radical change of course seven to eight hundred years ago. An Andean volcano pooping off or a landslide? Could be. But a climatic event could force a river to move too, and that would be interesting.

9

ON THE TRAIL OF
FRANCISCO DE ORELLANA

The National Science Foundation gave me the go-ahead to finish the job in Ecuador. Striking lucky to reach ice-age sediments would be wonderful, but go anyway. Even Holocene histories of climate change (or no change) at the Amazonian equator would be a plum, together with as detailed a limnology of these unknown lakes as we could manage. But I knew, and I am sure the NSF panel knew, that an ice-age record was the goal. My original quest, started at Yale so long before, was ice-age climate data to help with the diversity problem. And looking over everyone's shoulder, including mine, was the refuge theory. And increasingly I was losing my old conviction that the paradigm might hold truth. The soaking Amazon I was coming to know had never dried as the paradigm said, nor could the refuges have been small or discrete enough to serve evolution's purpose. Mostly I think it was the very weakness of the arguments, particularly on land forms and soils, in Prance's book that had set me to realizing the absurdity of a drying Amazon. I hungered for a glacial section that would settle the matter, even as I doubted I should get it in Ecuador.

What NSF gave me amounted to a license to explore the Oriente Province of Ecuador for ancient lakes with a team of limnologists and paleoecologists, rather as it had done for the Galapagos Archipelago fifteen-odd years earlier. The first target lakes to be reached from the Napo River already had Quechua names, for this was country long used by the ancient civilizations of Ecuador. So we set our course for Limoncocha, Garzacocha, Añangucocha, Taracoa, Zancudococha.

Conquistadors had been through the Napo country long before we did, and left records that told of its properties as we unwittingly followed a path hallowed by history. Francisco de Orellana and his gang of ruffian adventurers went down the Napo in 1541, eventually to force their way from the Andes to the Atlantic Ocean, 3,000 miles downstream as rivers wiggle. It was the first crossing of the continent, and it happened by accident.

Orellana was a lieutenant in the service of Gonzalo Pizarro, brother of Francisco Pizarro, self-appointed first governor of Peru and founder of Lima. In a nepotistic *coup d'état* Francisco Pizarro had made his brother Captain General of Quito in 1538. Gonzalo consolidated his power, then heeded the lure of riches in the Amazon jungles. "Gold was there in the east," lied the conquered people of the Inca empire. "El Dorado," imagined the Spaniards. So Gonzalo set out in 1540, taking Orellana and some 350 Spaniards with him, as well as horses, pigs, and a multitude of Indian slaves.[1]

Months of sickness, hunger, and dead men later they reached the jungles at the foot of the mountains and saw a river running east. They followed its banks for days, accompanied by more sickness, hunger and death. The jungle would have been like that at Lago Agrio on my first Amazon walk, for the route was at about the same altitude in riverain bush, not friendly to walk through at any time, but deadly stuff for men in bad shape who are out of grub. The river was the Napo, and where they finally stood knowing they could cut their way no further, the Napo was already bigger than any river they knew in Spain. This was probably between what are now the Ecuadorian settlements of Coca and Nuevo Roquefuerte, not far from Limoncocha, where we planned to begin.

A council of war decided that there was nothing for it but to build a boat, let whom it would hold go on by water, later to report back on such fortunes as they found. I use the phrase "council of war" deliberately, for these were men bent on conquest and loot. They went armed, humping through the jungle under steel helms, with broadswords at their sides, a few primitive and heavy firearms, perhaps with steel breastplates (the old accounts picture the conquistadors so equipped), dreadful impediments to travel in sweltering jungle. Like other war parties of the period, they marched without a formal commissariat, expecting to live "off the land," which is to say by plunder. But the jungle was not well supplied with plunder.

Those of us who have canoed the Napo and camped on its banks can wonder why Gonzalo's large expedition found so much suffering and death there. But they had no proper commissariat, and they had people by the hundred to feed. To be fair, they seemed to have tried by rounding up all the pigs and livestock they could from the settled lands of the inter-Andean plateau, using their Indian forced labor to herd the collection across and down the Andes. Scattered animals might live in the forest, but not large herds. My ecological training tells me that the animals died first, then the Indians, those who did not manage to run away.

The old accounts talk of sickness; this also we understand. When the sickness was not from starvation or bad water, it came from airborne carriers of disease, of which the Spaniards knew nothing. The region now has dengue fever and malaria, though the latter is now coming under control. Both are diseases of human settlements where infected people are the reservoir and mosquitoes the vectors. Native peoples lived all along the Spaniards' route. Even in the jungle the Shuars, Sionas and head-hunting Jivaros made a living. This jungle population was not so large as to yield much food for plunder but still sufficient to harbor dengue and malaria. It also showed fight; doubtless some Spaniards were killed by poisoned darts from blowpipes shot from ambush. If superior weapons kept these "enemy action" losses low, broadswords and breastplates were no defense against pathogen-loaded mosquitoes biting while men slept. The men had months on end of sleeping rough in which to contract dengue or malaria and, already weakened by hunger, to die.

Chances are Gonzalo's men were also drained of blood by vampire bats, for such was the lot of a later party that followed the same route. In 1897, F. W. Up de Graff, a young New Yorker, and a like-minded companion, set out by canoe from roughly the same place on the Napo. Like Orellana before them, they were heading for the sea. De Graff made it three years later, but by steamboat from the flourishing port of Iquitous in Peru, where the Napo enters the Rio Amazonas itself. From the first night on the Napo the vampire bats found both of them.[2]

The young men had local porters to set them on their way, who built them a sleeping platform under thatch, the type of Amazon shelter still in use. (I have slept on them, but with a mosquito net over me.) The young men's porters slept outside, wrapped in cotton sheets from head to foot, like a row of mummies. The young men laughed at the sight as they lay awake talking. Then, "*Suddenly I noticed that every now and then something would fly in at one end of the shelter, cross over our bodies, and disappear into the night at the other end. At times, as the night wore on, these spectres would pass over us so low that we could feel the air from their wings on our faces. 'Owls' we decided.*"[3] But they were not owls. The young men went to sleep regardless. In the morning a "great ugly clot of blood" was hanging from the back of Jack's head. De Graff's blood clot had seeped into his blanket from his toe.

De Graff notes he was "tapped" about twenty-five times during his days on the Napo, each time leading to a few days of weakness. Perhaps the average of strikes at Gonzalo's war party was less because his men outnumbered the available bats, thus distributing the burden. But for the unlucky ones, loss of blood cannot have helped the survival of sick and hungry men.

Perhaps we were spared vampire bats by zipping ourselves into our little sleeping tents and their mosquito nets; the same was probably true when we borrowed one of the many roofed sleeping platforms left at likely places on canoe journeys but still used our nets. I never gave a thought to vampire bats; perhaps they are all gone, though I do not see why.

Gonzalo's decision to build a boat came too late for many of his men, and the task itself was daunting. He determined on a real ship. Survivors called the result a "brigantine." They had to cut the living trees for wood. Presumably they had axes amongst the impedimenta that slowed their march, but adzes or saws? Surely not saws. Wedges to split planks they could chop with axes, and I can imagine adze blades and chisels forged from broadswords. The old records say they had saved the nails from the shoes of their horses as they died.[4] But a brigantine large enough to carry fifty men! That they built it on the muddy bank of the Napo is testimony to the human spirit whatever the motive of the expedition.

Boat building spanned more weeks of privation. Gonzalo gave command of the brigantine to Orellana, together with a detachment of fifty Spaniards, and sent him ahead to scout out a rumored settled land where they might be able to loot both food and gold, while he and the remainder made camp to await their return. Orellana and the brigantine never came back.

Not surprising, this failure to return. The Napo is a fast-running river, shallow in places, with rapids. Surely their crudely built ship could not breast that flood to return. Walking back was not an attractive possibility either, particularly since the expected settled land with food and riches did not appear. Orellana's men were as starving as the rest. Like a sensible commander he pressed on. Months later, the brigantine still serving him well but with very few survivors, he reached the mouth of the Amazon after perhaps the greatest exploring journey in history.

Orellana's passage was predatory, the only way he could hope to feed his people. He was resented by those he sought to loot, leading to skirmishes all the way. Many attackers had long hair after the fashion of jungle Indians, so the close-cropped Spanish soldiers thought they must be women. They talked of being attacked by "Amazons," using the name of the warrior women of Greek mythology. The world's greatest river had got its name.

Four hundred years later my field party boating down the Napo was just six, with an excellent commissariat also from the inter-Andean plateau (from a supermarket in Quito) with the meat in cans that would not run away, and rice and other staples dried and well bagged. The natives were friendly, including the one manning his outboard motor, driving the great freight dugout canoe. The coring party was five strong: Miriam Steinitz-Kannan, myself, Mike Miller, Ian Frost (a graduate student from Ohio State), and Chela Vasquez (an Ecuadorian student who attached herself to our party and whom I promised to take back to Ohio State as a graduate student). It was as fine a field party as any with which one could hope to serve. We six canoed down the Napo, heading for Limoncocha, not so much in the footsteps of conquistadors, or even New York adventurers of the 1890s, but of the missionaries and yes, the tourists, who came after.

Formidable craft, these great dugout freight canoes thirty or forty feet long and a yard wide, the simple shape of a tree trunk gracefully tapered at the ends. The hollowed-out interior is spanned by planks jammed between the sides to serve as thwarts for crew and passengers; hard seating for city-life buttocks but letting a paddler sit high to apply his weight. The sides and bottom are one to three inches thick, making for a craft of great strength and extreme rigidity. Heavy certainly, but I have seen one managed by four small girls who looked to be ten to twelve years old; they had no problem, giggling all the way, as they paddled the great canoe into the bank against the Napo stream. Dugouts last for years if kept wet. To store: fill with water and sink; to use: bail out and go.

When paddled by twenty strong men the result would be a war canoe, fast and dangerous as recorded in numerous historical anecdotes. Doubtless Orellana saw some in this mode. He would have been better off if he and Gonzalo had built a flotilla of these instead of the clumsy brigantine. If they felt they had to sail, pair them off in catamarans. To go upstream without benefit of motor, hug the bank and punt. In the 1980s you could still see families without the luxury of an expensive motor doing this, the man with the pole standing four square in the bottom of the canoe, driving from the shoulders with body erect in the approved Oxford manner.[5]

Limoncocha is one of the larger lakes in the Napo drainage, only four hours or so downstream from the roadhead at Coca by motorized canoe. "Coca" is the ancient name of the settlement where the Coca River enters the Napo, making the joint stream a more navigable proposition. Gonzalo and Orellana must have passed that way. The growing settlement of Coca, complete with airstrip and bridge, is now known officially as *Puerto Francisco de Orellana*, but you only discover this name by map-reading. Ecuadorians still call it "Coca."

Limoncocha itself drains by a narrow stream winding into the Napo, a stream too small for the passage of a freight canoe. Instead, you unload your canoe onto the muddy riverbank and walk to Limoncocha, or, if you have Miriam organizing things, you are met with a farm tractor and trailer to help with the baggage while you stroll along behind. The end of our stroll was not a camp but houses.

American Protestant missionaries had built the houses, now deserted. The story was that the Ecuadorian government of the time had ordered the builders to leave the country because of complaints that the mission was offering medicines for conversion. The truth of this tale I do not know, but it was at least clear that this was not an *African Queen*–type mission as shown in the Bogart-Hepburn movie. The buildings were standard American-style suburban houses of the period, properly fitted out with furniture and electrical appliances. I am almost sure I ran my clothes through a washing machine before we left.

This suburban retreat was beside the lake, beside the river where Gonzalo's men had starved and built their brigantine. We would not even be worried by vampire bats (if they were still around) because of wire screens on the windows. Our only bat adventure was associated with toilets.

There were no flush toilets, the one concession to frontier living I could find in the ex-missionary compound. So we used the missionaries' privy, well built in traditional American rural style: a little hut and a can with a seat, over a voluminous dark hole. Long unvisited, a colony of bats had found the dank cavern to be an ideal cave in which to pass the day. Our scientific ingenuity was adequate to meet this bat problem. You began your visit by banging the can with a stick; the startled bats flew out through the hole and away. Only when the procession ended did you take your seat.

First sound the lake. We now had a recording sonar run from a 12 volt car battery instead of a weighted string, also a tiny Seagull outboard motor to drive the rubber dinghy. Early in the morning a crew set out to chart the bottom of the lake, looking for the best place to core.

Was Limoncocha to be like Lago Agrio and Santa Cecilia, a young bit of abandoned river channel, useless for an ice-age history? It did look as if it was, despite its greater size and rounder shape. If, once more, all we got was a bit of fairly recent river history, then the likelihood was that the rest of the Napo valley lakes would be the same; not a comforting thought for one increasingly bent on an ice-age section for the health (good or bad) of the refuge theory.

From the middle of the lake came a shout, "Forty feet," followed by the jubilant comment, "We have found a basin." Oh joy!

But the survey crew came back to shore subdued. They had misread the scale of an unfamiliar instrument the first time round. The true depth was a little over three meters, not the ten meters that "forty feet" would suggest. What was more, they had completed enough traverses to draw a rough chart of the bottom. Limoncocha occupied a u-shaped channel, as obvious an old river bed as one could find. This was once the path of the Napo itself.

Nevertheless, we cored in two places, finding three meters of gyttja over river sand. The question of origin was settled: Limoncocha, a largish lake in the Napo drainage, was definitely a remnant of an old river channel abandoned some centuries ago, probably a cut-off meander rather than the shift of a whole stream, like that of the Aguarico, which made the smaller lakes Agrio and Santa Cecilia.

Now face the fact that probably all the lowland lakes of Ecuadorian Amazonia would turn out to be young relics of river dynamics. Like the Andes before it, this was a young and active landscape, poured on by some of the heaviest rainfall on earth, with water running onto it down the sides of the Andes themselves to stir the muddy mess. No place to be if I wanted the mud of 20,000 years ago, lying quiet and undisturbed.

It was then that I knew that my constant thought must be, "Get above the rivers, work only above the rivers." Not too high; stay in the rain forest, but keep your feet dry; and do your boating well above the flood plain. But where in the barely mapped vastness of the Amazon was I to find places like that? No computerized images from satellites to which to appeal in those days of the early '80s.

Meanwhile, I had a team of skilled limnologists, a comfortable work place, and some of the least-known lakes in the world in its region of highest diversity. An opportunity not to be wasted. Mike Miller and Miriam led the lake study effort. M. C. Miller, the "Mike" of virtually all of our expeditions for the next ten years, is a limnologist and professor at the University of Cincinnati. We first met at Toolik camp on the Arctic slope of Alaska back in the '70s. Toolik is now an important Arctic research facility with buildings, but in those days the station was no more than the sleeping tents of individual scientists pitched on the tundra, a cookhouse, and some trailers beside the haul road and airstrip, a place where a little science rode piggy-back on the building of the Alaskan pipeline. Sunday relaxation was to climb something in the Brooks Range. Mike led a climb that he advertised as "six thousand feet before going home," and in that long Arctic Sunday we did indeed stand on a nameless summit above 6,000 feet. I dubbed the peak in my notes "Peaceful People Mountain."

Riverain lakes of the Napo valley tumbled into our notebooks: bathymetry; temperature, oxygen, and light profiles; plankton; rooted aquatic plants; carbon reduced by photosynthesis (i.e., productivity); and water chemistry. Mike had brought a battery-powered spectrophotometer, a sophisticated instrument, to the remotest of field sites. Operating a thing like that in the mud beside a lake in a mixed cloud of evaporating sweat and flying insects is not everyone's idea of fun. Mike apparently thrives on it.

Limoncocha, Garzacocha, Taracoa, Añangucocha, all were collected into our notebooks and vials. At Garzacocha, which we first reached on foot and without a dinghy I got mud samples by diving naked and went ashore with a fringe of leaches dangling from one arm, a bit proudly I think, like Humphrey

Bogart in *The African Queen*. This was the only Amazon lake in which I was to encounter leaches.

Añangucocha seemed worth coring, so we did, or rather Ian Frost did. He needed a thesis out of this expedition, and Añangucocha seemed the best bet. The lake was more remote from the parent river of any we could find, giving some hope that it had not been picked up by the wandering river recently. But it took three attempts by different routes to reach it. In the first attempt a boy, conveniently named "Darwin," from the local village was tipped as a guide; only a boy, but fair enough: those of us who have been boys once devoutly believe that boys reckon to know the land through which they range. But Darwin apparently didn't, and he managed to get Mike and Ian lost in the forest. But old hands as they were they had kept their own checks on position. They navigated themselves (and the boy guide) back to the Napo and their canoe by dead reckoning and a compass, but only after three and a half hours in the forest.

Next we tried to trace the mile-long outlet stream in two small dugout canoes. The going seemed as much dragging as paddling, the stream being both narrow and shallow. A dugout canoe filled with drill rod and stores is a heavy object to drag over obstacles, but the skills of the Indian canoe men were enough to overcome even fallen tree trunks across the way. They cut bark off a local tree conveniently growing in the wet bottoms by the river, I think a species of *Cecropia*; the inner side of the bark of which is incredibly slippery. Place a slab of the bark, slippery side up, on the blocking tree trunk, lift the bow of the dugout onto the bark, then have us all, waist deep in the stream, heave the laden canoe over. A good enough heave and a heavy laden canoe slid up and over the frictionless surface in a smooth rush, landing with a splash on the other side. It was a grand lesson in watermanship in the jungle. But after half a mile we came on a fallen tree too large for even this acrobatic feat and had to quit.

In the third attempt we got a better guide to Añangucocha and walked: just an hour and a half on a marshy trail, though in places up to our waists in water. It took us three days to pack in the coring rig, drill the lake, and pack out again. Añangucocha turned out to be a wetter place in the soggy lowland across which we had splashed to get to it: a four-meter-deep hollow filled with water, its edges a ring of swamp vegetation sliding into underwater plants, an uncomfortable place to build and launch a platform of rubber boats.

I suspect this basin of water within a very large basin of wet marsh that is Añangucocha is a model for that "lake" of the coastal hacienda to which the owner said he could take Miriam and me with ten or twenty horses in the dry season. Glad we did not spend the time or money, though it would have been fun.

But Añangucocha had an important feature of Amazon waterways that I had read about but never seen: a floating mat of grasses. Perhaps a third of the total area of what should have been open water looked like a grassland, a very bright green grassland, as verdant as a putting green but nicer because studded with flowers tended by clouds of butterflies. It was a floating meadow, literally floating because the water underneath was four meters deep (say twelve feet and deeper than most swimming pools). As a raft, it would almost support the weight of a cautious man, but only "almost"; it was as good a proposition for a walk as cat ice (or as bad). A lovely curiosity of the Amazon but important to us because it was a potential source of grass pollen within the rain forest.

Ian Frost got the best core from a flood plain lake that we were to get in that expedition, 3,000 years of it and not without signs of climate change either. We came away convinced there was more to be got that we could not reach. The cause of this failure was that the raft drifted badly. Eventually, keeping the raft on station became impossible; we lost the hole and had to give up. Our anchors were inadequate, which was my fault.

Typically we use three large rocks, at least 25 lbs each (ten kilograms), to hold the raft steady between its Mercedes star of ropes. But there were no rocks in the muddy mess of this flood plain. Not fun to pack in a hundred pounds or so of rock on a long jungle walk: go to plan B. My plan B was to use folding metal anchors advertised for small boats. No good, the lightweight things merely dragged through the soft, peaty gyttja in which they were embedded. Even so, with the gentle handling of one used to boating, Ian got 5 m of interesting history: gyttja over peat, over clay, over sand, over a sequence of alternating layers of peat and gyttja. He was working through a massive thickness of peat when we finally lost the hole. How much more is under there, perhaps the whole of the Holocene? But I am not going back to find out.

Añangucocha had given us enough to show that there were environmental changes in centennial or millennial timescales in the Amazon lowlands, even in the Holocene times in which we live; also the material for Ian to write a fine thesis.[6]

So much happening in the Holocene in all our flood plain sites whets the appetite still more to know of the ice age.

When packing out, Mike and I helped each other on with our packs, perched as we were on a slippery mound of mud on the swampy shore. I remember brushing off about fifty small bees from the back of Mike's shirt to make room for his pack, then knocking a scorpion off the top of the pack. The large spider on Mike's shoulder yielded to an oblique punch aimed to remove, not squash. I felt the impact on my knuckles as the spider yielded to my blow. At the end of the hour and a half walk out 'twas blessed to strip to the skin and swim out into the Napo River.

One thing we did have. The ages of the lakes were roughly the same, whether in the Aguarico or Napo river systems; even 3,000 year old Añangucocha ended its column of gyttja at about a thousand years ago, the older peats and clays suggesting earlier, pre-river adventures. Interesting: to us it suggested climate change on a millennial timescale, a subject much talked about at the time. At least we had something of interest to publish, a sop to self-respect and, more important, a nugget of results to report to the granting agencies.[7]

Now what to do? An ancient record was becoming a burning necessity, and I was not going to get it by punching trivial holes in this soused, muddy field of a forest stretching round me for hundreds of kilometers. I badly wanted that ice-age mud, the goal I had been set at Yale so many years ago and, more important, the only way I could test that increasingly irritating paradigm of refuge theory. I measure my own rejection of the paradigm from those days in the soggy reality of the western Amazon forest. Had lowlands like these really dried up? I began to doubt it. Even the idea that they might have remained as a "forest refuge" while all about them dried began to seem as absurd, if not more so. They were the size of a nation-state, not the geographic isolates of vicariance evolutionary theory. Although not conscious of the fact, I was finding for myself the truth of Agassiz's ancient dictum: "Study nature, not books."

But I still had the problem of what to do next. No help to be had from professional geographers. I still possess the 1/50,000 scale mapsheets of Oriente province provided by the Instituto Geografico Militar, the best maps that then existed. Many sheets are plain white expanses of paper surrounded by coordinates, with the channels of rivers drawn as black lines and the names of rare settlements printed beside them. These charts are a small boy's dream-of-adventure types of map, on the clear white spaces of which he is tempted to scrawl, "Here be griffins."

Not the Instituto's fault: it takes huge resources of money to make the air photographs of great jungles on which contemporary mapping was then based, added to which the clouds and rain storms over this forest, one of the wettest places on earth, greatly hinder the necessary low flying and photography. Flying the jungle for the subtle contours of Colinvaux's lakes might well be regarded as low priority!

So: nothing on the maps and no air photographs or satellite imagery to be had. There remained only word of mouth, and myth. I had the best possible word-of-mouth collector in Miriam, but she had drawn a blank. One myth, however, did nag, though I knew perfectly well that it had to be just a myth.

In government offices I had seen large, wall-sized, hanging Ecuadorian maps of ancient vintage that showed a big lake in the center of the jungle. The maps were varnished paper mounted on cloth between wooden rollers, like the teach-

ing charts and maps employed in the late nineteenth century. That they reflected some long dead entrepreneur cartographer's imagination was obvious. Remarks by me pointing out the lake were met with embarrassed laughter by occupants of the offices.

But I remembered an Ecuadorian engineer telling me of his many flights
across Oriente Province. Of course there was no big lake there, but he often saw
water glinting through the trees; earnestly he said it, his eyes screwed with re-
membering, "When the sun is right, you can often see water glinting through
the trees." Here was a myth as good (or as bad) as Orellana's report by Indians of
gold in the jungle.

When hopelessly stuck near this same spot, Gonzalo Pizarro and Francisco de
Orellana had built a ship to follow the river and search the land ahead of them.
Mobility and "See-for-yourself" was the answer to "What to do next" for me too.
I should charter an aircraft to search for lakes above the rivers, meanwhile pass-
ing over the land of the mythical central lake. Counsel of desperation for con-
quistadors balanced by my own counsel of desperation. But I had the better luck.
My blundering about in the air was the end of the beginning of my hunt.

At Limoncocha I had an airstrip; all that was needed was an aircraft and a pilot, both of which were to be found at Coca. Thus back to Coca, with Miriam to
work her Ecuadorian magic and rustle up a bush plane. Driving a motorized
dugout upstream to Coca against the strong flow of the Napo was not like the
restful journey down. It took our canoe man eight hours to cover the distance,
hours made memorable by the fact that it rained the whole time. Cold misery in
a hot jungle, icy water falling as from a fire hose, pounding through rubberized
jackets, soaking skin and clothes, our only activity the vigorous bailing out of
rainwater filling the canoe, learning that teeth can chatter even in equatorial
jungles.

The gods of Coca smiled and sent us a Pilatus Porter flown by the Ruales
brothers; smiled because there could be no better combination for our purpose.
The Pilatus Porter was (I suspect still is) the finest thing in bush transport. A muscular "porter," the Pilatus version cheerfully carries heavy loads. It is a single-
engined, high-winged aircraft, vaguely like the Cessna 180 workhorses of 1960s
Alaska, though bigger. And the single engine is a huge turboprop, projecting far
in front of the cabin and wing as if a snout. The power of the jet turbine hidden
in that snout gives the aircraft its giant muscle. We were to find that a Pilatus
Porter could take a modest coring rig, together with crew, food, and camp supplies, delivering same to a tiny grass runway in the jungle in a single trip.

Better still were the pilots. The Ruales family had been flying the Ecuadorian
forests for more than forty years, a family tradition passed from father to son.

Knowing the skies over their native jungle, modest and careful, I sensed in them that quiet competence you find in the older Alaskan bush pilots. Flying with the Ruales was going to be safe.

We flew a five-hour circuit of Oriente, flying low over the forest most of the way. We did see "water glinting through the trees" at several places, but however eagerly I looked, the water glint would not resolve itself into an ancient marsh, let alone a lake. The truth is that Ecuadorian Oriente is a soggy place, a bottom land laced with innumerable waterways that flood, and pounded with rainstorms that can cause temporary overload for any drainage system. I dare say that the forest through which we walked to reach Añangucocha would have shown a "glint of water through the trees" to a passing aircraft.

We saw no undiscovered lakes. The water that was not just sogginess "glinting through the trees" was all in the riverain lakes of each river valley that we crossed in our broad circuit, none of them worth a closer look. We wound up by flying low over the whole parts of the courses of the rivers Aguarico and Napo that were under Ecuadorian airspace, seeing the riverain pattern of the lakes we had visited. It was on this flight that we saw the trace of the old riverbed snaking under the forest between Lago Agrio and Santa Cecilia.

Only Zancudococha, in the angle where the Aguarico meets the Napo, might warrant a visit. The lake lay large, rounded, black, and still, with no river valley through it that we could see. We crossed and recrossed this broad lake, so low that the wind of our passing blew a trace across the surface. It looked wonderfully enigmatic in its silent forest, in the most remote corner of northeastern Ecuador. Its outlet stream was tiny, and the course to open water meant tracing a few kilometers through a wetland of exotic plants. It would take a canoe trip of several days down the Aguarico to reach it. I think we all hungered for a try. But it was not "above the rivers," "No."

In truth, I had found no lakes above the rivers, but I talked about it constantly. The Ruales brothers heard and understood. They spent the night at our camp, and we talked. They told of lakes that really were "above the rivers," two of them, in the farthest southern portion of Ecuadorian Amazonia where we had never thought to penetrate. The airmen were quite sure of their facts and had remarked the oddity themselves. These lakes were on the flight paths to remote airstrips that served frontier operations in the jungle, one to a mission school, the other to a military post. The Ruales regularly flew supplies into each place, crossing a lake at each when flying low on their approach to land. They were quite sure that each lake was "above the rivers."

My budget was already taking a beating by the venture into the air, but this was no time for penny-pinching. If it came down to using my savings, so be it. I signed on with the Ruales brothers to load coring rig, supplies for three days, and crew. We left in the early morning.

The two lakes above the river were at roughly 2° and 3° S latitude respectively, all but dammit on the equator itself, neither marked on any map, both unknown to geography and nameless. Both, according to the pilots, lay in the forested foothills at comparatively low elevation. Both were close to the no-man's-land of disputed frontier between Ecuador and Peru, where hostility, even occasional firefights, were to linger on for several more years before international mediation finally settled the dispute.

We flew flanking the Andes all the way, the peaks high above us, the tall forest of the foothills beside us as well as below, the forest diversity wonderfully obvious in the subtle changes in shape or color of trees, with occasional shouts of Gauguin color when a tree was in full flower.

The Ruales brothers' lakes were real. The first we saw we now call "Lake Ayauch[i]," and the second "Lake Kumpak[a]." Both lakes really were unknown to geography and shown on no known maps. They were, in a sense, "nameless," but only nameless to modern geography. They could not be "unknown" or "nameless" to the ancient inhabitants of the forest, for a lake is always a magnet for humanity.[8] They must, therefore, have names. Miriam talked, as she always did, with those we met and found the names given to them in the language of the Shuar people of the forest. I have no idea what "Ayauch[i]" and "Kumpak[a]" mean. But those are the correct transliterations of the Shuar words, complete with the elevated final letter in each name. We have published the names in the scientific literature in the hope that the Shuar heritage will be preserved.[9]

Ayauch[i] lies at 2° 5′ S, 78° 1′ W and at about 500 m elevation. Being in splendid rain forest, and 200 m above the Amazon lowlands at that longitude, it exactly answers the impossible dream of being in the lowland rain forest while also "above the rivers." What is more, it is lovely, a lake man's lake. Circling above Ayauch you look down on an unheard of sight in the Amazon, a blue-water lake. In fact, it is not quite blue, not the Tyndall blue of a mountain tarn, more a greeny blue. All the other lakes of the Amazon that I have ever seen are in shades of dirty brown of bog or turbulence, like Agrio, or blackest bog water, like Zancudococha. Ayauch[i] is close enough to a blue-water lake to pump the blood of a professional lake man from the cold north. And quite unknown to science.

Kumpak[a] is a degree of latitude further south, at 3° 2′ S, 77° 49′ W; like Ayauch[i] in the jumbled eastern foothills of the Andes. A somewhat larger lake on a significantly larger crater hill, with broad slopes perhaps a kilometer wide, converging on the lake on all sides from a weathered crater rim. Part of the forest cover had been cleared for farming so that drainage streams could be seen converging on the lake from several directions. This lake, although a true crater lake of the maare variety, like El Junco, was not closed, having sundry small inlet streams, and we could see a clear outlet where a low point in the crater rim accommodated a small but turbulent rivulet. Streams pouring in on all sides and a single outlet made the lake a giant sediment trap, as we were to learn to our cost. At 700 meters, Kumpak[a] was higher than Ayauch[i], though still well within the rain forest.

The pilots looked at me, "Now what?" All we had time and money for on this trip was one lake, and that had to be cored in a hurry. Ayauch[i] was a little more exciting (because it was blue!) but here we were at Kumpak[a], looking down on a classical maare crater lake like those that had yielded ice-age histories in other parts of the world. And Ayauch[i] might take a day or two to reach from its airstrip. An easy decision. "Please put us down here." So they did.

They gently brought us down between the great trees of the long jungle clearing, and we unloaded our mound of coring rig and supplies onto the grass. Then they flew away, promising to return "the day after tomorrow."

Silence in the rain forest after the last drone of your departing aircraft has died away is a profound silence. We were on grass between trees, green on green, the long clearing itself the only sign that people had ever been here. It was not clear what to do next. But before we could set out to explore we saw something white moving through the trees. The moving something resolved itself into a woman. She walked into the clearing, starched white and crisp against the greens and browns of the forest, the skirts of her habit swishing against the grass, her stride the bold stride of one in command. In the full regalia of her Salesian Order, Sister Rosa Vargas, the mother superior of the Yaupi mission school, had come out of the forest to welcome whatever visitors the aircraft might have brought.

And welcome us she did. Excited talk between her and Miriam ensued. Soon two or three sisters of the order arrived with a column of teenaged boys of the Shuar forest people. A flurry of orders, and the boys grabbed loads. "Food first," Miriam said they had been told. We followed the parade of good sisters and Indian boys along a winding path into the forest.

The Salesian mission of the Shuar people at Yaupi is as different from the abandoned suburban houses of the American missionaries' camp at Limoncocha as can easily be imagined. Yaupi was the real thing: a small clearing, with simple, bare wood structures made from trees cut from the forest. And hospitable! A room was cleared for us to spread our sleeping bags. The reason the boys had been told to "bring the food first" was made clear as the sisters set about cooking us dinner. A tired knight errant of old arriving at a remote monastery or convent might have known hospitality like this.

A posse of the Indian boys guided us on their trail to Lake Kumpak[a], where they sometimes fished for the small piranhas that appeared, deep fried, at breakfast one day. In the morning a larger posse shared our loads of drilling equipment as we started for the lake in the dawning light, a walk of about an hour through the forest.

Mike and Miriam made their usual survey of the lake bottom with their sonar, finding that Lake Kumpak[a] had the shape of a flattened "U" in cross section. The bathymetric map they made has nicely concentric depth contours at the edges and a broad, featureless middle for the flat lake bottom. This is the classic shape of an explosion crater lake, one more bit of evidence that this lake "above the rivers" was volcanic in origin. It should be as old as the explosion that made it, making the lake literally "as old as the hills." My hopes for an ice-age record now depended on how old the hills were. My goal was in sight.

Or was it? "As old as those hills" might not be old enough. I had been drilling in these hills for many years and had found nothing old. Instead I had found a restless landscape, erupting, exploding, burying with ash, the scene of the giant smashup between the irresistible force of the moving sea floor and the immovable continent. Why should the land round Kumpak[a] be any different?

The water in the middle was 19 m (60 feet) deep. Unlike Limoncocha, this time we really had found a basin. Good, but I had also found a problem. I had brought barely enough drilling equipment to deal with so deep a lake because I had cut down on the amount of drill rod and casing to get everything into the Pilatus Porter for the reconnaissance. I had just 20 meters of casing and 30 meters of rod. The casing was just enough to reach the mud and project through the raft as required and the rods could reach no more than 10 m further for the core. Deep water makes coring slow and hammering hard. We got the full ten meters that my rods allowed in the single, twelve-hour tropical day, our guides afterwards leading us back through the forest in the dark. But it was very clear that much more sediment remained to be drilled. And drill we ever so deep, we might reach the age of the local hills and still find it not enough.

Next day came the Ruales brothers to start us on our journey home from the hospitable Salesian mission. We left with the blessing of Sister Rosa Vargas for our return. And return we should.

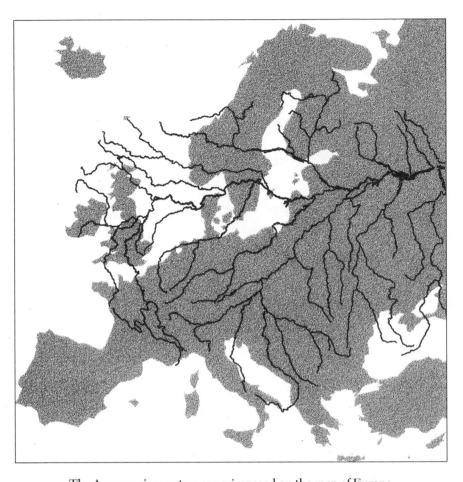

The Amazon river system superimposed on the map of Europe.
The wide use of the Mercator projection disguises from our consciousness the
real size of the Amazon drainage. The river and Europe are drawn on approximately
the same scale as taken from separate Mercator projections of the two areas in
standard atlases.

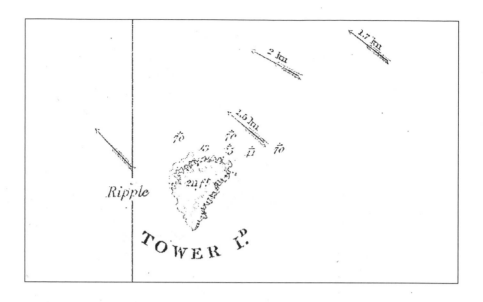

Ripple

TOWER Iᴰ.

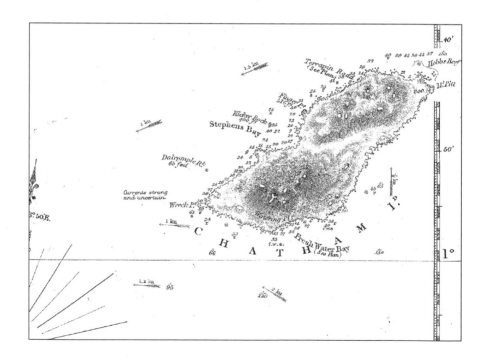

Terrapin Rgᵈˢ
(See Plan)

Hobbs Bay
Mᵗ Pitt

Kicker Rock
Stephens Bay

Dalrymple Rk.
65 feet

Currents strong
and uncertain.
Wreck Pᵗ

3° 50′.E.

Fishing Pᵗ

Fresh Water Bay
(See Plan)

C H A T H A M I.

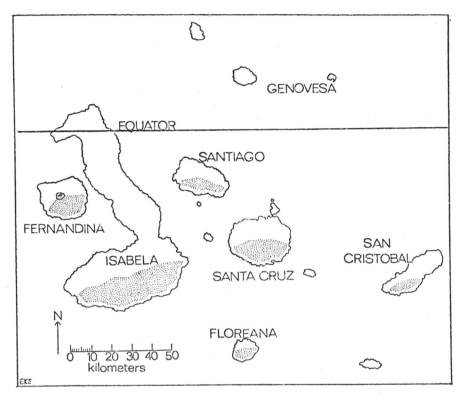

Sketch map of Galapagos Islands based on the modern Bowditch survey (1942).
Apart from the tiny island of Genovesa, this sketch map could have been drawn with
equal accuracy from FitzRoy's chart. The shaded areas of the larger, higher islands
denote the wetter sides facing the southeast trade winds. The northern sides of these
islands are semi-deserts.

Opposite: **Details of the Galapagos chart made by Captain FitzRoy**
of the *Beagle* on the voyage that took Charles Darwin there.
FitzRoy worked by sextant, compass, and sounding lead from a sailing ship in
difficult, essentially uncharted waters, and yet his chart is almost identical to that
produced by the survey ship *USS Bowditch* in the 1940s with such aids as air
photographs and sonar. The only serious difference is the shape of Tower Island
(Genovesa) where FitzRoy failed to discover Darwin Bay on what can be a dangerous
lee shore. *Top:* FitzRoy's erroneous chart of Tower Island (compare Genovesa map
in gallery). *Bottom:* FitzRoy's highly accurate chart of Chatham (compare San
Cristóbal on sketch map in gallery). All the islands other than Genovesa were as
accurately portrayed as was San Cristóbal.

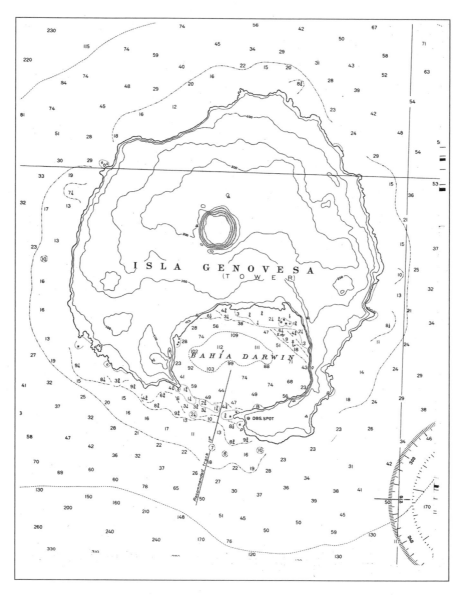

Bowditch survey chart of Isla Genovesa.
Darwin Bay on the south side is a large caldera, discovered and named in the 1920s
by William Beebe, who went round the island in a small power boat from a mother
ship at anchor on the other side of the island, something not possible in FitzRoy's day.
The map also shows the central lake, another caldera, that we cored.

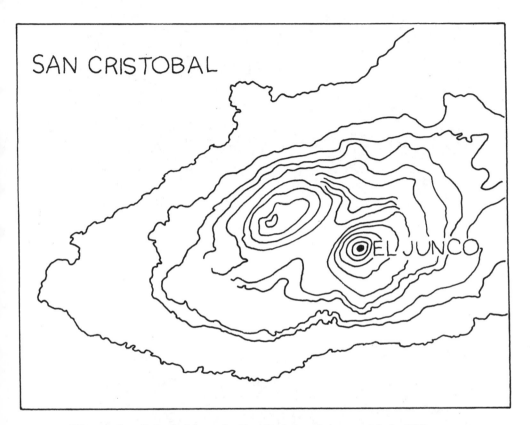

Western San Cristóbal from the Bowditch (1942) survey with the El Junco contours added (1966).

The Bowditch charts have the cloud-shrouded highlands as white *terra incognita*, with El Junco still unknown to geography in 1942. FitzRoy got a little closer to the truth with his ring of volcanic peaks. Summit contours sketched by P. Colinvaux from the rim of the El Junco crater and are only approximate.

Lake El Junco and the author, who was present at a recoring in 2005.
Crews from the Massachusetts Institute of Technology and the University of Arizona
were able to position their coring rafts close to the original positions of the 1966 expedition with the aid of sketch maps from the original work. (Photo M. Steinitz-Kannan)

Coring raft as used on most of our lakes.
This one is in use on the Manacapuru lake-like embayment of the Rio Solimoẽs in the central Brazilian Amazon. The basic design is two small inflatable yacht tenders (Avon Redstarts pictured) lashed together to form a catamaran and bridged by a wood platform with an access hole set between the hulls. The top of the casing pipe is visible center. The man on the left holds a core tube ready to insert into the casing and to push it down with connecting rods. Crew are, left to right, Paulo de Oliveira, the author, and Mike Miller. (Photo N. Carter)

Opposite, below: **Forty years on (2005) in a church on San Cristóbal.**
The beautiful church was built to meet the needs of the burgeoning population (now 12,000 souls). It has a mural of Laguna El Junco with a legend that the lake is sacred. The author is seen admiring the painting of the lake that he introduced to geography and gave to it the name taken from a local marsh by which it is now known. (Photo M. Steinitz-Kannan)

The author rowing what is believed to be the first boat ever to float on the great saline crater lake on Galapagos Genovesa (Tower Island).
The cliffs ringing the crater are between 30 and 40 meters high (some hundred feet).
We roped down it, and cored the 6 meters of sediment under 30 meters of water to bedrock. (Photo D. Greegor)

The author at the rail of the Beagle II in 1966.
In the background is the coast of Fernandina (5,000 feet), which we shall climb in the
morning, find a Galapagos Shangri-La of virgin forest and a large colony of land
iguanas at the summit before descending the steep 2,000 foot face to the bottom of
the great crater to spend the night. The lake in a small subsidiary cone within the
crater yielded a new species of diatom now named *Amphora paulii*.
(Photo D. Greegor)

The Haffer ice-age forest refuges as they appear in recent refugialist papers, from which these boundaries are drawn.
The postulate was that forest was fragmented because of drought that left only disconnected patches of forest separated by dry vegetation such as savanna.

Mark Scheutzow, the author, and Kam-biu Liu in a triumphant mood.
We had just raised a core from the glacial fjord lake San Marcos at an elevation of
3,414 meters (11,198 feet), on the flank of snow-capped Mount Cayambe, from under
37.5 meters (120 feet) of water (our deepest yet). We had high hopes of at least
reaching the last glaciation when the glacier began its retreat but were to be
disappointed. (Photo by M. Steinitz-Kannan)

Scene on the bank of the Rio Napo.
Women preparing a monkey for dinner. They are lifting the corpse out of a pot of
boiling water, the boiling either to ease skinning or to kill fleas (perhaps both).

On the trail of Francisco de Orellana.
The author paddling a well-laden 2-person dugout canoe up a tributary of the Rio
Napo with Miriam Steinitz-Kannan (who took the photograph) at the other paddle.
We were trying to keep pace with a similar canoe paddled by two river men whose
seemingly effortless paddling drove their canoe at a pace that could be matched only
by exhausting labor; hence the expression of near collapse on the author's face.

***Above and opposite*:** **The author on the march to Añangucocha.**
Packing in the coring platform on this "trail" was an experience that set the author to
design a platform that could be broken down into pieces small enough to fit into a pack
for future expeditions. Much of the floor of the Amazon forest of Ecuador (Oriente
Province) is flooded like this, such that the glint of water can be seen under the trees
from a low-flying aircraft. (Photo I. Frost)

Soaked in sweat, the return from Añangucocha.
The author after packing out. With relative humidity high and air temperature close
to blood heat we were often in this condition. (Photo I. Frost)

Sister Rosa Vargas at the Salesian jungle mission at Yaupi, Amazonian Ecuador.
It was Sister Rosa's white shape that we saw advancing through the trees after our
aircraft had left us in the silence of the grass airstrip in the forest. Directing the
Shuar boys, who lived and were taught at the mission, the Sisters took instant charge
of our lives as would have a church foundation far away in feudal times. Poor in the
world's goods but hospitable as one dreams of hospitality. Bless them.
(Photo M. Steinitz-Kannan)

The Salesian Mission to the Shuar people at Yaupi.
The mission structures are built from simple materials culled from the surrounding
rain forest. Here youngsters for whom the mission serves as a board school are at their
daily convocation. Here it was that we were welcomed as honored guests while we
cored Lake Kumpak[a].

Miriam Steinitz-Kannan in the field.
Miriam took the lead in all our expeditions in her native Ecuador. Her speciality is
diatoms, of which she is now the authority for the whole region.

The Chief (Haki) of the Sionas and his wife (Hako).
Haki is best known by his Ecuadorian name "Victoriano." It is he who keeps alive the
legend of the anacondas of *Lake Sọcora* (Zancudococha), and of the settling
of the once far more numerous Siona people. He captained our expedition of three
great canoes crewed by his young men while Hako took charge of the commissariat.
(Photo M. Steinitz-Kannan)

Mark Bush hauling one of the great canoes through the marsh at Zancudococha.
The broad marsh choked the outlet stream of Zancudococha and had to be crossed
to reach the open lake. Living and dead vegetation up to waist or armpits must be
penetrated; men chopping with machetes, others pulling, all taking their turn. It took
three hours to cross. The sacred lake of the Sionas is not easily reached, though an
undertaking for cheerful exuberance. (Photo M. Steinitz-Kannan)

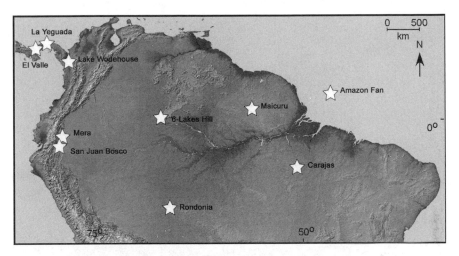

Relief map of northern South America showing principal research sites.
The three Panamanian sites include Lake Wodehouse in the Darien and the two
deposits (El Valle and La Yeguada) that together provide a long, continuous section
from glacial times to the present. El Valle is the name of a town built on the floor of
an extinct caldera that held a lake until some 12,000 years ago. La Yeguada is an
existing lake with sediments going back at least 13,000 years. Sediments from the late
ice age in both yield pollen and phytolith data showing the descent of high altitude
plants to mingle with rain forest species at low elevation during times of glacial
cooling. This confirms the chronology of the Mera and San Juan Bosco records and
validates the conclusion that the Amazon lowlands were modestly cooled in the
glacial period. Our Brazilian sites of Six Lakes Hill, Maicuru, and the Amazon fan
together provide the transglacial record of the persistence of unfragmented lowland
Amazon rain forest. The two rival sites claimed by the Van der Hammen team to
demonstrate savanna in glacial times (wrongly in our analysis) are Katira in Rondonia
and a small lake on the Carajas Plateau. (Map from public sources by M. Bush)

Hostile chase up the Rio Negro.

A river boat giving chase at maximum speed resolves into one owned by a skipper
with whom we had a preliminary arrangement for the four-day journey to Six Lakes
Hill but which we cancelled when displaced by a supercargo of a squad of soldiers.
A game of chicken between riverboats followed. We proceeded unhindered, though
finding the civil law waiting for us when we reached journey's end.

Lake Pata on Six Lakes Hill.

Pata is the oldest lake known from the Amazon lowlands from which we were to extract 170,000 years of forest history spanning two glacial cycles. But the core was so short, only 6+ meters, that we feared failure at the time. The lake occupies a rare kind of solution basin in siliceous rock that has been dissolved by percolating water with exquisite slowness. These lakes have the property that solution persists at the lake bottom, providing them with slow leaks. Accordingly, water level can change markedly with changes in rainfall. (Air photo T. Leme)

Lake Verde on Six Lakes Hill.
Verde was the second of the pseudokarst lakes of Six Lakes Hill that gave us a good transglacial record. The lake now occupies what appears to be the bottom of a much deeper solution basin, the steep sides making a slow labor of the descent to the water for the crew carrying a complete coring rig. (Photo T. Leme)

Paulo de Oliveira and Mike Miller making a sonar map of the bottom of Lake Manacapuru.

Manacapuru is called a lake out of courtesy, being large, open, and with minimum water flow. It really is a large embayment of the Rio Solimoēs near Manaus. The river flows down one edge of it, leaving the water more than half a kilometer away seemingly as motionless as a lake. In fact the sonar shows that the bottom is dead flat under 5 m of water, a demonstration that the river floods silt across the whole lake so that the sediments are in effect stable remnants of river sediment and thus a river history that could be cored. Our two cores to the bottom of the sediment body are thus long records of sediments passing down the river for the last 5,000 years, as measured by the radiocarbon age of the oldest sediment. The record, therefore, is sufficient to show us the sediment and pollen flowing down the river before modern settlement. This was important later to compare with the long records from the Amazon fan.

(Photo N. Carter)

The author coring Lake Manacapuru.

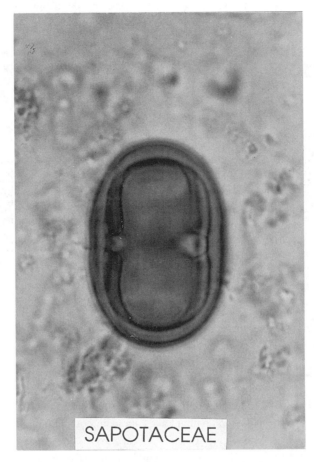

Pollen grain of forest tree from sediments of Lake Pata.
This pollen grain is identified only to family rank (*Sapotaceae*), all that usually can
be achieved with a light microscope in tropical forests, as are most of the 400 pollen
types identified in the Lake Pata sediments and in our pollen atlas. But Sapotaceae
is a family so largely (perhaps entirely) composed of forest trees that we are justified
in using the pollen as evidence of forest in the watershed. The pollen is about 30μ
(30/1000ths mm) long. (Photomicrograph E. Moreno)

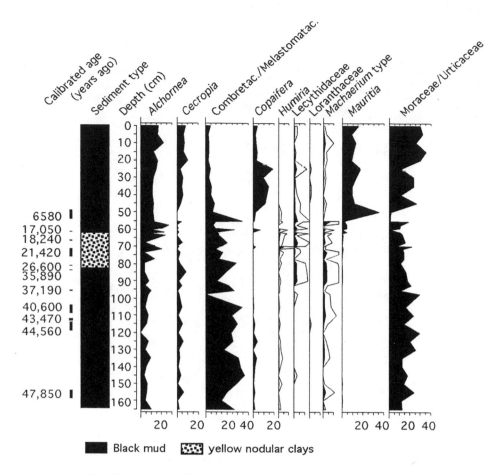

Calibrated age (years ago) — Sediment type — Depth (cm) — Alchornea — Cecropia — Combretac./Melastomatac. — Copaifera — Humiria — Lecythidaceae — Loranthaceae — Machaerium type — Mauritia — Moraceae/Urticaceae

6580
17,050
18,240
21,420
26,600
35,890
37,190
40,600
43,470
44,560

47,850

■ Black mud   ▒ yellow nodular clays

The first ice-age pollen diagram from the Amazon rain forest.

Pollen diagram from the top 1.6 m of Lake Pata sediment spanning the last 40,000 years of the lowland Amazon forest history at near zero latitude: percentages based on pollen counts (sums) of 300 grains at each interval. A total of 400 pollen taxa (types) were found, but the large number found at concentrations of no more than one or two grains in each count of 300 are shown only as cumulative sums on the diagram, their identities being listed in tables that accompanied the published diagram (*Sapotaceae* among them). This diagram presents in slightly different form the data in the paper that was rejected by *Nature*, apparently because the data led to conclusions not acceptable to the refuge paradigm, but was immediately afterwards published by *Science* (248:85-88, 1996). The data are arranged youngest at the top, oldest at the bottom. In this rearranged version the radiocarbon age determinations shown at the left have been corrected to show true calendar ages. Three observations should be evident. The percentage sum of tree pollen at the right of the diagram is always high, and the sum of herb pollen, especially grass (*Poaceae*), always low. This is direct evidence that the forest was not replaced by savanna but rather persisted

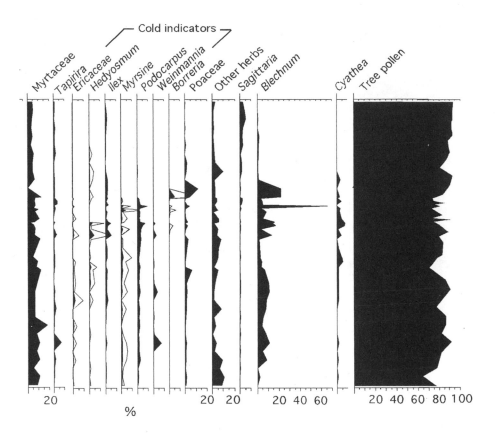

Cold indicators

Myrtaceae Tapirira Ericaceae Hedyosmum Ilex Myrsine Podocarpus Weinmannia Borreria Poaceae Other herbs Sagittaria Blechnum Cyathea Tree pollen

20   20 20   20 40 60   20 40 60 80 100

%

throughout the glacial period, and is thus a direct falsification of the refuge hypothesis and paradigm. Second, the core section shows that the most intense glacial period (the Last Glacial Maximum) between 60 and 80 cms, from 17,000 to more than 26,000 years ago, deposited different sediment from the rest of the core, described as yellowish nodular clay. A good inference is that water was often very low, or even absent, from the lake in these many years of the glacial period. It follows that rainfall was reduced. But continuation of the pollen record of forest in this sediment shows the climate did not become arid, fit only for grassy savanna as postulated by the refuge theory. It is our contention that what happened is that rainfall was reduced just enough to be unable to keep up with the leak rate of a lake type that always has a leaking basin: the lake lost its standing water in several months of most years but the forest still had what water it needed. The third remarkable observation in the glacial period was an upsurge in pollen of cool adapted plants living within the rain forest (cold indicators) similar to the pattern we had found earlier in the Ecuadorian sites of Mera and San Juan Bosco. This was the first evidence of glacial cooling in the central Amazon lowlands.

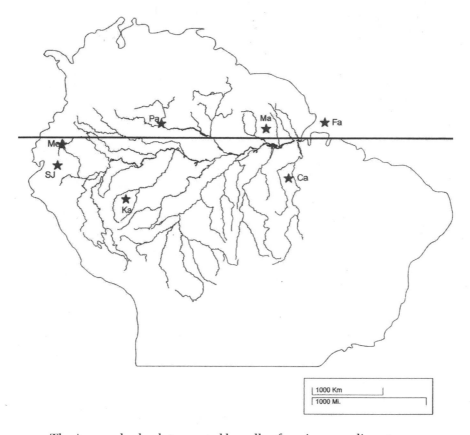

The Amazon lowlands transected by pollen from ice age sediment cores.
Me = Mera; SJ = San Juan Bosco; Pa = Pata, Six Lakes Hill; Ma = Maicuru
Inselberg; Fa = Amazon Fan. The first four of these are lacustrine deposits from within
the lowland Amazon rain forest. The Amazon fan sediments collect from the entire
river system draining that forest. All five sedimentary records are from the author's
laboratory, with the pollen analyses performed to the protocols and taxonomy of our
published pollen atlas. The two more southerly sites, Ca = Carajas and Ka = Katira
(Rondonia) are from high altitude savanna (Carajas) or from marginal woodland at
the edge of the forest (Katira). Both are from the Van der Hammen laboratory, famed
for the quality of its high altitude records in Colombia (the longest continuous pollen
records yet obtained from South America). The different Amazon climate
histories deduced by these two laboratories from their different data have been a
continuing subject for controversy.

Expedition helicopter landing on Maicuru Inselberg.
The helicopter was able to land only where a flat outcrop of rock was almost clear of vegetation, although partly covered with ephemeral shallow lakes. A disadvantage was that we must penetrate three kilometers of dense forest with our equipment to reach our target lake at the other end of the inselberg. An advantage was that we could study first-hand in this barren region vegetation like that described round the Carajas plateau lake.

Shallow ephemeral pseudokarst lake beside the helicopter landing place at Maicuru.
Note the emergent grasses and other weeds growing out of the water, there was even an aquatic species of the classical tropical weed of waste places, *Borreria*, the pollen of which the Carajas team had assigned to wind transport from distant savanna of the lowlands. It is our contention that the weed episodes in the Carajas pollen diagram merely represent the vegetation occupying the draining lake or the completely drained surfaces like that on which our helicopter landed.

Grass marsh near helicopter landing place at Maicuru.
These tufts of grass growing on very wet ground on the floor of another pseudokarst depression clearly are not savanna. The bushes in the wet and ponded ground at the edge of the basin included thriving *Byrsonima*, again clearly not part of a dry savanna. We take this site to be a replica of the Carajas lake in its low water period of glacial times. The plants of the shallow lake and the grass marsh, together with their surrounding bushes, nicely match the glacial pollen record at Carajas.

The "house" built in the forest at Maicuru.
One of the two living trees forming the end gables can be seen in the foreground, as well as the forked pole supporting the ridge pole against it. Three of the men from Recreio who built it are taking their ease on a comfortable bench made of chopped branches. They slept in hammocks slung inside the house. (Photo J. Curtis)

Opposite and above: The author launching the first boat
on Bean Lake, Maicuru Inselberg.

An inflatable without wooden parts like the Avon Redstart can be dived into as
comfortably and safely as onto an inflatable mattress floating in a swimming pool.
This is the easiest way to get in the boat for a row in any remote lake without a
developed bank. Getting raft and equipment down there afterwards means work,
certainly chopping, possible digging. (Photos J. Curtis)

The author coring Bean Lake on the Maicuru Inselberg. (Photo J. Curtis)

10

Ice-Age Forest Found

Kam-biu Liu was from Hong Kong, with the English language as much a native tongue as Mandarin. His Ph.D. in geology from the University of Toronto was based on Arctic pollen analysis, just as mine was, thus giving him technical qualifications for the Amazon project fully equal to my own; two Arctic pollen analysts blundering into the Amazon, perfect. But he lacked one essential qualification: he could not swim. A boyhood in the Hong Kong of those days apparently did not include swimming. I have made it an iron rule that members of my lake expeditions are swimmers. I do not use life jackets on my tiny drilling rafts, where they would hamper freedom of movement close to the point of being dangerous. Kam-biu joined me as a postdoctoral fellow for what I hoped to be the decisive coring of lakes Ayauch[i] and Kumpak[a]. But I stipulated that he learn to swim, and swim well, before we went. The university pools and their instructors were available. He did learn to swim, competently.

TAO, the family-run airline of the Ruales brothers, had its Amazon base at a tiny settlement with an airstrip and hanger called Mera, at the very foot of the Andes in the northeastern corner of Oriente Province. Walk westward from the graveled runway and the trail, or dirt road, starts at once on the long climb that ends only at the pass through the Cordillera Oriental above 3,000 m.

We drove to Mera from Quito in a battered rental van: Miriam, Mike, Kambiu, me, and another new recruit: Mark Scheutzow, a six-foot-four freshman graduate student at Ohio State out to see more of the world.

I had overreached in my estimate of the huge loads a Pilatus Porter aircraft can carry, having brought every meter of drill rod, casing, and sample tube I possessed. Neither Kumpak[a] nor the blue-water Ayauch[i] should get away because I had too little equipment. Their role was to yield up their ice-age interiors, all other lakes in Ecuador having failed me. Would all this kit go in a Pilatus Porter?

Probably. Would the aircraft then fly safely to a tiny airstrip with my five men and supplies? No. TAO decreed that we use two aircraft, the other being a smaller Cessna 152, then away on other duties. Wait a couple of days to assemble our little air fleet.

We had time to walk the trails and the dirt road beside the Pastaza River in that forest at the foot of the Andes, five naturalists in wonderland. And there we found what we sought by pure chance, the sedimentary section that would be shown to hold the first ice-age record of the Amazon forest. It was first manifest as a tree stump out of place, evidently fallen from a layer of stumps buried in place high on the face of the road and river cut, ten meters above our heads and out of reach. Forget natural history, we watched that face with devoted intensity as we walked. Three kilometers on we found the sediment section of our dreams, conveniently at head height, neatly sliced through by blade or bucket of bulldozer or backhoe.

Here the cliff face was layered with mud and organic debris that shouted, "quiet water deposit." We had collectively seen too much lake mud, too many sediments, to mistake the signs. These deposits held twigs and small branches, evidently ponding close to shore. A pond of sorts, perhaps the edge of an ancient pool in the bend of a stream. Very likely the pond deposit and the forest bed three kilometers away were relics of the same ancient period and had a common burial.

Exciting? yes, fossil finds are always exciting. I was less excited than I should have been, too hooked on coring lakes, too used to thinking of visible fossils in cliffs as things to do with geology far more ancient than the ice age. Not so Kambiu Liu who set himself at the exposure as if this was the holy grail of an ice-age deposit, as indeed radiocarbon was to show it to be. He sampled everything, organic fragments, sodden wood from the pond sediments, and chunks of tree stump from the forest bed for radiocarbon, together with plastic bags and vials of mud for a complete vertical record of the exposure, with sketches in his notebook and photographs. At the end of the day Kam-biu had collected all that we would need to date the fossils and do the pollen analysis. In his first day in the Amazon, he had found what had defied our years of searching: the first ice-age record of the Amazon forests. The find has become known as "the Mera section."

Now for lakes Ayauch[i] and Kumpak[a], where we could hope for even better records, running from ice-age time right up to the present. Our little air fleet flew the very next day, and landed on the grass airstrip serving the military base at the little village of Santiago. South America is rife with settlements called "Santiago," from the cosmopolitan capital of Chile to, well, this tiny place where the military had made an airstrip.

We deplaned, unloaded, and sat on our baggage as our air fleet departed. As with the departure of the aircraft down the long green strip at Yaupi the year before, silence fell, with not a living soul in sight. But the ambience was not the same. More cleared land, and a muddy path led away across a plank bridge to where wooden buildings were visible through trees.

The huts had to be the military base. I remember a tinge of anxiety: this was, after all, an active duty site on a frontier that had already seen some shooting the year before. Our unheralded arrival just might not be all that popular with the military command. So once again Miriam to the fore. She set off down the track and over the bridge and vanished among the trees. The rest of us had an hour to ourselves, sitting on our pile of baggage. Had Miriam been arrested?

No, she had triumphed once more. She came back with a file of soldiers, small, wiry men in fatigue uniforms and rubber Wellington boots, each pushing a wheelbarrow. The soldiers were to help, not to arrest. Miriam had quickly reached El Commandante, a Captain in the Ecuadorian army, who had welcomed her with something like the courtesy I remembered from that naval commander of the Galapagos twenty years earlier. He had assigned us a hut in the barracks. The soldiers with the wheelbarrows had the fatigue duty of bringing all our kit from the airstrip.

More good news followed. The trail to Ayauch[i] was good and took only an hour and a half to walk. The river round Ayauch[i]'s hill was not as formidable an obstacle as it seemed from the air and could easily be forded. And, finally, an impossible dream, a man with a packhorse could be hired to carry the heavy equipment to the lake. We moved into the Santiago base forthwith. That same day guides led us to the lake. The next day we cored it, occasionally watched from a military helicopter by an observer who stared down at our anchored raft of rubber boats through binoculars.

Again the survey team of Mike and Miriam led off with their sonar while the other three of us built the raft. The sonar results were a shock. There was not one crater here but two, perhaps made by twin explosions of the same eruption, perhaps one older than the other. However it had happened, the two craters overlap, the result being an obloid lake of two intersecting basins. The deeper of the two basins was 45 m (150 feet) deep, its bottom mud out of range of the coring technology I was using, and indeed of the 37 m of casing that was all I owned.

But the second crater had a flat bottom at 25 m. Seen in cross section, the combined basins would have looked like a 45 m deep hole with very steep sides, and a broad ledge 25 m deep jutting out halfway across. It is a strange piece of geology that surely has tales to tell to a volcanologist. Not for me, though. I was content that it offered a stock of possibly ice-age mud on the 25 m shelf that I could core.

I cored it with Kam-biu and Mark Scheutzow while Mike and Miriam had an exciting day exploring the workings of this strange lake. The secchi disk transparency was nearly 5 m, an utterly different result from all the other Amazon lakes where transparency tended to be less than two meters. It was this unusual transparency that made possible the extraordinary blue color: the Tyndall effect was working.[1] A side effect of clear water was that the water held free oxygen down to 20 m, a quite extraordinary result, making the oxygen reservoir in Ayauch[i] comparable to that of a mountain trout lake in the Rockies. By comparison, Kumpak[a] water runs out of oxygen in the top meter. This is not to say that trout could live in Ayauch[i], even at 20 m the water is much too warm for trout.

Ayauch[i] achieves the clever trick of being a deep, deeply oxygenated, blue-water lake without cold bottom water, something I take to be unique in the experience of limnologists. The lake is unproductive, what we call in the trade *oligotrophic*; which fits with the idea of blue water and high light penetration, but equally unproductive are the other Amazon lakes with their muddy brown water and poor light penetration. Mike measured the productivity of Ayauch[i] by the radiocarbon uptake method, finding it similar to other lakes of the region. It is not, therefore, an unusual absence of the respiration associated with organic death that lets the light and oxygen in; some other unique property must be at work.

Having driven the argument that far, the answer is easy. Ayauch[i] is unique in having almost no input of sedimentary matter from outside, despite sitting plumb in the middle of a rain forest. Looking around us the reality of near-zero input of debris, both vegetable and mineral, was self-evident. The lake is on the summit of its cone-shaped hill; the communal rim of its paired craters is close by, with little slope down to the edge of the lake; and there are no inlet streams at all. Little erosion, little chance of detritus or debris being washed in: no wonder the water is clear, letting light pierce down far into the depths to come out again blue. And because light can penetrate deeply, in amounts sufficient for photosynthesis far deeper than shown by the crude secchi disk measure of 5 m, algae of the phytoplankton pump oxygen into the deep water. Miriam found these algae down there with her plankton tows; they were desmids. Their presence gave the slight greenish cast to the blue of Ayauch[i] water. As Sherlock Holmes would have said, "My case is complete."

This discovery was vital to explaining what the coring effort found. We got only three meters of mud before we reached volcanic sand and pumice that were the debris from the original eruption. Of course I would have preferred more mud, because long cores spark dreams of great age. But this was peculiar mud, black, slimy, fine-grained, with no visible particles, nothing mineral, clearly the well-rotted remains of the lake's inhabitants, with probably little more than pollen-sized material blown in from outside. Without significant inputs from erosion, this residue of life in an oligotrophic lake should collect extremely slowly, making an ice-age reach quite plausible.

A detail of soldiers helped us lug our equipment back to the airstrip in the morning, and a cheerful crew took leave of the commandante. The Pilatus Porter arrived as promised and ferried people and equipment to the Yaupi airstrip for the coring of Kumpak[a], making two trips of it rather than bringing the Cessna all the way down from Mera. Sister Rosa Vargas welcomed us once again to the mission, and we settled back into that hospitable outpost. They killed a pig so that we could have fried pork rinds with our meals of manioc.

We could eat with their pupils from the forest, learning what they could tell us. Not for the last time, I had reason to curse my lack of language skills. Our rendezvous with the aircraft gave us four clear days to stay at this poor but happy mission while we cored Kumpak[a] properly, bent on reaching the bottom of that huge mass of sediments under its cover of twenty meters of water.

Again a crowd of Shuar boys made light of hauling our gear over the miles of jungle trail to Kumpak[a], happy boys excited by this strange visitation. Once at the lake, many remained, eager to see, wanting to help. One task we could easily surrender to them: inflating rubber boats with bellows. The boys quickly mastered the art and took over.

I indulged in my usual conceit of launching the first boat and going for a fast row on the lake. I have given myself this childish pleasure in virtually every remote lake we have reached. The boys looked on. Some were eager to try their hands at boating in objects so different from the dugout canoes of their homeland, in which I was willing to bet they were already web-footed. I let them have their well-earned fun.

But a rubber dinghy is not a canoe. It has a tendency to spin in circles, and it is made for rowing by those used to the crazy notion of driving a boat backwards. The people of the Amazon like to see where they are going; they paddle their boats facing forward, as sensible canoe and kayak people do. The Shuar boys proved to be not immune to spinning syndrome when they tried to row. Their solution was to turn round and paddle, which to those who know rubber dinghies makes the spin all the more likely. But these were Shuar boys, canoeists as soon as they could walk. Quickly they paired off, two to a dinghy, and paddling on both sides with one beat, they made the wakes bubble.

The first coring crew was afloat on our raft in no time at all, with three days to beat our way down to bedrock, if we could. We did not make it, though we got the longest core of all my efforts from the Galapagos and across the Amazon: 20 m. At the end we needed a 40 meter string of drill rod to span 19 m of water and a sediment hole 20 m deep. Slow business taking up a 40 meter drill string, meter by meter; changing sample tubes; resetting piston; then putting down the 41 meter replacement. Hard to apply pressure too, whether by hand or drop hammer, to a flexible rod forty meters long. Sticky clay sediment all the way, occasionally with vexing sandy patches of volcanic ash, made for a nasty mess of mud to core. I omit details of the gripping saga of three days of hammering through sixty-odd feet of sticky mud, but there were moments.

We were pelted by a few heavy rain showers. Once or twice we went over the side to keep warm in the lake for the few minutes it took the icy sluice to pass. A grand memory is of Kam-biu Liu swimming cheerfully on his back as the raindrops machine-gunned the water around him. This was the very first time in his life that he had been swimming since he had learned in the Ohio State pools. So Kam-biu took his first real swim in a lake unknown to geography. What is more, the lake was said to hold piranhas, the small ones that the good sisters had cooked for our breakfast.

The rain showers also showed me something very germane to the research. After the heaviest of them, great deltas of sediment-heavy water billowed in from all sides of the lake, puffing out like dense, pink clouds. This was the effect of sheet erosion on a large, weathered watershed. Whatever the vegetation, the heavy deluge had mobilized the surface layer of soil into a muddy goo and washed it down in sheets or trickles into the lake. This was the reason that the sediments and color of Kumpak[a] were so different from Ayauch[i].

Kumpak[a] might very well be no older than Ayauch[i], even though our Kumpak[a] core was seven times as long, the difference being due to the huge influx of clay and rapid increase in sediment that came with storm rains at Kumpak[a] but not Ayauch[i]. An exciting possibility was that the banding I had seen in the shorter Kumpak[a] core from the year before just might yield a record of individual rainstorms that might provide a nicely detailed history of storms over thousands of years. It was only the very heaviest of rains that would leave such a record, because we saw the clouds of sediment form only once. A history of extreme storms would be an interesting nugget for climatology. However, my personal holy grail was not a history of rainstorms. The goal remained the discovery of the ice-age forests, and this required long histories, continuous histories, perhaps far back beyond the start of these cycles of rainstorms. I held to the idea an ancient record needs a long core. We pressed on to get ever deeper into Kumpak[a]'s heart.

The night before the aircraft was due we were on the twentieth meter. We were stuck, unable to force the core barrel down to the bottom of the old hole, the sides of which must have collapsed. The mud had been getting decidedly stiffer down there, which hinted that we might be nearing the bottom. But as yet we had brought up no rock or gravel, nor had we got that beautiful, decisive evidence of dents in a steel cutting shoe that tells of striking rock. Agony: the ice age might be in the next few meters that we could not reach. There must be one last push in the morning.

We were there at first light; the aircraft was due that afternoon; we had an arrangement with the local headman, through Sister Rosa, for Shuar boys and young men to come and get us at high noon itself. And we pounded away all morning to cut out the debris that blocked the bottom of the hole. We did not make it. At twelve o'clock by my watch I gave the signal to haul up casing and anchors and head for shore. As we did so, the bank came alive with Indians. The Shuars had had a nice sense of timing. They rushed into the water, meeting us waist deep, and dragged everything out. They packed and hauled that mound of wet boats and heavy tubes of mud over the long trail to the airstrip. We walked along behind, feeling blessed.

Two last memories of that happy mission at Yaupi. A youth, about fifteen years old, was sitting next to me at the long refectory table. He managed to explain that his people had once been headhunters and declared, "We have been colonized, but over there they are still real."

The second memory is of Miriam working with Sister Rosa, giving penicillin shots to sick children. Sister Rosa was a dab hand at it.

The laboratory news from those first lakes above the rivers was not good. The oldest radiocarbon date from the bottom of the Ayauch[i] core was 7,010 +/− 130 years BP (uncorrected radiocarbon years before present). On the latest calibration that works out to about 8,000 calendar years ago. Nice, but not good enough. I needed at least another 15,000 years before that, preferably more. The date from the bottom of the long Kumpak core, nearly 19 m down, was not even nice: 5,120 +/−100 years BP. I had failed to reach the ice age yet again.

Did this mean that a policy of "above-the-rivers" as a talisman for finding ancient lakes was a failure too? Not so. In Brazil, "above the rivers" was to prove to be the guide to our eventual triumphs. The policy had failed in Ecuador because "above the rivers" was possible only when the lakes were raised up from the jungle lowlands by the ever-climbing Andes. Volcanically active Ecuador was just not a good place to look for ancient lakes. I had found this out the dumb way by searching the whole country before I learned the truth. Too late to remind myself that the beautifully ancient volcanic lake El Junco was on an island long since volcanically dead.

Yet the Republic of the Equator had relented already. Mera mud and fossil trees in those two exposures so carefully sampled by Kam-biu Liu proved to be a true time-slice of the ice age. Wood from the tree stumps yielded a radiocarbon age of 26,530 +/− 250 years BP, a date squarely in the last glacial advance.

This was a believable age with a tightly constrained measurement error of only +/− 250 years. Many age determinations approaching 30,000 years BP made by the old method of radiocarbon dating of counting the decay rate (beta decay) were of samples whose radioactivity had already decayed to so low during those 30,000 years that the slow decay of the remainder was hard and tedious to count. The profession became used to treating many a 30,000-year date with the proverbial pinch of salt, particularly if the stated age had a large stated error, say 1,000 to 2,000 years, suggesting that the counting was difficult or that a truly ancient sample had been contaminated with a trace of modern carbon. This could be from something so ordinary as groundwater moving carbon picked up from freshly rotten material and losing it to the old samples when it percolates through the soil, or from the unmentionable possibility that someone had been careless in the handling of the sample. For all these reasons we habitually treat old-style decay count radiocarbon estimates in the 30,000-year range with caution, letting them mean no more than "very old," rather in the way that horse-copers report the last wear-class of a horse's teeth as "aged." We even compound this feeling of "hopelessly old" by using the grossly inaccurate and dismissive term "infinite" to describe our oldest age class.

This inbuilt uncertainty of 30,000-year dates is particularly infuriating for ice-age studies because the glaciers were at their greatest extent 22,000–30,000 calendar years ago. The very ages needed to demonstrate an ice-age sample are the ages where the dangers of mismeasurement or contamination are at their greatest. If someone's age estimate is uncomfortable for your theory, dismiss it as probably contaminated. Things are a little safer now with AMS (accelerator mass spectroscopy) dating, which does not need to worry about decay rates but counts instead the relative mass of the ^{14}C in the carbon mixture, measuring the whole amount of ^{14}C instead of just the tiny amount that happens to decay while you are watching. But AMS was not available when we dated the Mera section.

A particular reason for caution is not so much because of the difficulty of measuring the vestigial radioactivity of an old sample near the end of its decay history (probably fairly represented by the stated large error) but for fear of contamination likely in such a site. The minutest addition of modern organic carbon to an ancient specimen will be detected by the Geiger counters and fed into the calculations to give a date in the 30,000-year range when the true answer should have been "No measurable radioactive decay; sample of indeterminate but great age." Wood cut out of a tree trunk was the safest possible candidate to avoid these pitfalls.

Kam-biu's Mera section was, by accepting the validity of the date on wood cut out of a tree stump, the first discovery of an ice-age record of the Amazon rain forest. Radiocarbon dating of an amalgam of twiggy bits buried in the wet lacustrine mud of his cliff samples yielded an age consistent with that of the tree stumps 33,520 +/− 1,010 BP. This result would be legitimately challenged on suspicion of contamination if it were all we had, perhaps contamination by forest water sluicing down the face of the road cut. The large stated "error" might then be consistent with an ancient sample of radioactively "dead" carbon mixed with a tiny addition of modern humic acid to give a low but detectable total count-rate. But the good wood date on the tree stumps gave credence to the more dubious date in the wet sediment.

We proceeded on the working hypothesis that the mud section and the forest bed three kilometers away were related deposits from the same ancient forest of the ice age. They would, of course, have had to have been buried at roughly the same time, but this was no problem in a steep landscape chronically afflicted with volcanic upheavals: a volcanic mud flow (*lahar* is the technical term) or landslip would have done the trick. The landscape had been transformed, such is the violence of volcanism, but the pond sediments and a section of forest had been preserved as in a grave. The grave goods left with this ancient burial were tree stumps and pollen grains from the ice age. These revealed that the forest of those times differed from the forests we knew in rather startling ways.

Our Amazon pollen skills were still in their infancy so that we could not name a disturbingly large proportion of the pollen grains we saw, 50 percent in some samples. But the diversity of pollen types was readily evident, and we could compare percentages of both known and unknown pollen types from the ancient Mera pond mud with our young forest lakes from the Napo drainage, and our collection of surface samples gathered from all parts of the Andes, including all the lakes we had cored. It was immediately evident that the Mera pollen array was a unique mixture of Amazon and Andean pollen types.

Rain forest trees clearly had left their signatures on the pollen array, probably a large collective signature if most of the types still unknown to us in 1984 (that 50 percent of unknowns) were from the immensely diverse Amazon forest, as they probably were. But our fascination was for the Andean types that were included. These were present in quantities too large for a convincing case to be made that they had blown down the mountain. Among these were pollen grains of the tropical coniferous tree *Podocarpus*.

Coniferous softwoods, like the pines, spruce, cedars, and fir-trees whose wood is used to build most American houses, do not grow naturally in tropical countries. Only two specialized genera make it to tropical South America: *Araucaria* (odd-looking trees that include the monkey puzzle) and half a dozen species of *Podocarpus*. Both trees are distinctive, requiring no refined botanical skills to spot. Their pollen is equally distinctive, particularly *Podocarpus* whose pollen grains are of the bladder-winged type, superficially like pine pollen, though the difference is distinctive when you know it. Both look like the head of Mickey Mouse, with the bladders as Mickey's ears, but there are no native pines in tropical South America. *Podocarpus* is common in forests of the high Andes, where we do find measurable amounts of its pollen. But we never found a high percentage of *Podocarpus* pollen, even in our Andean surface samples where the trees grow. *Podocarpus* does not produce the huge clouds of pollen that pine trees do, although both kinds of pollen are made on the same plan with bladders. Pine (*Pinus*) pollen blows far and wide and is so copious that a black car parked in a pine wood can turn yellow with pollen over night. Not so *Podocarpus*. Our surface samples showed that a grove of *Podocarpus* trees in the immediate neighborhood contribute only a few percent to the total pollen sum.

The "head with big ears" grains of *Podocarpus* were present in the glacial pollen assemblies of Mera in numbers that we could duplicate only from a high Andean site with the parent trees growing in the immediate vicinity. We never found the pollen in our modern lowland samples. The clincher to showing that *Podocarpus* had lived at Mera in the ice age came when an expert at the Wisconsin laboratory of the forest service, R. B. Miller, identified the Mera wood samples as a coniferous softwood. He could not specify the genus, the preservation being not good enough. But there was nothing else for it to be. Pollen and wood were the same, the two deposits were indeed correlated. Together they demonstrated that a vigorous population of *Podocarpus* trees grew at Mera in full glacial time.

This was not evidence of forest replacement, of the tropical forest gone to be replaced by a temperate forest come down from the Andes. It was rather an invasion of the tropical forest by trees that cannot live there now, at least not in large numbers. Something in the environment had changed, and the most likely "something" was temperature. Our hypothesis was glacial cooling.

Confirmation for this was readily found. The Mera mud yielded high proportions of other pollen types that were common in the Andes samples but less so in the modern Amazon lake mud. Cooling, therefore, is an acceptable hypothesis to explain the modest difference of the lowland forest of ice-age Mera.

Cooling was also agreeable to geological evidence that glaciers on the highest

Andes had expanded in glacial times. The glacial expansion was a certainty, with end moraines mapped a thousand meters below the present glacial limit.[2] From a glaciologist's perspective, therefore, cooling was respectable.

But it was not draconian cooling. The lowland tropical forest was still there, even at the foot of the mountains, where the Mera airstrip lay. This was clear from our analysis of the entire pollen mixture, in which a large proportion of individual pollen types matched what we had found in our many lake cores from the modern forest. The best explanation of this is that the environment in ice-age times of the western Amazon lowlands was still suitable for most, perhaps all, of the living things of the lowland forest as we know it. The tropical lowland forest was still there, but there were changes in that forest. Cooling in the ice age had allowed *Podocarpus*, and a few other plants, to muscle their way in. A well-trained Amazon botanist, if allowed to walk through the Oriente of Ecuador 30,000 years ago, would have realized that something was odd, "Hey, Bob, that stand by the creek, aren't they podocarps?" Nobody else would have seen the difference. It was, to use the ecological abstraction, "the same biome," albeit with minor modification.[3]

The cooling hypothesis explained this shuffle in the forest nicely. All the hypothesis requires is that relative success of species is determined, among many other factors, by ambient temperature. Success in competition could be a simple function of mean annual temperature, or it could be a critical temperature at some particular time of year, perhaps as a trigger for flowering, or seed set, or germination. Or the critical temperature might be the temperature of the coldest month in a hundred years. If the plants were recording the temperature change that brought the mountain glaciers down, however, the cooling was probably continuous. The betting was that the cooling would show up as lowered mean annual temperature. How to measure this from the pollen data?

We followed a well-established tradition in paleoecology by estimating temperature from the vertical displacement of plants on a mountainside. The method is loaded with assumptions. It requires that the fossils used (*Podocarpus* for us) really do have their distribution set by temperature and that the lapse rate of temperature with altitude is known for that time and place. These requirements make further stipulations in their turn, that the fossil populations had the same tolerance of temperature as their modern counterparts, for instance, and that the ancient air had the same moisture content and hence the same lapse rate. As a paleothermometer, therefore, vertical displacement of plants has problems. Nevertheless, the method has a respectable track record. Prudence suggested that we be conservative in our calculations, and we were. Our answer was a lowering of mean annual temperature of 4.5° C.

Conservative or not, this was a provocative conclusion, running foul of the then prevailing view in paleoclimatology that cooling of tropical regions in glacial times was minimal, as well as having unpopular implications for the forests of the Amazon lowlands. Perhaps that was why *Nature* published our paper in 1985.[4]

One immediate result of that note in *Nature* was a personal letter from Al Gentry himself (see chapter 1). I quote, "My reason for writing to you is . . . to question one of your critical data. *Podocarpus* is very definitely *not* restricted to the high altitude Andes, as you suggest. I have collected *Podocarpus* at 140 m altitude on white sand soils near Iquitos."

Gentry's datum of *Podocarpus* at 140 m in the Amazon lowlands would turn out to be of more interest and use to our reconstruction of the ice-age forests than he could have imagined. *Podocarpus* is in the forest, and from one side to the other, though I have never heard of large population; small groups of trees in sub-standard habitats like the "white sand soils near Iquitos." These are all of one species of *Podocarpus*, whereas there are many species at higher elevations. That these small colonies of *Podocarpus* are in place, ready for an environmental change, like the Fifth Column waiting inside Madrid was to feed my later thinking about the Amazon forest as a great reshuffler of species populations throughout the course of an ice age, while maintaining an intact forest cover throughout.

On the face of it, Gentry's datum makes nonsense of our calculation of temperature depression. If the spread of *Podocarpus* did not come from population descent of the mountains but from the spread of plants that were already in the lowlands, the temperature reduction calculated from lapse rates with altitude has no validity. There was, however, an immediate way round that objection using the pollen data. These demonstrated that many shrubs other than *Podocarpus* now known only from high elevations also thrived in the lowland forest. We let our calculation stand.

The cooling we saw in the ice-age Mera pollen data was strongly in accord with the undoubted descent of mountain glaciers, and I remained comfortable with it. Yet here also we ran into opposition, head on to another belief of near-paradigm status. This was that glacial cooling was minimal at the equator, a belief predicated on a grand program to map the ice-age surface temperature of the world's oceans by a group of paleooceanographers calling themselves CLIMAP.[5]

Paleoclimatic data for reconstructing ancient climates on a global scale came mainly from the oceans. It still does. The surface of our planet has far more ocean than it has dry land, and the most powerful heat-engines driving climate are ocean currents. Furthermore, the oceans everywhere leave records of their past in their sediments or carbonate reefs. Oceans are everywhere, in all latitudes. A global library of sediment cores from the world oceans now exists, scattered among the great oceanographic institutions in several countries admittedly, but all accessible to researchers. It is from this great library that a large part of the history of global climate has been written.

The paleothermometers applied to ocean sediments are multiplying, but up to twenty years ago, when we published the Mera record, the temperature map of the world oceans at the last glacial maximum was based on the species composition of fossils of foraminifera, tiny planktonic animals that graze on the even tinier plants of the phytoplankton. The method was to compare species lists of fossils to species lists of foraminifera now living in water of known temperature, a method having much in common with our correlation of tree fossils with altitudes of known temperature and subject to a similar range of debatable assumptions.

The consensus from this work was that surface oceans of north and south were perhaps as much as nine degrees cooler in glacial times but that equatorial oceans cooled hardly at all. The consensus had not, perhaps, quite the authority of a paradigm, but was powerful enough to constrain most climatic reconstructions. It was this consensus that our Mera claim of more than four degrees of surface cooling at near zero latitude confronted. If the surface ocean cooled only 2°C how come the Amazon lowlands could cool >4°C? Worse, how come mountain glaciers had descended 1,000 m? These were serious issues of contention when Kam-biu and I published our first Mera record.

More or less ingenious arguments were found in the popular hypothesis that the tropics were arid in an ice age. The temperature lapse rate with altitude is less severe in dry air than wet. Our calculations assumed moist air lapse rates, and were backed up with measurements of actual temperatures on a transect of the Andes in modern times, when of course the air is moist. Dry air in ice-age times was a requirement of the refuge hypothesis, the proponents of which immediately dismissed the Mera imputation that the major climatic change of ice-age times was cooling. With their assumption of dry air, the measurement of cooling could be dismissed. For them all was well.

More difficult, though, to dismiss the descent of glaciers, but here again a true believer that tropical cooling was minimal could, with a little ingenuity, latch onto the tropical drying paradigm to explain glacial descent too. The models were a little too clever by half for me, and although I sat through passionate seminars in which glacial descent without cooling was explained away, I cannot recapitulate the arguments here. They had to do with melt rates, cloud cover, and mass balance models. They were a lesson in, "Do not try to dismiss data with a clever argument if said data conflict with your hypothesis."

Kam-biu and I left our paper to speak for itself, not rising to the bait of these criticisms. An unexpected reward came in 2005, twenty years after publication, when an anonymous reviewer of a grant proposal for climate research in which I was heavily involved noted approvingly that, "Colinvaux was the first to argue for glacial cooling at the tropics."

But all my associates were vividly aware of one massive failing in the Mera paper. The pollen statistics were still frighteningly weak in that we had fifty percent "unknowns," pollen types to which we could not attach a name. Most we "knew" to the extent that we had seen them before in the lakes of the Napo, and had assigned them numbers and descriptions. But name them we could not. We assigned them to rain forest plants because we found them only in rain forest lakes, a fragile reed on which to hang mighty conclusions. This was the moment of truth when we knew we must deal with the problems of the 80,000 species and the predicted silence in the pollen record of the rain forest trees that were pollinated by animals.

So we did.

11

POLLEN, ANACONDAS, AND THE
COOL, DAMP BREEZE OF DOUBT

I had met Mark Bush while in England on a professional visit, when academic friends had knocked our heads together as likely collaborators. We clicked. Kam-biu Liu had left me for his first tenure-track appointment, and Mark joined me in Columbus as postdoctoral fellow in Kam-biu's place shortly thereafter. This opportune arrival started the long journey that solved the pollen problem. Mark's doctoral thesis involved pollen analysis in Britain of an unusually precise kind, letting him redesign an old pollen trap to one gorgeously preadapted to our needs.

Mark's traps were simple and cheap, such that we could set them out by the hundred and absorb the cost of the inevitable losses. They are essentially 10 cms plastic funnels filled with rayon floss (to hold the pollen), covered with a wire screen (to keep out debris), and perched on a plastic bottle with a hole in its side (to let rain water out). They sample all airborne pollen, both the true windblown stuff knocked by rain out of the clouds of pollen from wind-pollinated plants, but also any animal-carried pollen that happened to fall from trees directly overhead, perhaps dislodged by messy feeders among the insects, birds, and bats.

Our plan was to seek out places in the Amazon forests where botanists had set out their plots, identifying all the trees. There are such places; we found two in the Amazon lowland itself that we could use, others in Central American forests. These were enough. In each set of plots we should put up to a hundred traps, reload them every six months, and keep them running for at least one complete year. One lot we ran for three years.

From the botanists we had a complete list of the local plant species, which we could add to our pollen reference collection by raiding herbaria, if we did not already have them.

The traps worked as intended, collecting both airborne pollen and that accidentally dropped. The big surprise was how many of the accidents we got, some having made surprisingly long airborne journeys, the longest being 45 m from the parent tree.

When Mark had finished this simple exercise in a few places, we knew how to do Amazon pollen analysis. It is an art quite different from the classical pollen analysis of Europe and North America.

Amazon pollen may be divided into three classes. First, as the doomsayers had predicted, the kind from plants that never let their pollen get away, never to reach our traps or lakes, and thus to be silent in the pollen rain. This class was very large, fortunately accounting for most of those terrifying 80,000 species. Included were all the orchids and passion flowers, and most of the smaller plants and swamp plants that so delight tropical naturalists. With very few trees in this class, the omission of all these plants from the sedimentary records was no loss to our reconstruction of the general kind of vegetation present, whether tall forest, scrub, savanna, etc. And none of the pollen in this silent class did we need to know, thus cutting the dread 80,000 down to size. The north-temperate belt has this class too, plenty of pretty flowers in Europe. But in the Amazon they are present in legions beyond imagination. Classical pollen analysts usually do not bother to learn the European versions either. But in Europe this class does not include trees.

Secondly the Amazon has its own component of wind-pollinated trees, but unlike in the north temperate regions, only a small minority of tropical trees use this method. The ones that do are almost always understory bushes or small trees of disturbed sites. In forest lakes this wind-pollinated component typically makes up no more than half the total pollen. The third class are the animal-dispersed forms which fell to the ground, reaching our lakes, not through the air, but by being washed across the soil surface in runoff water. Finding their pollen in sediments, therefore, is direct evidence that their parent trees grew on the watershed itself, often a most useful bit of evidence.

Far from being "silent" as European doubters claimed, therefore, Amazon forests leave pollen records that are more informative than the crude snapshots of regional forests that we can get from airborne pollen clouds of Europe, where pollen analysis was invented. We do have to learn more pollen types though, even after discarding all the truly "silent" ones never found in our traps. Again circumstance comes to our help, because tropical trees belong to a large number of families, all with distinctive pollen. With a light microscope we cannot iden-

tify most of these more closely than family rank, simplifying our pollen vocabulary even further. Even so, we must learn some 400 pollen types well enough to
be certain when we see one. In some analyses we have found nearly all 400 of
them in a single section.[1]

It was the 50% unknown pollen types in the Mera glacial-age deposits that had
brought the pollen trapping program to the fore. Few specialists would take seriously a study reporting 50% unknowns, even when published in *Nature*. Until
we had licked this problem, pollen trapping took precedence over all. And we
had a good site still within Ecuador, where our connections and local knowledge
were good. This was the Cuyabeno station on lakes by the Rio Aguarico, run by
a young couple of Ecuadorian naturalists, Eduardo Asanza and Ana Cristina
Sosa-Asanza, where a team of European botanists had been active: naming
plants and mapping quadrats.[2] It was easy to set out our first array of traps in the
few days we were at the station. Ana Cristina and Eduardo had a firm friendship
and alliance with the local Siona Indians, so our traps would not be without
guardians.

When planning the venture, it had not escaped us that from Cuyabeno it was
just two or three days by canoe to Zancudococha, the remote black lake that we
had looked at longingly when crossing at nearly zero feet with the Pilatus Porter
during my "reconnaissance of the last resort" two years before. This appeared by
far the most promising lake in the bottom lands of the Ecuadorian Amazon; not
perhaps worth a special expedition now I had resolved to work only above the
rivers, but since we should be so close anyway, too good to miss before starting
the hunt for another Mera.

For as well as making us name the pollen, Mera had added a second urgent requirement into our planning. To find another one.

The believability of our published conclusions could not safely rest on this
one site. These conclusions were just too provocative: lowland forest to persist
despite prevailing views to the contrary; a reshuffle of species that included
plants that thrive in cooler places; and that extra provocation of a general cooling
of the lowlands, contradicting the prevailing opinions of paleoclimatologists.
Conclusions of such importance could not stand for long without more data.
We badly needed another Mera, another such relic of ancient lake mud buried
at the foot of the Andes that would back up the first.

My next grant proposal duly promised that second Mera. This was a rash
promise. I was promising to go exploring again, and to come back with the Holy
Grail. Many before me have made like promises to their sponsors, only to fail. An
aging Sir Walter Raleigh tried to repair his declining political fortunes with a
shipful of gold for his sovereign, found iron pyrites (fool's gold) instead, was sent

to the Tower, and eventually beheaded. The beheading of a modern scientific explorer takes a different form. It means loss of grants and standing in the profession, a less messy form of decapitation, but little to choose between the methods in effect.

To find a second Mera we had one lead. We had talked at the Yaupi mission with a traveling Jesuit priest who had shown the usual interest of literate churchmen in the work of groups like ours. He had told us of seeing forest beds in road cuts elsewhere as he made his way along the foot of the Andes. If we drove all the likely dirt roads between the Andes and the Amazon lowland in Ecuador (perhaps two or three hundred kilometers of them), a second Mera was not too forlorn a hope. The payoff could be huge.

And so once more to Quito, bound for Lago Agrio and an overland journey before we could take to the water again en route to pollen trapping at Cuyabeno and coring of Zancudococha, before the hunt for a new Mera. No oil company hospitality at Lago Agrio this time; their accommodation was full, and there were too many of us: Miriam, Mike, Mark, me, and two graduate students: Melanie Riedinger, apparently as hooked on diatoms as Miriam herself, and Paulo de Oliveira, a Brazilian transferring from a completed master's degree in Mike's lab to join my Brazilian Amazon project for his Ph.D. Texaco's "No room at the Inn" was completely understood.

We flew commercial to Lago Agrio, split up, and checked into several small and filthy hotels, the worst I have encountered anywhere on Earth. Afterwards I promised my people that if we were ever stranded in Lago Agrio again, we should make our way out of town to pitch our tents in the clean jungle.

Morning bus to a landing at the mouth of the Tarapoa River, hire a freight canoe, and travel four hours down the Aguarico to the Cuyabeno station.

Cuyabeno station was peace. The Asanzas had built it with their own hands, simple huts from local materials. They were both field naturalists of exceptional caliber. Ana Cristina was a virtuoso performer with a tiny, one-person dugout canoe, in which she navigated with precision for miles through the flooded forests (*igapo* it is called). Among her extensive knowledge of the forest life was familiarity with eighteen species of parrot. Eduardo had made himself expert on the natural history of the largest crocodilian of the region, the black caiman (*Melanosuchus niger*). He could call them to him. I invited Ana and Eduardo to come to Ohio State with me to study. They both hold doctorates now, by an odd coincidence, one on parrots and the other on the black caiman.

Pollen traps set, some coring in deeper basins within the flooded forest, and we were ready for Zancudococha. Half a dozen of the Sionas would go with us. We needed help, and it was their lake, their sacred lake in fact. We were honored, even to being led by the chief of the tribe, *Haki,* usually addressed by his Spanish name "Victoriano," a giant of a man, strong, infallible source of good cheer, and in his seventieth year.

The Sionas have their own name for the lake *"socora,"* a name embedded deeply in the traditions of the tribe. Here is the version that I got of the legend from Victoriano, sitting across a table in a Cuyabeno hut one evening by the light of a lantern. The full story is said to take five hours to tell. I got the shortened version. It explained how the lake could be sacred to anacondas, and why it got its Siona name.

> *First the forest burned twice, then there was a great flood, then it was dry again, and then the people came. A shaman made magic there and caused a big hole. Actually it was the Queen of the wild pigs who made the hole under the people. Then the hole filled with water. When people went back to live, their houses were sucked into the water; when they swam, they were sucked under. And the anacondas were as common as the shoals of tiny fish (soco). Some shamans took hallucinogenic drugs to see what the Queen of wild pigs wanted, and where they spilt the drug was made the island that is in the lake now. The lake got its name because the anacondas were as many as the tiny soco.*

I liked that story. It said there was something different about this lake, not at all just one more of the river embayments, the meanders and cutoffs that litter this drenched land. The Sionas make their living by canoe, and know well the kinds of lakes that rivers make. Their story said that Socora (Zancudococha to follow our craven use of the name on Ecuadorian maps) was like none of these. Why have legends of shamans and Queens of the wild pigs making a big hole for this one lake alone, when they knew perfectly well that rivers made all the other lakes they knew? Good enough evidence for me that Socora was not just another lake made by rivers. A lake not made by the rivers was still my dream, and one not above the rivers should be a prize indeed.

Three days by canoe to reach Socora, and three of the big freight canoes to carry our whole party, half a dozen Sionas led by Victoriano, coring rig with all our other gadgetry, our tented camp, and food for us all. We overnighted in empty huts, and we called at other huts not empty where we found sick people asking for help. Little help we could give beyond dispensing aspirin, and they had no doctor to "call in the morning."

The last lap of the journey was tracing the outlet stream. There was one, a vaguely defined overflow, as much marsh as stream, at first a garden of great white lilies, acres of them to gladden any florist's heart. Not if you had to push a forty-foot canoe through them, though. It was over the side and cut our way through, ruthlessly chopping at the lovely blooms with your machete. Then came the "Elephant's ear," stout plants with pithy, semi-hollow stems 4–6 inches in diameter, rising out of the water high above your head. Wade on, seldom more than waist deep, chopping your way through and dragging the heavy laden canoes for half a day. A thin tree grows out of the water and a Siona young man climbs it to check direction. OK, and open water ahead.

And then we were out on the open water, back in our canoes, and driving across to the island where the shamans had dropped their magic herbs. It was nice, firm ground. All hands to unload and make camp. Not me though. I wanted to see what the bottom was like and, being already soaked, why not remove the outer layers and go down to have a look? So I did.

I swam out some thirty meters or so from the island shore, then down I went, about 4 m to the bottom with, strange to say, enough light filtering through to see by. Sample the mud by hand, filling a plastic bag, have a look round as well as I could without a face mask until my lungs gave out, then shot to the surface. I doubt I had been down more than a minute, being a modest underwater swimmer. But I was greeted by all the Sionas in our party, lining the bank and staring at where I had gone down. They were sure the great anaconda had had me for supper.

The Siona legend that this was no ordinary riverine lake was certainly correct. I traced the circumference in one of the Avon Redstarts with a seagull outboard mounted and found no sign of any entry or exit streams other than our route through the swamp.

We cored the lake in seven places, a complete transect, always getting something over 5 m of good lake sediment, under 3–4 m of water, before reaching gray basement clay. A set of consistent radiocarbon dates ended with one at about 5,420 +/− 90 yrs. BP on each of the seven cores. So now we knew when the Queen of the wild pigs cast her spell. Zancudococha was undoubtedly the best lake we had found hidden in the riverine mess of Ecuador's section of the Amazon lowlands. Interesting, but of no value to the ice-age study. I conclude that the lake is held by a biogenic dam of slowly rotting plant remains in the position of the swamp. James Lyons-Weiler worked a beautiful master's thesis in my laboratory from the cores.[3]

Back to Cuyabeno by canoe. An advance party of Miriam and Eduardo canoed on to Tarapoa to arrange overland transport to Lago Agrio. This time Miriam's magic was in full flower as she had arranged with Texaco that there would be "room at the inn." When the flotilla of canoes took the rest of us to Tarapoa in the morning, we were met by Miriam and Eduardo in an ancient bus chartered by Miriam to take the heavy kit to Quito, an overnight climb of the mountains. The people would spend the night at Texaco accommodations and go on to Quito with Texaco's morning flight, only possible if our heavy equipment went overland in the bus. Many an army could use Miriam as its chief of staff. By general agreement someone should stay with the bus and our kit. Mark and Paulo drew the short straw for this unenviable task, I forget why. Doubtless I shall be reminded when they read this. Meanwhile I had a wonderful hot shower in the Texaco compound, afterwards to attend a party.

Mark and Paulo also took on that almost forlorn task of exploring the foot of the Andes for another Mera when, in Quito, it was apparent that I was falling sick from some unknown cause. We suspected I might have picked up something from those visits to dispense aspirin to sick people in their huts on the Aguarico. I also had an infected foot, with the butt end of a large thorn visible below the skin. We rented a truck for Mark and Paulo and wished them luck.

Miriam took me to a clinic to have the thorn extracted, but the sole of my foot being lined with pus by this time the medico made a lengthy incision and scraped out the pus over a large area. There was no anesthetic. An interesting experience that I am not anxious to repeat. But I still have my foot.

The rest had their sights set, in the interest of diatoms, on some of the highest lakes in the Ecuadorian Andes, the Mojanda lakes, which lie above 3,700 m (in the 12,000-foot range). Melanie and Miriam had found aberrant forms of some diatoms in plankton collections from these lakes and wanted cores to see how consistent were these changes through time, a matter of some theoretical importance. Nothing to do with my research on the ice-age climates of the region, but we had the opportunity and took it.

Came the day when Mark and Paulo returned to our base at the hospitable Steinitz home in Quito to report. No Tower of London for us, and our heads would remain safely attached to our necks. They had driven for days through muddy forest roads, forever scanning road cuts that were nothing but mud and rock, I suspect getting heartily fed up at so fruitless an undertaking. The end came when they stopped for a call of nature where the road ran beside a jungle stream. They looked up, and saw across the stream a cliff with layered sands, clays, and organic debris on the facing high bank. This was near the village of San Juan Bosco, just over 160 km from Mera. We had the second "Mera," in some ways better than the first.

At San Juan Bosco the radiocarbon measures were all on buried wood, of which there were large, sodden pieces throughout the section. The ages ran from 30,990 +/− 350 years BP near the bottom of the five-meter-thick section to 26,020 +/− 300 years BP near the top: dates deliciously and squarely comparable to those at Mera itself. Remaining, of course, was the reality that the dates were close to the limit of beta decay radiocarbon dating, leaving room for skeptics to cry, "Contamination!" But they were dates on wood with small quoted errors, as good as you could get.

We now had independent histories of the lowland Amazon forest in Ecuador during the great glacial advance from two sites 160 km apart. From the first efforts of Miriam and myself to find an ancient lake in Ecuador to the completed publication of these two records in 1990 had taken more than ten years and a search of the entire country. To use an oilman's language, we had brought in a lot of dry holes before we struck lucky.

Back in Ohio, we took stock. The harsh truth was that these two records were all we were going to get from Ecuador. Until we ventured into the vastness of Brazil, the Mera and San Juan Bosco sections gave us our only handle on the ice-age history of the Amazon. We had best get whatever was to be found from both of them.

Mark Bush counted San Juan Bosco pollen to the standards that our trapping program was making possible, for good measure counting some samples from Mera to ensure consistency. Unknowns were reduced from 50 percent to under 5 percent; many of the new names being eye-openers. Mark found pollen that the more we puzzled over it, the more it had to be *Weinmannia*, which is now known as a principal shrub of the elfin forest that grows just below the paramo.

We found the idea of elfin forest species growing down in the jungle so hard to believe that we had to struggle to accept even our own evidence. Still, there was also lots of alder (*Alnus*), and, a noted pollen analyst of the Colombian Andes was on the record as stating that alder did not occur below 3,000 m. Then again, we had Ericaceae pollen, a family known from the region by the single genus *Drimys* a shrub of the high Andes. *Drimys* is a bush related to rhododendrons and azaleas (all Ericaceae), like them a plant of the uplands, another link with communities of cooler altitudes.

The wood stumps at the original find at Mera had been characterized by experts only as coniferous softwood. We needed something more precise, particularly at San Juan Bosco, where Mark and Paulo had found and dated wood of many kinds throughout the section. We needed a specialist in neotropical wood structures. Kam-biu Liu found him for us from his new position at Louisiana State University. Michael Wiemann was a graduate student there, setting a course that would make him an expert on identifying woods. By then we were in the second half of the 1980s, when computerized databases were already becoming available; one on the microscopic structure of wood from 2,500 genera of trees was already in existence, based on the accumulated experience of forest research laboratories the world over. Because of this, it was not necessary to begin virtually from scratch, as those of us trying to learn Amazon trees from their pollen rather than their wood had had to do. The wood technology labs had been there first.

Wiemann confirmed that the softwoods from Mera were indeed *Podocarpus*, adding that one of the Mera twigs from the sediment face was probably of the high-altitude bush *Weinmannia* (thus nicely validating our conclusion that we had found *Weinmannia* pollen). He found no *Podocarpus* at all from San Juan Bosco, though he did find *Weinmannia*. More exciting was his find of *Drimys* (the azalea-like plant that we had guessed at from the pollen) at two places in the long San Juan Bosco section, unmistakably identified to genus. He also found wood from the magnolia family, a tree that was 1,000 m below its modern limit in Ecuador. Finally he also found undoubted rain forest wood. The wood samples, like the pollen and the phytoliths, demonstrated a remarkable mixture of plants that in modern times are divided between Andean slopes and the lowland rain forest.

We had one more independent line of attack, phytoliths. Phytoliths are bits of magic left by good fairies to gladden the hearts of paleoecologists. Their name means "plant stones," but the "liths" themselves are fine grains of silica, essentially identical to the opaline silica in the glassy cases of diatoms. Like the diatom frustules, phytoliths are concretions of silica secreted by a plant system, but the plants that make them are the multicellular higher plants. Phytoliths are secreted inside cells so that they end up as minute casts of the cell's interior. Whereas diatom frustules are the delicate ornamental "box" that held a cell, phytoliths are more or less solid chunks in the shape of the parent cell. They tend to come in shapes characteristic for the plant that made them. And in acidic sediments they last virtually forever.

One of the oddities of the Mera pollen diagram (but not that of San Juan Bosco) was that the oldest samples, those dated beyond 30,000 years ago, had up to 30 percent grass pollen. There might well be some who would suggest that this was the savanna expected by the paradigm of a dry Amazon with refuges, despite all the other evidence suggesting a rather catholic tropical forest that was absorbing some more cold-tolerant plants. We cannot separate grass genera by their pollen, but it so happens that experts can separate them with phytoliths. Fortunately we knew just the right expert.

Dolores Piperno had in press a manual of phytolith analysis that has since become the standard reference on the subject.[4] An archaeologist by training, Piperno had developed the method to investigate the dirt of archaeological digs, Were big "grass" pollen grains in the spoil-earth from maize (corn), or were they just grass? And what else were the people eating? Archaeological digs yield better preserved phytoliths than they do pollen. Piperno then spread her knowledge, particularly of the tropical sites where she worked, to a knowledge of phytolith types from the main groups of tropical plants, not just the grasses. Essentially, she was creating a new tool for paleoecologists.

Piperno settled the grass pollen puzzle well before an advocate of the refuge paradigm could get hold of it. Bamboos were there, though only 2 percent of the total, and other grass groups also were consistent with lowland forest. But 14 percent of the grass phytoliths were from fescues, grasses of temperate grasslands, good for feeding sheep and making lawns. In Ecuador, fescues are plentiful in the high Andes, where you can indeed pasture sheep and from where farmers bring temperate crops to the Quito market. Fescues are vanishingly scarce in the jungle. Like all the other plants represented by pollen, the grasses were either rare inhabitants of tropical forest or cool-adapted invaders.

The other phytoliths in the Mera section were proper denizens of the tropical rain forest, Chrysobalanaceae, Marantaceae, Palmae, thus providing some of the clearest evidence that modern tropical rain forest trees persisted despite the cooling signaled by some of their new neighbors. Piperno found no phytoliths in the San Juan Bosco deposits, probably because the deposit was from faster flowing water where the minute casts of cells could not settle out. But we had that fine wood record from San Juan Bosco; our luck had balanced out.

These data from Mera and San Juan Bosco, both severally and separately, require three properties of the environment and vegetation at the region of contact between equatorial Amazon lowlands and the Andes at the last glacial maximum. The vegetation, then as now, was tall, evergreen, moist tropical forest. This forest, however, experienced significant changes in the relative abundance of many of its populations, in particular having much increased populations of trees like *Podocarpus*, whose largest populations are now in cooler regions or elevations, probably together with actual invasions of more cool-adapted shrubs like *Weinmannia*.

A blunt botanical summary of these conclusions is that the rain forest had had its species populations reshuffled as modern ecological theory predicts it should over so long a span of time and environmental change. Rainfall remained sufficient for tropical rain forest to be maintained. Mean annual temperature was reduced, though still clearly within the tolerance of characteristic trees of the Amazon rain forest.

"That rainfall was sufficient for rainforest trees" was the only measure we had for precipitation. We could not put numbers on this, no inches or meters per year. But we could do a little better with temperature, though only with the traditional thermometer of the descent of plants from higher elevations. Kam-biu and I had been cautious in our assumptions in the 1985 Mera paper, estimating this cooling at 4.5°C. But we now had two sites and many more data on what modern botanists think of as mostly Andean plants formerly thriving in the Amazon. Taken together, this evidence suggested both larger vertical displacements of plant populations than we had first thought and actual population descents, like that of *Weinmannia*. We dared to increase our estimate of cooling to 7.5°C to account for it. This was a conclusion bound to cause friction with a paleoclimate community that still cited the CLIMAP project estimates of minimal cooling of the tropics.

Our conclusions, then, were that a form of lowland rain forest had persisted in the western Amazon basin throughout the ice age, with subtly changing species composition as the northern glaciers came and went. Times were cooler in the ice age, or cold snaps were more frequent, or both. Nothing in this history hints at the truly dramatic climate change required by the aridity and refuge hypotheses.

But did the history of the western Amazon rain forest really have anything to tell us about the larger forest expanse in Brazil to the east? Obviously a defender of the refuge hypothesis could say of the Ecuador results, "So what, all you found was the biggest of the forest refuges where the rising face of the Andes provided rainfall even in the dry times: we always said that this was the site of a forest refuge." Well, yes. But the picture painted by our fossil data is not really that of a forest refugium. Rather, we had found a changed forest that had made room for more cold-adapted plants even as the other lowland species lived on.

Furthermore, Mera and San Juan Bosco raised the possibility that glacial cycles in the Amazon were felt as cycles of temperature, with or without changes in precipitation. The disruptions to plant life possible from low temperature should be greatest as you climbed out of the bottom land. Perhaps the most important property of those elevated places expected to have more rain was actually that they were colder. In a cooler Amazon, the places least affected by cooling would be the bottom lands. Refuge theory required the bottom lands to be too dry for rain forest trees. It seemed to us that rain forest trees of the ice-age Amazon should actually be in those bottom lands, where it was warmest, and that even there they might have lived in association with more cold-adapted species as we had found in Ecuador.

Mera and San Juan Bosco finally tipped my thinking from its early sneaking feeling that I should probably find that the Amazon had been drier as the refugialists claimed, despite their blatantly unsatisfactory arguments for aridity. After all, I had been a student of the scientist, Dan Livingstone, who had made the original discovery of ice-age aridity in Africa, the discovery that gave plausibility to Haffer's original hypothesis. But Mera changed this. From Mera on my working hypothesis was that the lowlands remained moist enough for rain forest at all stages of glacial cycles.

In 1992 I described these results, and the implications I saw for climate history, to a session of the Geological Society of America at its annual meeting. My remarks about the cool continent apparently drew blood, as it were, and a noisy argument broke out at question time. The consensus of the ocean sciences was still that the surface of the equatorial oceans had not cooled significantly, which required that the adjacent land had not cooled either. There was almost a shouting match between me and a very senior climatologist in the audience. Unknown to both of us, a reporter from the *Economist* newspaper was in the audience, delighted by the sounds of professors in dispute. The next issue of the *Economist* carried the story as its scientific lead under the title *A Cool Damp Breeze of Doubt.*[5]

12

THE PARADIGM STRIKES BACK

Mera changed things. Seeing the Liu and Colinvaux paper in *Nature*, I felt like a card player laying down a "full house," and quietly sitting back in his chair. Rain forest in glacial times in the Amazon lowlands, the very first record. Rain forest, familiar enough, but made all the more provocative by expanding its generous and catholic list of species to find room for a few trees and shrubs of cooler places.

San Juan Bosco was to spread the find down the length of Ecuador, for the two sections were but a single find in an Amazon scheme of things. Sections with the same wet ice-age forest at both places became known simply as "Mera."

Ice-age forest at Mera, was, by itself, not a direct affront to the refuge paradigm, whose authors had long realized the ways in which the Andes mountains would promote rain, thus a likely place for a forest to take "refuge" from the aridity of savannas in the central lowlands. Defenders of the paradigm could actually be jubilant that we had found the predicted refuge. I have seen letters arguing that Mera "proved" refuge theory to be correct.

Perhaps they found less congenial the evidence of cooling as a principal vector of glacial climate change. Not a far-out idea that, to associate cooling with glacial times. But it must mean that the forest "refuges" were not so pure a refuge as first thought. Our description of forests that let the heat-intolerant minority in, made forests like nothing on earth.

This did two potentially deadly things to established opinion. To the refugial paradigm it opened the possibility of a glacial-age climate quite different to the Haffer-Prance model, one in which the principal vector of climate change in the ice-age Amazon was cooling rather than drying. I stressed this possibility in our papers, arguing that the drying associated with cooling would be far too little to fragment the forest. This was not a strong answer to those who believed in savannas and sand dunes. Until we actually had glacial-age forest pollen from Amazon lowlands designated for savanna in the refugial paradigm, there was no chance of carrying the community with us. We did the best we could with Mera cooling.

The second deadly thrust was at those plant ecologists who still thought of plants as living in visibly distinct associations, refugialists among them. To their way of thinking rain forest trees lived in rainforests, recognizable by species composition as well as by shape and ambience, while mountain trees lived in belts on mountainsides in recognizable array. Changing the temperature should force the mountain belts to move up or down together, in rank order, each a single, separate association. Likewise the rain forest, though with local variations, was an association not to be mixed up with mountain belts. But this mixing is exactly what our pollen, phytoliths, and fossil wood declared to have happened. The ice-age forest housed populations of plants from the elfin forest, the highest belt of shrubs below the paramo. With them were plants from different mountain "belts" such as the azalea-like *Drimys*, magnolias, alders, and fescue grasses. *Podocarpus* had built significant rain forest populations, either from rare individuals on specialist habitats already in the forest, or by migration from on high. No permanent belts or associations here!

This heresy of permanent communities had already been slain by pollen analysts Margaret Davis and Tom Webb, who plotted the history of vegetation in eastern North America with the advance and retreat of the glaciers (chapter 7). Plants did not come in permanent associations or communities, but were built by individual species populations capable of living in various conditions and with different associates. This idea, a paradigm shift itself, had not yet achieved universal acceptance outside North America. Permanent associations, or belts on mountains, were still a convenient approximation where vegetation was simple, useful as a tool of pollen analytical reconstruction in much European work where finding of characteristic or dominant species could be clues to the whole forest association.

The strongest hold on this old association idea was the concept of belts of vegetation on mountains. It happened that European pollen analysts working in the Andes of Colombia used the "belts of vegetation" concept, expecting these to move up and down the Andes with the coming and going of mountain glaciers. Some have even published diagrams (probably still do) of all the belts they recognize being compressed, but never mixed, as the mountain glaciers made their descents.[1] For this way of thinking the Mera demonstration that assorted plants from these "belts" could be accommodated in a preexisting rain forest when temperature fell a few degrees was unsettling.

It was for these reasons that a hostile response from paradigm supporters was inevitable, for all that others affected to rejoice that we had "proved" the existence of forest refuges. One line of attack made possible by our data was obvious, and I waited with interest to see how it would be used. This was our special vulnerability to the perennial problem of radiocarbon dating in the 30,000 year range. We had the best dates (on wood) then available, with 26,000 and 30,000 + years at both sites, but the fact remained that these might, just might, represent no more than far older wood impregnated in some way with just a trace of modern carbon.

If so, our sections were older than we thought. But so what? Older was still ice age, which lasted a hundred thousand years. More, except for interglacials like the present it was ice-age all the way back for at least half a million years. What magical dating could airbrush away our conclusion that the tropical rain forest in the Amazon lowlands of Ecuador was invaded by a few heat intolerant plants in cooler times?

My interest turned out to be well fed when a German geologist, K. Heine, spent two field seasons looking at the geological setting of the Mera site, concluding that geological evidence demonstrated the burial of the Mera forest to be far older than the dates we claimed for the forest itself.

Heine's paper[2] came out some ten years after the Mera *Nature* paper. It was also well after the publication of our second site from San Juan Bosco 160 km south of Mera, a paper that included Mark Bush's more refined pollen analyses from both sites. Heine visited only Mera but spends the first two columns of his article confusing our descriptions of the two sites. I can only conclude that he thought San Juan Bosco was part of Mera, at most just down the road.

Next, Heine called the lacustrine deposits of the section "lignite." Now "lignite" is a good geological term. It refers to old organic deposits, half-way between peat and coal. Often called "brown coal," lignite has been used to run power stations in central Europe. It was grossly inaccurate as a way to refer to Pleistocene pond mud, but Heine offered neither reason nor analysis for his curious designation. He just used "lignite" as a descriptive term; it probably came from his field habits, a geologist at work. I shrugged it off as a minor matter.

Kam-biu Liu did not shrug at "lignite," he was incensed. I have a vivid memory of several years later when, both of us being in London attending a conference, I took Kam-biu to dinner at Simpsons in the Strand. We sat at one of the long tables in Simpsons' "Great room," our bellies well-filled with English meat and John Barleycorn, reminiscing on our days in the Ecuadorian forests. We talked of Kumpak[a], of Mera, and inevitably of Heine's "nonsense." Kam-biu's indignation about "lignite" was still raw, with its propaganda effect of lending the deposit great geological age. Kam-biu was quite right. "Lignite" may well have been an innocent mistake by a geologist expecting to find ancient deposits as I had dismissed it, but it was not to be shrugged off as I had done. It was potentially lethal. In truth, it suggested that Heine's "revisit" to Mera was not impartial but rather infused with the expectation of demonstrating great antiquity.

These calumnies were merely preliminary, setting the tone for Heine's real thrust, which was to build the case that the mud flow that had buried the Mera deposits was geologically ancient. Heine identified the mud-flow as a *lahar*. We had not been so specific, being cautious about volcanic geology, in which subject we were not experts, but Heine is almost certainly correct in this. A *lahar* is a mud flow directly induced by a volcanic eruption, a viscous slurry of volcanic ash and water carrying larger volcanic chunks and accumulating assorted debris from the preexisting hillside as it goes. The volcanic ash weathers to an array of clay minerals, rapidly in a wet tropical climate like that of Mera.

Heine's two field seasons, 1990 and 1991, working the lahar at Mera, noted that the surface was heavily dissected but also well supplied with wood fragments and organically enriched clay. Heine's radiocarbon dates on organic matter (consistently called "lignite") were varied, but positive in the 30,000 to 40,000 year range, quite consistent with our wood dates of from 26,000 to 33,000 BP. That his dates were variable within this range is scarcely remarkable, the upper deposit on which he concentrated being jumbled by the lahar. The deep dissection of the surface was equally to be expected at a site, initially bare of vegetation, with five meters of rain a year. But for Heine these things meant that the deposits, all of them, were much older than we had claimed. For verification he went to uranium/thorium dating of the ash-derived clays in the lahar.

Thorium is one of the "daughter" products of the radioactive decay of uranium 238. Thorium itself then decays into further daughter products, the end of the sequence being lead. Because the decay rates are known, measuring the ratio of uranium and thorium in suitable minerals lets the age of the rock in which they are held be calculated. The accuracy of the conclusion obviously depends on the assumption that there has been no loss or addition of uranium or thorium since the rock was formed, meaning when the ash blew out of the volcano. The original rock or lahar must also have been without thorium or if not the original thorium must be calculated independently.

Heine applied this method to a number of samples from the clay and organic clay layers in the lahar, making every reasonable allowance for the obvious possibilities for error. His calculated ages varied widely, with corrected estimates ranging over an order of magnitude from 8,000 to 80,000 years BP (the oldest), all with very large estimated "errors."

What Heine failed to emphasize, even perhaps to realize, was that these were all good glacial ages, any one of which validated our observation that the glacial rain forest at Mera accommodated minority populations of heat-intolerant plants now mostly living at high elevations. Instead, Heine concluded that the lahar was older than the 30,000 or so years suggested in our papers, that our radiocarbon dates were the worthless results of contamination (like his), and that our claim for an ice-age forest could accordingly be rejected.

Heine's paper, however, did not stop there. Convinced that he had demonstrated an error in our dating, he cried "Gotcha." Two pages of conventional wisdom follow his sections on dating. Citing the pillar papers of refuge theory. These pages show how others have described the ice-age Amazon quite differently from Liu and Colinvaux. This is Heine's real argument, his *prima facie* case that we must be wrong. Then comes Heine's formal conclusions section, often the only part of such a paper likely to be read. I quote, "The Mera forest beds seem to be of at least middle Pleistocene age according to the *geomorphological and paleopedological setting* [my emphasis]. According to ^{14}C in combination with U/Th age determinations, the age of the organic layers near Mera are evidently older than ca. 50 ka BP."[3]

Why stop at ca. 50 ka BP if he really thinks the deposits to be of "middle Pleistocene age" and thus half a million years old? Not even the most eccentric of his uranium/thorium "dates" come close to that. The reason given for this remarkable conclusion is the *"geomorphological and paleopedological setting,"* which translated means, "It looks old to me." I prefer our radiocarbon age of 26,000 years from good wood samples.

Flawed though his paper is, it did its work. "Heine" was to be cited in the opening paragraphs of many a refugialist's paper in the years to come as disposing of the inconvenient "Mera."

The practice of science has its own response mechanism, and I set out to use it. I wrote a formal reply to *Quaternary International*, cataloguing Heine's grosser errors, pointing out that even his dating let the Mera deposits stand as of glacial age, and noting that he had apparently not seen, nor realized the significance, of the second site at far away San Juan Bosco. I sent this response to the journal's editor of record, a scientist of high repute and the president of a solid international organization. We knew each other so that my covering letter could be in "first name" form, and I had no doubt that my response would be published quickly; for that is the way of good science. I was in for a shock.

The editor wrote saying that he had no authority to accept my paper and that all such matters must go before the full board. Strange; indeed unprecedented in my experience. Surely this was a routine editorial matter for a scientific journal? — not, apparently, for *Quaternary International*.

The board's official response, in writing, came after the usual time-lapse known as "due course." It was against the board's policy to accept critiques of papers published in *Quaternary International*. End of story.

Belatedly I took to reading the available "fine print" describing the purpose and policy of the journal. To paraphrase, this is to give scientists from outside the rich countries a chance to be read by the global audience of science. The method chosen is to appoint scientists from the target countries to edit single issues of the journal, recruiting their authors by invitation. The papers would get read because *Quaternary International*, an English language journal, should, from the prestige of the organization backing it, be taken by all the major libraries. Admirable idea.

Yet sensitivity persuaded the inventors of this alternative publishing route that third world scientists using it should not be subjected to the harsh possibility of rebuttal. An experienced German scientist ought not to need this protection, but he had been invited to contribute to *Quaternary International* by its then South American editor and had accepted, as would have most of our profession. But accepting means, in all ignorance, you have bypassed science's system of checks and balances. I had been undone by affirmative action in science.

But the paradigm had something more potent than Heine waiting for us. A team of Brazilian and French scientists was about to release the results of a study of a lake core from the eastern Amazon, which penetrated at least fifty thousand years, and provided what was claimed as hard data for both aridity and savanna in glacial times. Not only had a Franco-Brazilian team beaten me in the race to find an Amazon lake with sediments going back to glacial times, but they were also claiming a positive, and decisive, test of the aridity predictions of refuge theory.

Paulo and I had no advanced hint of this momentous discovery, having no grapevine in Brazil to tell us that "something was up." What we got were invitations to a Quaternary conference in Brazil where, "The latest work on the Amazon would be discussed." Obviously it was in our interest to go. We accepted.

What was "up" was "Carajas." Carajas is an immense plateau towards the southeastern edge of the Amazon, to this non-geologist (who has never been allowed to see it despite trying) a giant version of the inselbergs of erosion-resistant rocks that rise out of the Amazon plainsland like castles in a moat of river-washed bottom lands.[4] The sides of the plateau are steep, and much of the hard, rocky surface at 700–900 m above sea level. The rocks, as with true inselbergs, are mineral rich and were being actively mined and explored for iron, manganese, gold, and copper.

On the southern end where the Franco-Brazilian team worked, soils were thin and vegetation sparse, a barren sort of place even though precipitation at 1.5–2 m was just enough to support forest, had there been any soil for the forest to grow in.

The actual vegetation is described as an open mixture of shrubs, scattered small trees, herbs, bromeliads like the wild pineapples whose spines rip your legs in passing, and the ubiquitous coarse grasses of open ground in the tropics. This "altitudinal savanna," or *campos rupestre* as Brazilians call it, is a Brazilian peculiarity that few botanists have seen. It is molded by a combination of thin soils, hostile rock, sufficient moisture, and altitude.

But the exciting, and somewhat amazing, feature were the small lakes that dotted the plateau. Those in the altitudinal savanna or *campo rupestre* of the southern part of the plateau where the ORSTOM[5] team worked were apparently unconnected to any river system and generally without serious inlets or outlets. They were the closed basin lakes of a limnologist's dreams.

It was to one of these lakes that Brazilian geologists took a French government team from the ORSTOM organization by helicopter to go a'coring in the Amazon. They had chosen well. From the very start, the Franco-Brazilian team were to do their lake coring "above the rivers," no years wasted punching holes in soggy riverain mush of the lowlands as I had done. And at their very first try they raised a core that passed easily through the magic 30,000 year limit of accurate radiocarbon measurement, well into ice-age time.

The Carajas lake was what the ORSTOM team described as a "pseudokarst" lake. This meant that it was similar to lakes common in the limestone region of the Balkans in Eastern Europe known to science as "karst lakes," after the local name for the limestone region. "Karst" lakes are solution basins. They form as slightly acid rain or groundwater washes away the easily soluble limestone. Running water being what it is, a percolation trickle can quickly expand into a cavity, eventually into a deep basin. A more revealing name for karst lakes is "sink holes," well known to dwellers in limestone regions the world over.

But Carajas and its kind are not formed in limestone, hence the "pseudo" part of the lake name. The lakes lie in ancient rocks on ironstone formations that have resisted erosion, a strange place to find a solution basin. No good looking for limestone, there isn't any. But solution basins they are, with all the overt signs: pudding basin shape, whether shallow or deep, in a hollow or even on the flat, no inlet or outlet, typically steep-sided. Nothing else makes water-filled holes like that. They are not craters, whether volcanic or impact, which have their own stigmata, as it were. But no soluble rock? very well, leaching away of the least insoluble part of this hard resistant rock must be the answer. Current opinion credits silicates as the most likely suspects. This leaching cannot have been anything but a very slow process, even allowing for Amazon heat and rains. The lakes, and their sediments, just might be magnificently old.

And old indeed they are. The five meter Carajas core easily passed through the thirty thousand year radiocarbon mark, yielding an extrapolated age for the whole core of more than fifty thousand years. So short a core, so ancient a lake, but this, too, fitted. No inlet streams to bring silt, a rocky landscape with thin soils, comparatively sparse vegetation, and unproductive water made sources for sediment scarce. Thus Carajas lake mud gave a true record of ice-age time on a plateau in the south-eastern Amazon basin.

Of this record, the interesting bits were evidence of fluctuating lake level, even to something like total drying at two stages in the glacial cycle. The evidence was mineralogical, coupled with gross stratigraphy that interrupted thick black layers of organic mud with light brown sandy clay, rich in siderite. Siderite is a crystalline form of iron (feric) carbonate, the presence of its crystals good evidence that the lake had lost its water. On the other hand, siderite is a mineral easily oxidized, suggesting that the anoxic state of a lake bottom and lake mud prevailed while the siderite crystals were formed, and had probably remained in force throughout the period of lowered lake level.

That this was never a sun-baked arid basin is shown by the maintenance of the siderite crystals themselves, and is further supported by the published observation of plant debris included in these layers. There can be no doubt that the ORSTOM team had discovered evidence of lowered lake level, almost certainly of lowered precipitation, at intervals in the last glaciation. But with no estimate of its magnitude, always an extremely hard thing to do. And no evidence for actual aridity.

The best evidence that can be gleaned from the published data is that what is now standing water was then marsh, firm but vegetated. This is not a large environmental shift in that the modern lake is described as shallow with rooted aquatic grasses emergent over large areas. This is not the sort of gross climatic shift required by the refuge paradigm, with its predicted loss of rain so grievous as to fragment the forest.

But then, of course, the Carajas lake was not in the forest, it was perched high in an altitudinal savanna which, despite its name, is not a grassland but a place where shrubs, weeds (including grasses of a sort), plants like bromeliads and cacti that can survive on short commons for water, and small trees grow scattered out of fissures and other places where thin soils collect. The real forest grows below the edge of the plateau, creeping up the plateau flanks from the tall tropical forest on the Amazon plain 500 m down.

It was in the tropical forest of the bottom lands that the paradigm allowed true savanna grasses to flourish in place of the modern forest. This was the event that the ORSTOM team needed to demonstrate as happening during their low water times. They called in their pollen analysts to do the rest of the job. These did their best, though the pollen record was not helpful.

They published what they called "a partial pollen diagram." "Partial" to me means something left out, though it is legitimate in the trade if, and only if, a complete list of other taxa found, even if not identified, together with their collective sum, is given. There was no such list. Nor was there a statement of the pollen sum, that critical number of pollen grains counted for each sample without which it is impossible to assess the validity of percentage measurements or of any other statistics used. In short, and to be kind, an uncommonly obscure account of results.

The pollen diagram presented on those terms gave but thirteen plant names, most of them weeds, the rest prominent among the shrubby plants of the altitudinal savanna. This is in itself remarkable, even allowing for the barren place in which they worked. In our experience, if the lists did not run into the hundreds, they certainly ran into the scores at all Amazon sites. Perhaps that stipulation "partial" had something to do with it, though impossible to think that some had been left out deliberately.

The names of weeds, whether herbs, shrubs, or small trees, that made up most of the thirteen were all of plants common in the altitudinal savanna/*campo rupestre*. Their histories told us nothing of ecological note. The "action" was concentrated into a separate diagram of a kind once common in European publications, a plot of the ratio of total arboreal (tree) pollen against total herb pollen. The most abundant herb was grass, and the total herb plot had by far its largest percentage in relation to tree pollen at the two glacial times when the lake dried down to a marsh, or even a muddy field. For pollen analysts of the ORSTOM team these were the *eureka* data. The forest was being fragmented by savanna in glacial times.

None dared to say that a real savanna-grassland grew on the Carajas Plateau itself, too little change in the makeup of the rest of the pollen diagram. Besides they did not want that. What was needed was replacement of the rainforest below the plateau. And it was this that they proclaimed. In dry glacial times the lowland forest had been replaced by savanna as predicted by the paradigm, resulting in clouds of grass pollen. Wind had blown these clouds high over the plateau where some had settled on the marsh that filled the Carajas basin. Mission accomplished.

To scholars already convinced of the beauty of the paradigm, this was the clinching evidence sought so long. The very accomplished Brazilian geologist who had first pointed the coring team towards Carajas, Professor K. Suguio of the University of São Paulo, said to me, "At last we have got a real record."

I regret to say that I did not think they had. Certainly something happened to rainfall on the Carajas in at least part of the glacial cycle, but there was nothing to show a change to actual aridity, just a very shallow lake losing open water to become a muddy marsh. And the simultaneous increase in the grass to tree pollen ratio was far too modest for belief in the great pollen clouds inevitable from grasslands tens or hundreds of kilometers wide sweeping up to the plateau sides. So for me, a pollen analyst by training, the "partial" pollen data we had been shown did no more than suggest continuity of vegetation on and around the Carajas Plateau throughout the glacial cycle, probably with weeds and grass growing on the bog surface itself. I tried to keep these views to myself, though I fear I was not entirely successful.

But I learned, absorbed, and inwardly digested Professor Suguio's vital message, inherent in his organization of the Carajas project. "Go to the inselbergs," those strange steep-sided, flat-topped hills of resistant rock that stand out of the Amazon lowlands like miniatures of Carajas, with the great rivers washing past, unable to tear them down. On top you will find sinkholes, pseudokarst lakes of great antiquity. "Drill there." When we followed this unspoken Suguio prescription in the years to come we were to succeed in getting actual histories of the rain forest.

Carajas was a brilliant stroke, but like many a pioneering effort fated to direct others to snatch the real rewards later. One of our inselberg lakes in the years to come was to prove to reveal lowland forest holding its ground for two whole glacial cycles even as fluctuations in rainfall came and went.

Meanwhile, Carajas for what might be called "the paradigm community" was the blade that cut down the pretensions of the Colinvaux team. Here was decisive evidence, well dated by radiocarbon in continuous lake deposits, for the two environmental requirements of refuge theory, drying and the replacement of lowland forest with savanna. As frosting on the cake, the pollen showed no evidence for cooling. The supposed lack of Carajas cooling, allied to the Heine attack on Mera, served to debunk all thought of glacial cooling in the Amazon. No more need for conflict with the paleoclimate community at large who still based their understanding of tropical climates in ice-age time on CLIMAP, whose foraminiferal data suggested little cooling at the equator. This it was that had prompted the clash at the geological society meeting reported on by the *Economist* (chapter 11). Carajas killed the cooling debate among refugialists too.

The São Paulo conference ended with a banquet, a real one, not part of the rubber chicken circuit so depressingly familiar in the United States. There were speeches from the floor. I grabbed my chance, because the city had always been a distant echo in my life. The father whom I had never known had lived there, and died there not so long before. I found his name, and that of the company he founded, still in the São Paulo telephone directory, a publication well-known for its continued ability to be years out of date (in this case fourteen years). The separation of my parents had begun in the 1930s soon after my birth in England, its effectiveness then made paramount by an ocean and the war that filled the years of my English boyhood. I knew only the few things that I had been told about my father, among which that he had been president of the foreign chamber of commerce in São Paulo. So here I was in a city that might well have been my home, and which my father had served all his working life. Might my family's long service to Brazil help my own endeavors there? anyway an excuse to make an after dinner speech which, strange to say, I like doing. I had once proposed a well-received toast at the dining table of the Novotny News Agency in Moscow in the old USSR days. If I could get away with that, some remarks in a city where my father had headed a chamber of commerce ought to win friends. No go. A year or so later, a senior Brazilian scientist told me he thought I had made the whole story up.

So ended the conference discussing "the latest work on the Amazon." As it happened, it also used the last of the more explosive ammunition in the paradigm's limbers, leaving it little else with which to "strike back" beyond repetition. But it also served us well by telling us where to look in Brazil. Choose your inselberg.

But Paulo and I did not return directly to the U.S. It just so happened that we had a complete coring rig, rubber boats, casing, coring rods, sample tubes, drop hammer, the works, in our airline baggage, for which we had paid a very modest excess baggage fee. And here we were in Brazil. How nice if we happened to come upon an Amazon lake. We had not learned the possibilities of inselbergs when we packed or such would have been mightily in our minds. But the great lowlands of the Brazilian Amazon were surely more fertile ground for us than the mountain-base lowlands of Ecuador. We certainly had to go there sometime, with this conference we should steal an early start.

The National Geographic Society had seen the point, providing one of those grants, modest by the standards of the big government foundations, but vital when opportunity knocks. Paulo is a native of São Paulo and a graduate of São Paulo University, we could feel sure of finding what else we needed. With one of us a Brazilian citizen, and the other with an official invitation, we expected no problems with officialdom.

Big mistake. The Brazilian Amazon and officialdom are linked together in a tight embrace that "let no man put asunder." The word got through to me that if I cored an Amazon lake while on a tourist visa there would be trouble. In the words of one adviser, "You will never get back to the Amazon as long as you live."

But the National Geographic Society had already given me the money. What to do? Obvious, give it back. But before I telephoned the Society, Paulo offered a counter proposal. He knew of likely lakes in the states of São Paulo and Minas Gerais, outside the edge of the Amazon basin, certainly, though near enough, and well clear of Amazon administrative regulation. He could write his Ph.D. thesis on these lakes. I "bought" the plan, and only then made my sad confession to the National Geographic.

This meant talking to Dr. Snyder, the longtime director of the society's research office. Feeling a little cool about the feet, I told him my tale, "They won't let me core an Amazon lake, I can send back the grant money at once? Or we could core lakes outside the edge of the Amazon basin that would escape hostile administration but still be of use to the research." Silence on the wire. Then Snyder's well-known gravelly voice, "You are not getting any cores staying here, are you?" This was the authentic voice of an outfit set to explore the earth. Bless them.

So after the conference we went coring in civilized Brazil, traveling on black-topped highways in a borrowed truck, eating our meals in what must be the best truck stops in the world, and sleeping like normal folk in good hotels. And we got to the "bottom" of three lakes *Lagoa Santa, dos Olhos,* and *Serra Negra.*

Lagoa Santa is a lovely stretch of water, possibly five hectares in area, and with sandy beaches. A well known place for a swim, and deservedly so. Because of the land forms around it, local geological opinion was that it might be very old. But our radiocarbon dates were all in the last 10,000 years of the Holocene. Nice try, but no.

We had our rewards in the other two lakes, particularly *Serra Negra* in Minas Gerais State. It lies in a cultivated area where only traces of the original vegetation remain, vegetation known in Brazil as cerrado and cerradão. What this suggests is largely open vegetation ranging from savanna to open deciduous woodland. The lake was reduced and shallow when we cored it, although it occupies what looks like a large crater and the lake is known to be larger and deeper at times. Our best core was slightly less that 2m long but a basal age of 19,520+/−180 yr. BP was consistent with another 25 cms above it of 15,630+/−140 BP. We had reached glacial times, while still in the range of believable dates.

Paulo's pollen diagram from *Serra Negra* shows sixty-nine taxa though many others are named as rare grains. The woodland savanna mix shows up nicely in presettlement times, but our eyes were fixed on the glacial-age samples at the bottom of the diagrams and the truly extraordinary changes revealed. A distinctive tree of southeastern Brazil, *Araucaria angustifolia,* left its pollen there in such numbers as to suggest flourishing local populations. This is 600 km north of its present limits.

The conclusion to which to jump is that this means cooling, letting a tree from Brazil's deep south advance to cross the tropic of Capricorn and head for the Amazon basin because the weather cooled. For once the crime of jumping to a conclusion got the right answer. Because we know, perhaps, more about the natural history and tolerances of *Araucaria* than of any other tree in Brazil.

Araucaria is the state tree of Estado do Parana, the next state down the coast from São Paulo. *Araucaria* has the odd shape of a very tall umbrella, and it stamps its appearance on the Parana countryside. In towns, like the beautiful capital city, Curitiba, the umbrella silhouette of *Araucaria* is everywhere. Because of this notoriety, *Araucaria angustifolia* is one of the best studied trees of all the vast Brazilian forests. One datum in particular served our purpose: the tree needs frost to break seed dormancy. Paulo's *Araucaria* pollen record, therefore, moved the frost line 600 km to the north in glacial times, suggesting a fall in mean annual temperature for the region of about 6° C. This is as close as measurement allows to our estimates from Mera and San Juan Bosco on the Andean side of the Amazon basin.

You believe your own data. Ours told us that both sides of the Amazon basin, the West in Ecuador and now the East as well, were cooler in glacial times. Not a very hard idea to absorb that, "colder in the ice age." That some drying should be associated with this cooling seemed reasonable enough, though the only satisfactory, clearly dated evidence for drying was from this one lake site of Carajas, where the persistence of the native vegetation said to us that the drying was restrained.

It was a little hard, though, that the Carajas pollen analysts had found no evidence for cooling in their sediments. I did not want to think that it was there and they missed it, science does not work that way. Another of those times when all one can do is shrug. But Paulo's finding strong evidence for cooling on the western edge of the Amazon basin to set against ours for cooling on the eastern edge did blunt the Paradigm's hitting back.

13

THE ADVENTURE OF THE DARIEN GAP

> Then felt I like some watcher of the skies
> When a new planet swims into his ken;
> Or like stout Cortez, when with eagle eyes
> He stared at the Pacific — and all his men
> Look'd at each other with a wild surmise —
> Silent upon a peak in Darien.
> — John Keats, "On first looking into Chapman's Homer"

Ever since the early 19th century when John Keats wrote those lines, English-speaking people have felt the magic of Darien. It has been my wretched fate to have been coached by the inhabitants of those parts that it was not Cortez who first saw the Pacific but Balboa. I admit that this is what the history books say too. Yet Keats must have been right, because "Balboa" does not scan. With the ultimate proof of great poetry Keats made Cortez a symbol of Darien adventure, to add to less poetical exploits of Cortez leading to the sack of Mexico and the death of Montezuma in their world of plot and counter plot.

A century later Keats's symbol became not a peak but a gap: the Darien "Gap"; the last broad patch of wild country some 150 miles long that connects the Panama isthmus to South America at the mountainous edge of Colombia. To the land traveler, just jungle, almost impenetrable. It earned the name "gap" because all roads stopped at each end of it. Even the so-called pan-American highway stops short on either side. If you must cross the Darien Gap, go round by sea.

The story of conquistadors, whether Cortez or Balboa, casting their eagle eyes over the Pacific from a look-out in Darien, however, tells us that the Darien was not always a "gap." The conquistadors reached their lookout point along trails, good enough for horsemen as well as foot soldiers. Accounts from the time tell of an inhabited land with clearings and villages, all now long gone from rapine, slavery, disease and flight, leaving the land to its wonderful forest.

To a naturalist the result is close to marvelous, one of the last true bits of un-spoilt tropical rain forest left on earth, a dream to go there to be made true if pos-sible. As it happens an accident of history has left a way in. At Cana, a place halfway along, the conquistadors found gold. The deposit was small, but never-theless was worked intermittently, even after the Second World War when some-one built a dirt airstrip for access. A few simple huts, a local jeep trail, and a muddy tunnel into the hillside were all that was left in the 1980s. The dirt airstrip has been kept up, a unique port of entry to the Darien rain forest.

Bird men and Botanists use the airstrip at Cana, the sole easy point of entry to Darien. Look into any of the great museum collections of plants or birds from Central America and you will find the 'Cana' label in the Panama drawers. One of these visitors told me about a swamp near the airstrip. It sounded good and deep. It is true that my luck with swamps in Ecuador had been spectacularly lacking, but a thick swamp deposit could yet yield an ancient history, had done so in other climates. Moreover inquiry turned up accounts of open water seen from the air not far away, perhaps a decent lake? Cana was little more than 7° north latitude, still relevant to my Amazon quest (just); and another region where people had argued that the land might have been arid in the ice age. This was enough; try it.

The National Geographic Society gave me a grant to cover the field costs. I al-ready owned the equipment. Mark Bush and I set off for Panama, met with Do-lores Piperno at the Smithsonian Tropical Research Institute, and the three of us chartered a small plane to fly to Cana. The swamp was so-so, but we got ten me-ters of sediment out of it, a result not without hope. But the better news came from the two men who, with their families, were the sole remaining inhabitants of Cana. Yes, there was a lake, half a day's march away through the forest. We hired both men as guides and porters.

We set out at dawn; Mark, myself, and the two guides, a complete coring rig spread equally in our four pack-loads, plus what we hoped was enough water and food for the march, perhaps fifty pounds per man all told. One of the guides in-sisted on the extra burden of his shotgun.

The guides knew their forest. They said they rarely went so far, but they got us to the lake between dawn and noon through a forest without a definite trail, and home again, complete with core, after the setting of the sun. Memory chooses its own moments. Mine is of stumbling in the open of the last lap, a wet Avon Redstart balanced on top of my regular packload, and seeing far off the pinpoint of light which was the oil lantern the women of the huts had put out as a guide. Memory is good sometimes.

Air photographs and maps show we had covered about eighteen miles, through a rainforest and stream channels, sometimes on bad trails, other times with no trails at all. The only incident of note was an encounter with a pack of wild peccaries that led to some banging from the shotgun (without effect). And I remember one moment on the return walk when, badly dehydrated, I came across water bubbling out of the forest floor. I threw myself down and slurped it out of the ground.

Journey's end was a vast clearing in the forest, a rough meadowland of waist-high ferns and grass punctuated with spindly shrubs and small trees. There was certainly more than a mile or two of it before us. Not a lake. We looked at our guides, I suspect accusingly, but they pointed forward saying, "Agua, agua." So we cut and pushed our way on. We soon realized that we were walking over the dampish infill of an old basin, and after perhaps half a mile "agua" did hove into sight. Here was indeed a lake, although only two or three hundred yards across at its widest. Yet it was obvious that we had found something much better that another pond in the jungle. This small patch of water was the shrinking vestige of what had once been a much larger lake, now mostly filled in. Our guides said the whole thing flooded in the wettest seasons still. We salivated at the thought of the mud that might lie beneath us.

Mark and I cored it under the noonday sun, while our guides went about their own affairs, which for sensible people was bound to include rest in a patch of shade. I rowed across to sound the bottom, finding the water only 1–2 m deep and the bottom flat. It seemed clear that there should be as much old sediment under the soggy bank as in the patch of open water itself. We gave up on the chore of building a floating raft and cored at the water's edge from our wooden platform spread on the quaking infill where we stood.

The sediment was a limnologist's dream of happiness after our struggles in Ecuador, a compact gyttja of clay and organic matter with discernable layering, yet soft enough to require minimal use of the hammer. With just two or three hours work we got eleven meters of this splendid stuff before striking bottom, as neat a bit of coring as I have done in a lifetime drilling mud.

For those hours Mark and I were parboiled under relentless sun, entertaining each other with gallows humor. Humidity was high, and the shade temperature can hardly have been less than 90° F. Heat-haze shimmered. Vultures circled over us, black vultures mostly but once I was delighted as a huge bird with the black reduced to edgings on starkly white wings, far bigger than the circling black vultures, launched itself from trees at the forest edge and swept over us before starting its own circling climb. A king vulture, the first I had ever seen.

A crocodilian of some size, only its eyes and snout visible above the surface, watched us from the open water, probably a black caiman though an alligator was possible. Mark was wading in a collecting mission at the time, when my sense of gallows humor got out of hand. I reminded him in a loud cheerful voice that attacking crocodilians come unseen, under water. Blunderings in the marshy edges suggested largish rodents, one of which Mark saw from his offshore position, a capybara. Kara-karas (hawks, exclusively tropical) scolded us from perches on the taller spindles of shrubs round the lake, a constant background of complaint a bit like the cacophony of crows but shriller. What with the heat haze, the sweat, the thirst and the watching wild things, this was a memorable, and rather rum, place in which to work.

An eccentricity of mine is to carry in my pack a P. G. Wodehouse paperback, for peaceful reading in tent or sleeping bag, often by a candle's flickering light. I have read the Jeeves and Lord Emsworth novels so many times, that I always know what the next page will bring but, tired at the end of an expedition day, the master's prose is still the most restful medicine I know. I had a copy with me as usual. Mark knew the world of Wodehouse too, and we fitted the thought of Jeeves into our gallows humor, thinking of him looming out of the heat haze, the tall glasses on his tray frosting with moisture. We named the place "Lake Wodehouse," and so it appears in our published reports.[1]

Lake Wodehouse yielded the first ever pollen history of the Darien, and it turned out to be a story of man's impact on the land. The course of this history ran directly opposite to the usual sequence of events. In the topmost mud was pollen evidence of the rain forest we saw around us, the rich and glorious forest of the Darien Gap with its wild array of species. But then, less than two meters down came a stark change. Almost from one sample to the next, the pollen percentage of the true forest types fell away, to be replaced in the counts by rising percentages of the trees and shrubs of tropical disturbance, together with an influx of grass pollen. Our radiocarbon date for the transition was just 310 +/− 50 years BP, a nice enough stride into the sixteenth century to meet the conquistadors coming.

This different pollen regime persisted to nearly the bottom of the core nine meters of sediment below the transition. It took no great historical acumen to realize that we were seeing human history through the Spanish conquest. An agricultural people had lived for thousands of years in cleared parts of a forested land, until exterminated by conquest 350 years ago. Their fields had been re-claimed by forest only in the last three centuries to leave us the precious jungle of the Darien Gap.

We found plenty of verification for this history, both from Lake Wodehouse and from the Cana swamp fifteen kilometers away. Agriculture was confirmed by pollen and phytoliths of maize throughout the agricultural period at both sites, and by flecks of charcoal from people's fires. These traces vanished from the sediments at the time of the conquest, not even the fires remaining. Hard times.

The thriller for ecologists was the spectacular restoration of the forest once agriculture was stopped. Is this a message of hope for those who agonize about the destruction of rain forests now plaguing the world? Mark and I think not. Our reading of the pollen is that patches of forest were always there, if only as remnant woodlands between the agricultural fields. The trees did not have far to go, even to the middle of the small fields. This conclusion from the pollen fits common sense as the decimation of a rain forest by people armed only with stone axes, and without draft animals or tractors, does not compare with what chainsaws can do. The Darien people had not clear-cut their land because they could not.[2] When they "left," the forest could grow back.

Fun though it was to have got the first sediment history of Darien, it did noth-ing for our ambition to reconstruct the ice age. The core bottoms, both of Lake Wodehouse and the Cana swamp, were just four thousand radiocarbon years old. We needed at least another ten thousand years of history even to begin to talk about the ice age. To confirm the Mera history we should need longer still. But the luck was about to change. Dolores Piperno, resident of Panama at the Smithsonian as she was, used her local knowledge. We got first the minimal 14,000 years of history, then a whole ice-age worth, though of course not in the Darien itself.

Dolores's prime choice was Lake La Yeguada, the biggest natural lake in Panama. She and I had already cored it in a brief operation on a previous visit, as a side-line of my giving a seminar at the Institute, but this had been little more than a reconnaissance. This time Dolores had everything laid on: truck to the lake, launch the raft, core the lake, go home at the end of the day. Just like that, or more or less so (something nags memory; it might have taken two, or even three days). Mark later went back with Dolores and cored it again at another place. The cores were 13 m long. Our oldest radiocarbon date was just short of 14,000 years BP, thus reaching the last gasp of the ice age, what we call in the trade the "late glacial."

Lake La Yeguada lies at just 700 m elevation in Central Panama, well to the

West of the Canal and just north of the Azuero Peninsula. It fills a depression on a low tongue or ridge of land running east from cordilleras that rise above 1,500 m. For the lake purist, La Yeguada has the annoyance of being mucked about by engineering projects, but these do not seem to have interfered with the lie of sediments.

La Yeguada's comparatively low elevation of 700 m, with its precipitation of about 3 m a year, supports tropical forest in the watershed. Apart from recent human plantations, the pollen record shows that this tropical forest has persisted throughout postglacial time and even into the bottom sediments from the late glacial period.

But the payoff for us is that the bottom sediments also contain significant percentages of oak (*Quercus*) pollen. Panamanian oaks do not live in modern tropical forests but are confined to high altitudes, typically above 1,500 m. Oaks can be found now high on the cordillera near the frontier with Costa Rica to the west. Thus we have at La Yeguada a repeat of the Mera and San Juan Bosco story. Cooling in glacial times allows more thermophobic (heat-avoiding) plants to thrive in the lowland forest previously denied because of excessive warmth. And unlike Mera and San Juan Bosco, this illustration of the phenomenon is in well-dated continuous lake deposits. This one cannot be avoided by claiming contamination of dates or geological intuition that something is wrong. To the contrary, we can and do use it to confirm the Mera dating by correlating the same phenomenon of glacial cooling that allows invasion of the tropical forest from on high without a fundamental change in the forest type, extrapolating the safe dating from La Yeguada to the shaky dating at Mera.

And then there was the city of El Valle. This city is built on the floor of the caldera of an extinct volcano. It lies east of La Yeguada at the very tip of the tongue of land on which both are sited. But El Valle is 200 m lower (at just 500 m) and closer to the Pacific Ocean. High enough to escape the sweltering weather of Panama City and still in the path of sea breezes, it is a highly desirable place to live. El Valle to Panama City is the Hamptons to New York, or Cape Cod to Boston. The caldera is filled with nice homes, with fine gardens, on large lots. And they get their water from wells, drilled in my time by a grizzled American engineer named Dick Wharry.

Wharry turned up at the Smithsonian office in Panama City one day saying he had drilled a well in the town of El Valle and found what he described as "camel shit," and he proffered a sample. "Never seen anything like this before, what gives?" He was shuttled to Dolores, the most likely staffer to know the answer. And indeed she knew. The "camel shit" was gyttja, lake mud, beautiful. Every one knows the enormous flood of adrenalin that overwhelms body, and even speech, when some tremendous, and wholly good, thing happens. That was us with Wharry's "camel shit" which was really gyttja.

Most extinct calderas are taken over by lakes at first. Obviously this one had followed this conventional route, at first a lake gradually filling with sediment until, millennia later, it had overflowed a low point of the crater rim. Then the great lake continued to fill with gyttja until the rim cracked, perhaps from an earthquake (not rare events here), or simple down-cutting of the new outlet stream let the lake drain. The resulting perfectly flat surface of drying mud, ringed with low hills of the old caldera rim, was a real estate broker's dream of heaven. Dick Wharry had found us our ancient lake, hidden from a layman's view though it was.

We were now faced with a caldera the size of a city (indeed with a city actually in it), and underneath this, one version of our Holy Grail. Our elegant piston sampler in our airline baggage was scarcely adequate for the task before us. Dick Wharry's equipment was on a different scale. It was mounted on an eighteen wheeler (or it may have been sixteen, I forgot to count). So I hired Wharry to go back to the site of the "camel shit," and core the thing through to bedrock. Deal quickly done.

Dick showed us where to drill, and we negotiated with the unfortunate land-owner on whose lawn we proposed to drive the vast tractor-trailer combination (later I paid the owner compensation for damage, which was considerable).

Then I flew home to Columbus to teach my classes, leaving Mark to get the core. Mark Bush and Dick Wharry got 55 m of lovely lake deposit before striking gravel. The radiocarbon age of the top of the core was about 8,000 years BP, the latest date for the draining of the lake. The magic 30,000 years of radiocarbon believability was passed at a depth of 10 m, so that the bottom 45 m were sediments deposited earlier still, deep into ice-age time. Extrapolation and pollen stratigraphy let us claim either 70,000 years or 150,000 years for the bottom of the core, with the older date being more probable. Add the record from Lake La Yeguada to cover the missing younger bit and we had all or most of the 110,000-year span of the last ice age.[3]

Forgetting the enticement of the possible 110,000 years, we practiced the iron self-discipline of first analyzing the upper layers where the dating was sure. We did not (yet) need to be greedy for a full glacial cycle that we could not yet prove. Merely putting La Yeguada and El Valle together gave us a continuous and well-dated history of the last thirty thousand years that we could prove, a history of that last low tongue of the cordillera where it reached into the lowland tropical forest of Panama. For this was a "sure enough" history of the last glacial maximum, the time-slice on which the refuge paradigm was based. Not since Galapagos El Junco had I had such a prize.

Mark and Dolores, using both phytoliths and pollen as at Mera and San Juan Bosco, revealed the forest as it was in glacial times. No savanna, no aridity, the lowland tropical forest was still there. But so were invaders from on high. Apart from the oaks, abundant at El Valle as well as at La Yeguada, was the record of a company of invaders: *Drimys, Podocarpus, Magnolia, Gunnera, Ilex, Thalictrum, Symplocus*, to name only the most certain and striking.

The adventure of the Darien Gap had paid an adventurer's rewards. We had duplicated and dated the Ecuadorian invasion of the ice-age lowland forest by populations of plants that in modern times thrive only in cooler climates. By doing so, we had also confirmed the dating and observations of Mera and San Juan Bosco, putting them effectively out of reach of further calumny. Out of these observations came the climatic observation that the tropical lowlands of America did cool in glacial times, an observation that was to gather strength until now the reality of it having happened is generally accepted in the paleoclimate community. But the bigger reward for us ecologists was its illustration of how even tropical forests meet climate change, they do not dissolve nor fragment, neither are they replaced by different vegetation, as both the ecology of the 1930s and the refuge paradigm still holds. Rather they slowly shuffle their populations and diversity, obeying the relative competitive advantage of individual species. Species suffer population fates, good or bad, but the forest lives on.

The Darien Gap had also given up the secret of its existence. It was formed three and a half centuries ago by one of the most brutal acts of ethnic cleansing in history.

14

THE ADVENTURE OF THE
FLOATING FOREST

Now for Brazil, and the heart of the Amazon. A history of the vastness of the Amazon plain was critical for the prime purpose that sent me to the equator from Yale, by this date, all of twenty years earlier. Here was the astonishing diversity that drove a generation of ecologists. It was also the land where the paradigm of aridity and forest refuges had developed and must be tested. In Ecuador and Panama we had found tropical forest alive and well in glacial times, but with an unmistakable stamp of a great reshuffling of species populations signifying cooling, not drought. Whether this glacial cooling was basin-wide, and the principal vector of climate change, must now be tested across the great hinterland of the Amazon basin in Brazil.

A wild land, the interior of the Brazilian Amazon, even with some of the legendary properties of the North American West of nearly two centuries ago. A wilderness harder to penetrate than the North American model, because forest rather than prairie. Landless men and gold seekers encroaching on homelands of aboriginal peoples, who resist. And the added conflict between those who would preserve part of the forest for posterity, and the many who could win riches by selling its wood and cultivating its land.

Many a stretch of real frontier, the line between "the desert and the sown" may be found in all this: soon we should be working on this edge. But first we must penetrate the newest frontier of all, the one that Brazil had invented for defense against the likes of us. We did not want to clear forest for a farm to feed a family, we did not seek wealth, whether gold, or mahogany, or diamonds, or some more esoteric form of twenty-first-century loot. We were not armed. The only harm we should do Brazil was to publish our findings on climate change in the Amazon, aided and abetted by our desire to share state-of-the-art techniques with Brazilian colleagues. The newest frontier was an administrative barrier, built high and strong to prevent the foreign researcher.

Before we could think of applying for a pass to the gate of this new frontier we must identify our target lakes. No wandering footloose to find them as we had in Ecuador. This was definitely forbidden, even if it had a chance of working, which it had not. Lakes above the rivers, lakes on forested inselbergs, better still a scattering of both. How to find them hid away in a great continent?

The military regime that had governed Brazil not long before we started work there had mapped the whole Amazon region from the air, mostly by sidescan radar. Someone, somewhere, must have a map to show the lakes I needed, hidden though they be in that vast real estate. But I had no idea who that person was. The *radambrasil* maps are mostly on the web now, but in those ancient times of the 1980s the "web" was still just a "network," the erudite child of the U.S. Defense Advanced Research Projects Agency. I was not even sure that the military connection in Brazil might not classify as "secret" these "maps of the interior." So where to turn?

Paulo worked his native Brazil, with my blessing. I tried U.S. sources. Paulo had more success.

If the lakes I wanted figured on the existing U.S. satellite image system called "Landsat" no one seemed to know, and the problem of searching thousands of square miles' worth of imagery remained. A curator of maps or images would want to know the coordinates I was interested in, something I could not give him because I did not know myself. The questions I wanted to ask were, "Where in the Amazon are there closed basin lakes above the rivers?" or, "Can you supply pictures of all the inselbergs in the Amazon basin?" These are not reasonable questions to ask. Knowing nothing at all makes things difficult. In desperation I called offices at NASA, got a hint, made an appointment, and traveled to an institute with "Goddard Space Flight" in its title.

The natives were friendly. One of the Gemini orbital missions had crossed the Amazon on a clear day that let an astronaut get busy with his Haselblad camera. Among his take were oblique shots of landscape with a line of lakes perched high above the valley of the Upper Xingu River in the southeastern Amazon. The lakes, some rounded, each appeared to drain by its own narrow stream cutting its way down to the Xingu. They were "above the rivers," a true target. The location was not ideal: too far from the equator at between six and seven degrees south latitude, in the southeast, but they would do for a start. I walked out of the Goddard Space Flight Center with adrenalin in my blood, and prints from an astronaut's Haselblad in my briefcase. One of my better days about then.

The Xingu lakes were to get honorary mention in every grant proposal I wrote for the next ten years. Our plan was simple: chartered river boat up the Xingu as far as may be, motorized canoe to the mouth of the outlet stream, climb the escarpment and pack in from there; a week or so of boating and walking each way ought to do the trick. But we never did it: the Xingu crosses a remote Indian reservation with jealously guarded privileges.

At least three layers of permission were needed before we could even start: government regulation for all foreign expeditions, then the ministry responsible for Indian affairs, then the local Indian chiefs. A nested set of permissions like this is always a difficulty. This set was made even more intractable than usual by the sheer remoteness of the target. We could not talk to the chiefs without permission to go; but permission from government might be easier once we had got an OK from the chiefs. Catch twenty-two.

We kept trying, but we never beat the odds. I still treasure the Gemini photographs, perhaps my old age shall yet see me rubber-boating on those Xingu lakes first seen through an astronaut's camera. But we no longer need them for our research, even should they carry magically long records. The old age dream is probably just that, a dream. But it would be nice: just to feel the bubble of my wake as I rowed my little Avon Redstart across a lake first seen on an astronaut's photograph would be an old man's thrill.

Paulo went back to Brazil for four months to scout out our options "from the inside" as it were. And he found those "someones, somewhere" who knew where likely target lakes were to be found. These "someones" were Brazilian geologists who had followed the *radambrasil* teams with ground truth operations, flown in by helicopter in quest of likely minerals. Deeply weathered lowlands of the Amazon are bad places for hard rock geology, with the interesting rocks buried tens or hundreds of meters down under weathered soil and clay.

In this land of mud, the exciting outcrops were the *inselbergs* themselves: hard rock formations that jut up from the great Amazon plain as teeth of resistant rock, rearing up through the clays and forest. We wanted them for their pseudokarst lakes, the geologists wanted them for the mineral riches they might contain. Inselbergs proved rich in rare and valuable minerals like niobium. With such a lure, government pays for helicopters. And with helicopters, the inselbergs' time had come to be mapped and sampled. The Radambrasil teams had found some beauties, prime targets for the long forest histories. Inselbergs and their pseudokarst lakes were the best chance for the long records needed to test the paradigm.

One splendid inselberg in particular was being talked about in Brazil, this one on the Rio Negro, three days upstream from the meeting of the waters at Manaus. The surface was so well supplied with little pseudokarst lakes like the surface at Carajas that the geologists named the hill *"Morro dos Seis Lagos,"* the Hill of the Six Lakes. As far as I know it has no other name. *Seis Lagos*, rapidly translated to "Six Lakes" by Paulo's anglophone colleagues, was our obvious prime target. It was a low, well-forested inselberg with at least one lake less than 500 m above sea level, an elevation still in the richer layers of lowland forest but well "above the rivers," at that point flowing perhaps 150 to 200 m above sea level. This made it possible that we could find a pseudokarst, inselberg lake little more than 300 m above the forest of the lowland plainsland, in effect a lake actually within the lowland forest itself, a heady proposition.

But Paulo also found some Xingu look-alikes in the central lowlands of the Amazon, themselves real possibilities for a long record from the very center of the Amazon universe, logistically straightforward, and perhaps an opportunity to organize the collaboration we should need for more elaborate inselberg operations.

Instead of the Xingu River, these lakes were poised above the Amazon River itself, just a day by riverboat downstream from the Manaus and twenty miles or so inland from the *Rio Amazonas*. Paulo had photographs from low-flying aircraft to compare with the astronaut's long-range shots of the Xingu from space, but both sets showed suites of lakes, well above the flood plain of the river, either closed basins or drained by minor streams. The air photographs also showed that the real frontier between the desert and the sown was encroaching, with some lakes nearly approached by peasant farmers. The first dirt access roads were being chopped through the forest from the Amazon north bank settlements of Monte Alegre and Prainha. Try these lakes first, mount a more complex inselberg expedition later.

I had no idea how these Monte Alegre-Prainha lakes might have formed. Indeed, the obvious systems for antiquity could all be discarded: pseudokarst, crater (whether volcanic or impact), or geological fault such as a graben. None could apply. These were lake basins whose origins I could not guess. They were clearly an interesting start to Amazonian Brazil, and should give us practice at crossing that newest and most difficult of frontiers: the barrier of official permission.

Brazil asks for a specific code of conduct to be followed by foreign research teams in the Amazon, the underlying principles being eminently sensible. If you find something, tell us; if you collect something, get one for us too; use a little of your funding to take along a Brazilian scientist, or, better still, take Brazilian students as apprentices, one Brazilian for each of your own. Sensible rules? They are actually no more than the dictates of good manners. But good manners work better than rules.

The rules fail, as have regimes through the ages who try to dictate the details of behavior, inevitably making a nonsense when the behavior is impracticable, or even impossible. Get one for us too! of course, when collecting plants, or butterflies, or soil samples. But drill cores? where perseverance and luck might, or might not, get a single complete core in days or weeks of trying. Kumpak[a] and Yambo in Ecuador required the expeditions of several years to core. Our triumph in the Darien Gap could not have been attempted if the condition of duplicate samples was imposed before we set out.

And as for taking Brazilian colleagues or that modern form of apprentice known as a student! Who could fail to welcome the local lore they can bring to projects as ambitious as ours? This requirement is more than just good manners, it is often an essential prerequisite to success; I was taking Paulo (or Paulo was taking me). When the project is over, the service to students imparts a lasting feeling of self-worth greater than that from having successfully tested a thesis. My years in Ecuador did, in the end, open the first crack in our understanding of the ice-age climate of the Amazon. But what I value most is the five Ecuadorian students working with me in the field at different times whom I was able to bring back to the U.S. for graduate study, all of whom now hold the Ph.D. degree in science.

Work in Brazil soon strips the naivety of believing you can offer comparable service as payment in kind for hospitality. From the start my Brazilian work was a collaboration between my Brazilian student, Paulo de Oliveira, with his degree from Brazil's premier university, São Paulo, and a master's degree from my friend and colleague, Professor Michael Miller of the University of Cincinnati. A Brazilian citizen, his work with our pollen trapping program had already made him one of the most advanced pollen analysts of the Amazon in existence. He was the effective leader of our field parties. But this made no difference to the rules. He was even counted as an American to be balanced by a "real" Brazilian when counting heads.

The rules of the frontier start by saying that researchers must first have a collaborative agreement with a Brazilian institution. "Collaborative agreement" means to me that you both do some of the work. But where is the Brazilian institution with expertise in coring lakes and the paleoecology of the Amazon? The French part of the Franco-Brazilian team that cored Carajas could certainly be claimed as competent but is actually a part of the French civil service, which is to say it is French. And they are ensconced in the structure of the University of São Paulo by government agreement.

For our project, as for many before us, the solution to this problem should lie with INPA, the *Instituto Nacional de Pesquisas da Amazônia*. This is the Brazilian government's principal organization for Amazon research (*pesquisas*). INPA operates a research station in Manaus, properly equipped, with a significant library and research collections. This is a permanent, year-round facility with an extensive staff of career scientists who make their homes in Manaus and are a living library of Amazon lore.

If you are a biologist with an Amazon project in mind, INPA is the place to be, or at least from which to start. Not only is the basic knowledge you will need to be found there, but so is your permission to work. And the laboratory, though in the middle of the Amazon forest, is a mere taxi ride away from an international airport. What we needed was a collaborative agreement with INPA. We were not to get it.

Deep down, I knew it was hopeless to try. One of the several goals of our research was to test the predictions of refuge theory, to test particularly that the forest was replaced by savanna during a postulated period of aridity in glacial times, in short whether the paradigm should stand or fall. But INPA was a home from home for those who found the paradigm "beautiful."

Collectors had gone out from INPA to see if the distributions of other organisms matched those of Haffer's birds. More important still, the attention-catching soils claims were from Manaus, the stone lines that had acquired almost legendary status, and all the surface weathering patterns that had been fitted so neatly into the paradigm. Even sand dunes "beneath the forest" were to be reached, if at all, from INPA.

INPA was also intimately involved in the work at Carajas, having on its staff the pollen analyst Maria Lucia Absy who, under the guidance of Professor T. van der Hammen of Amsterdam, had produced that "partial" pollen diagram which was so important to the efforts of the Franco-Brazilian team to find or interpret data that should uphold the refugial paradigm. At the Carajas conference I had tried to keep my mouth shut about the deficiencies of the pollen evidence, but my general views were out. By 1989 I had published my doubts of all the data proffered to support the paradigm to that date, forcefully and thoroughly in *Scientific American*, the wide circulation in four languages of which had been hard to ignore.

A member of the board of a major American museum visiting Manaus on an official tour told me later that she had queried what she was being told about ice-age refuges by her tour guide, because she had read differently in *Scientific American*, and was told dismissively that "Paul Colinvaux likes to be contrary." INPA was a place that took pride in its paradigm, not likely to take kindly to those whose research might be skeptical.

So I was not surprised when Paulo reported that there was little or no chance of a formal collaboration with INPA. He had to find another way, and had duly done so. He met Dra. Elena Franzinelli, on the faculty of the *Universidad da Amazonas*, and was able to arrange a collaboration with her university. Dra. Franzinelli was a sedimentologist, trained in Italy, who returned to Italy frequently to take advantage of the laboratory facilities there.

Like INPA, the Amazon University was a federal institution in Manaus though without the grandeur and financial support of the research institute. With the agreement arranged by Dra. Franzinelli came a splendid document holding a facsimile signature of the President of the Republic of Brazil. This new document gave us special privileges over a simple tourist visa, notably that of having to report to the police everywhere we went. But it also let us bring in equipment, take it home again, collect, and take specimens out of the country.

There were four of us, Paulo, Mark, Mike and Paul. With the National Science Foundation once more our mentor, we flew by Varig to Manaus in September 1990, bound for Santarem and thence on to Monte Alegre and Prainha, and the lakes above the Amazon.

Although Manaus is a thousand miles inland it is a major seaport where ocean-going freighters dock. It has customs. Indeed, it has more than the usual batch of customs because it has customs against the rest of Brazil as well as against the rest of the world. Manaus is a free port within Brazil, the result of a law to boost development in the Amazon. The combined customs is a formidable barrier to those with strange and complex baggage, coring equipment and rubber boats, for instance. Even a paper signed by the President of the Republic is not enough, you need someone with serious clout to help you through in person. On this, our first run through, Dra. Franzinelli brought someone from the university who did the trick.

Check in at a *Pensão*, report to the police, change dollars for great wads of depreciated currency at the bank (cruzados, then issued in 10,000 cruzado notes), buy hammocks, and take ship for Santarem.

One tourist thing I did first was to look in at the opera house. Built with shipped-in stone in the time of the rubber boom, it stands a prudent distance from the river, a dome on an esplanade, grand as nineteenth-century Europe. Inside, a painted perspective of the Eiffel Tower lines the dome. In the empty theatre I lay on my back on the floor of the stalls to see the shape of the Eiffel Tower rising high above me, the view you would get of the real thing from below. I should love to attend the opera there, but performances with imported singers are rare events these days.

And so by river boat downstream to Santarem, wooden boats, locally built on the river bank, with open sides under a deck. The creature comforts they provide for passengers are lines of hooks to sling your hammock: bring your own, hence our purchase in Manaus. Fortunately for the knot-challenged the crew takes on the duty of slinging said hammock for you when you go aboard in the evening with the other twenty or thirty passengers. The sun sets, you bed down, and the boat departs for the all-night journey to Santarem.

The thousand-mile reach below Manaus is the Amazon proper, the only part of the great river for which Brazilians will accept the name *Amazonas.* Above Manaus the river is the *Solimões,* however much the Spanish-speaking countries upstream call their stretches "Amazonas." In Brazilian eyes the real Amazon starts only where the rivers Solimões and Negro join at Manaus in the "meeting of the waters."

Below Manaus the result of this union of mighty rivers is a vast, fresh-water seaway, the true Amazon. Away from the city at night it is dark; pitchy, inky dark when the customary clouds blanket the stars; featureless even in daylight as the great stretch of rippled and foam-flecked water merges in the distance into the forest of the muddy banks.

I had read Mark Twain's accounts of his early days as a pilot on the Mississippi, with his tales of mudbanks and strandings in the days before the U.S. Army Corps of Engineers turned the Mississippi into a canal. I remembered canoeing the Napo and the constant vigilance for the telltale dead-heads breaking the surface. And I looked at this massive flood, without channel markers or lights, and wondered at driving along it through the night. I asked the Captain a foolish question about hidden mud-banks and night navigation. He looked at me pityingly and said, with slow emphasis, "It is seventy meters deep."

Passengers sleep in two long rows, the hammocks two feet apart, almost touching. If the boat pitches they all swing together with a common rhythm; no problem unless you find a heavy fat man in the next hammock whose added momentum gives you a buffet with every swing and ruins the rhythm. Otherwise you sleep in peace in the cool river air of the night, with mosquitoes banished by the motion of the boat.

When I opened my eyes, the night had gone and we had docked in the bright light of early morning. Against a sky that Constable would have loved to paint but never saw, waves of egrets flew over the greenest of flood lands as they made their way from night roosting places to the feeding grounds of morning. Doubtless other birds flew with them but twenty years of fading memory leave only the white gleam of thousands of egrets.

Buy stores in Santarem, register with the police, and greet a Brazilian team who would join us, two or three students. These were the Brazilian academic community required by our permit. All had some degree of interest in limnology, not of course in the paleoecology of sediments, a subject still to take hold in Brazil. Then we took ship again for the short river crossing (just a mile or two of water to cross) and back upstream a bit to Monte Alegre. Rent a truck, and we were jolting our way to the frontier farm. We pitched our sleeping tents in the very last clearing. The lakes were within an hour's walk now.

This was the real frontier. The cabin at the frontier's edge was built in the local idiom of poles, split bamboo, and palm thatch, not the logs of the old North American West. But the result is the same one-room rectangle for most of the business of living. The Amazon version has the extra simplicity possible where there is no winter to face. The climate gives you just rain and heat, lots of both. Make the roof strong, pierce the walls with openings for light and air, fix structural poles strong enough to hold hammocks, a door you can latch at night, and you have a family house. What the house lacks is protection against mosquitoes, a potentially deadly failure. Being the last house at the edge of the frontier just might reduce the risk a trifle because the local human reservoir of the disease should be small.

A man with a sacão (a big chopping knife, called "machete" in Spanish, "panga" in West Africa, and I rather think "bill" in old England) can quickly erect this house from what grows nearby. If he is rich, he might have a chain saw to improve the place with planks cut from the trees he clears. Probably he only needs to know a man who has a chain saw to effect the miracle of planks. A sacão and a chain saw, all a well set-up man needs. Poor yes, as we now measure poverty. But better than a flour bag shack in a shanty town. He has land and space and water, and a few cows, and the hope they bring. With his choice I know what I should do.

What in English lore would be called a "bubbling brook" ran through a shallow valley fifty meters from the little house, clear water, probably spring fed. The family used it for drinking water and washing, both people and clothes. We wallowed in one of its pools at the ends of sweaty days on our coring rafts; heaven.

The stream was crossed by fallen trees, handy for hanging on to as the cool stream massaged your heated body. Once as I reached for a tree trunk a Brazilian student with our party was shouting at me; a warning? But what. He was struggling with his English but got no further than, "Insect, insect!" He was in earnest.

I took the trouble to look where I reached. I was about to put my hand on a spider as big as the hand that reached for it. Another day there was a small anaconda, no more than three inches thick, lying on the bottom which swam away when we appeared.

Dangling by a noose from a shade tree near the house was a small dead snake, its head triangular. Mike identified it as "some sort of viper." As we talked it over our host came up, "He bite you," and he put his hands together in an attitude of prayer. Definitely not a thing to tolerate where your children play.

But the idea that all wildlife is hostile extends beyond vipers. I have long had a dream of one day seeing a harpy eagle, the largest of the eagles. Harpies hunt monkeys through the canopy. Think of a bird large enough, and powerful enough, to pluck a monkey from the branches of a tree. I asked our host if he had seen any. He said, "We shot one last year and it was a big one." Certainly a "big one," they all are. Vulnerable to shotguns, the big eagles that live in the trees will soon all be gone. Killing this one seems wanton, an echo of the ethic as old as the oldest frontier, the ethic of fighting the wilderness. And yet: can you expect a man with young children to tolerate in his backyard a bird that hunts monkeys?

The woman of the family offered the hospitality of the house to the woman cook whom Paulo had hired. We have always done our own cooking, but there is no doubt that it is nice not to have to after a day on a drilling raft in the scorching sun, and the hour's walk back to camp over rough country.

The one male Brazilian student who came was a tough and willing young man, with sufficient respect for his country's wilderness to wear thick leather leggings against snakebite. He was a welcome addition to the party when it came to packing heavy packloads of drilling equipment and boats. It was he whose warning stopped my hand trying its span against the legs of a huge spider on the tree-trunk bridge.

As always on my expeditions, we had small individual sleeping tents for all, each mosquito-proof and with floors. A comfortable spot of privacy in the wilderness. I know of no finer moment than, having swam or soaked in a bubbling brook, and dried off somewhat, one rolls into that tiny home and zips in the mosquito defense. One can lie there naked in the wilderness with a sense of sanctuary.

The three lakes we walked to and cored in less than a week were nicely up to expectations, each about a kilometer across, well clear of the Amazon flood plain which lay some eighty meters below and thirty kilometers away. The land was covered with what had obviously been diverse lowland forest before the settlers started hacking out trees. The lakes were closed, that precious ingredient of isolation that a time-explorer loves, also deep enough at nine to ten meters to have nicely preserved sediment and yet be easy to core by hand. Each lay in its own bowl-shaped depression with banks sloping up through the forest until they reached a height of land some two hundred meters from the water's edge.

None of us had seen, or even imagined, lakes quite like these. Their likeness was certainly not in the textbooks. So the Amazon was living up to its reputation for wonder. The lake basins looked to me like explosion craters on the pattern of El Junco. Even the scattered rocks poking through the soil looked like the boulders on El Junco that I had foolishly dated by potassium/argon twenty years earlier. But these rocks were the wrong type for a volcano, the region was far from any likely source of volcanism, and the basins were embedded in great thicknesses of sand.

One of the lakes, *Lagoa Comprida*, was one-third covered with a floating forest, a wonder almost beyond comprehension. Thin floating mats of grasses and flowers, with their associated clouds of butterflies, are common in the Amazon, but a floating forest?

Forests on soggy soils are commonplace in the Amazon, often dominated by the palm tree *Mauritia flexuosa*, a fine, tall tree that makes extensive swamp forests. *Mauritia* woodland in swamps beside Amazon lakes are routine, but actually floating on the open water seemed absurd. But that is what has happened at L. Comprida.

As we worked our way through the trees to gain the open water of Comprida the land underfoot began to undulate in a gentle wave motion, like walking on the floor of a giant rubber boat. No water oozed through and we pressed on a matter of two hundred meters until we came out on the last cliff of trees at the water's edge. Here there was no doubt because we squelched with every step. This forest was floating.

Our soundings showed the water at the edge of the forest to be between five and six meters deep. Probing through the forest floor with rods found the forest mat to be up to three meters thick with empty water underneath. The trees were "rooted," if that is the right word, in a pile of floating debris laced together with the trees' own roots. Each tree was tied down so firmly with this lacing that its tall trunk could stand erect through the frequent thunderstorms of the Amazon. Why the organic raft does not become waterlogged and sink like more civilized wood I do not know. What I do know is that the *Mauritia* forest floating on L. Comprida is already four hundred meters long by two hundred meters wide; will the day come when it covers the whole lake?

The limnological wonder of Comprida did not end with its forest. When Mike Miller set out on his water measurements he found the bottom water to be noticeably cold, some six degrees centigrade colder than the surface water in fact. With no winter as a source of cooling, Amazon lakes are warm from top to bottom, a spread of no more than one or two degrees being the usual limit. Our experience at Lago Agrio found bottom water about as cold as you can get, but Comprida was over the edge. Bottom water this cold in the Amazon is as bizarre as a floating forest.

The best bet to explain the cold depths of Comprida is that the bottom water comes from a deep spring injecting colder water into the deeper part of the basin. The cold water would be trapped there under the warmer surface in the same way that winter cold is preserved in the depths of northern lakes in summer, insulated from the heated air by the floating "lid" of warm water.

Comprida had seven meters of good sediment over a sandy bottom. Two other lakes nearby, L. Tartaruga (turtle lake) and another without a name that I know, also yielded seven meters of mud over sand, though neither had anything as exciting as a floating forest. We did not even see any turtles in "turtle lake."

That sand formed the basins of all three lakes fitted the cold spring water idea nicely; all three lakes might be tapping the same cold aquifer below the sand. How, though, did the spring water get cold in the first place? There were no mountains that might feed cold water to the ground for hundreds of miles, and of course there were no winters. The loveliest explanation to imagine is that the lakes are tapping an ancient aquifer dating from the ice age, reflecting the 6° C cooling that our pollen data were suggesting, but we have not a scrap of evidence to support that idea, except its "beauty." And we had had our bellies full of "beautiful" hypotheses with the paradigm. No! More likely the aquifer was fed rainwater over the great plain of ancient sand in which the lake basins were made (how I still know not), and the brief passage of rainwater through the ground was too short to warm it up. Once underground it was nicely insulated from the sun. Probably true, the mundane explanation often is.

In these days of camping in the frontier pasture we all came down with an unpleasant skin trouble. We acquired little red bumps, particularly all over our backs. They seemed to multiply, because every day there were more. They spread. They itched remorselessly, we scratched where we could reach them, unable to stop scratching. Spots became infected blisters that were sore even as they went on itching.

Paulo got back into the riverside town on one of his diplomatic missions and found time to visit the local clinic. He reported that we were infested with spider mites, that he had seen one when the local doctor had put a scraping from one of Paulo's blisters on a microscope slide and shown him the mite, minute but alive and kicking. The lesson learned was that an Amazon pasture with zebu cattle is not a good place to pitch a sleeping tent.

The Brazilian frontier is used to spider mites. Every apothecary's shop has the cure, a tablet of soap, ordinary to look at but laced with dieldrin. Now dieldrin is one of the chlorohydrocarbon relatives of DDT, perhaps the most long-lasting and deadly of them all. Its use is banned in the U.S. and Europe, along with DDT itself. But we blessed that dieldrin-laced soap, environmentalists though we were; lather down with it, wait ten minutes, rinse it off. One treatment is all it takes to kill the spider-mites. The itching stops and the blisters subside overnight. I still own a cake of Brazilian dieldrin soap; no use going to a U.S. corner drug store or British chemist shop for a supply when you need it.

The second set of local lakes lay to the east, halfway back toward the river but still well clear of it by 15 km and 80 m of height. This was a place of older settlement, the edge of the frontier for at least fifty years. The largest lake there, *L. Geral*, three kilometers long by one kilometer wide, was used by people of the small Amazon river town of Prainha as a family resort. They could drive to it.

Lake Geral looks like what it surely is, a swollen portion of an ancient river valley, the old course of which can easily be traced as a drainage channel winding down to the Amazon itself. The lake water is six meters deep over a flat bottom for nearly all its three kilometers of length. Lake Geral then reaches the residual drainage channel through a wooded back-swamp, effectively serving as a dam. The elevation of both lake and dam above the Amazon suggests that the original blocking of the channel reflects tilting of the land surface long ago, giving hope of the long history we sought. But our two cores half a kilometer apart were each only 5 m long, ordinary lake mud with no suggestion of great antiquity.

Lake Geral was big enough that we put an outboard motor on the Avon Redstart dinghy we used as a tender, a tiny 2 hp Seagull from a company now long since defunct. One windy day I noticed the dinghy had drifted away from the raft (by common consent we never asked who had tied it). I swam for it, promptly discovering what others have before me how fast a small boat can be blown along, but I was spurred on by something big splashing in the distance. I caught up with the runaway tender in the end, exhausted and knowing I did not have the strength left to haul myself over the side. But I did, and that right speedily.

It was at Geral, too, that Mike Miller acted strangely. He ran down the bank, slapping at himself, hurled himself into the lake, and swam. He had collided with a nest of wasps hanging from a tree. I got the impression that he did not want any humorous remarks from the rest of us.

Back in Manaus we did our best to divide the core collection between the Franzinelli team and ourselves as required by law. We had cored four lakes, three at the first site and one (Geral) at the second, with no more than two cores from any one lake, not enough to share. The only possible way out was to hand over two lakes. Without giving any hint of my preferences, I asked the Brazilian team to decide which two lakes they should keep. Their choice left us with Geral and Comprida, an allocation with which we were well content. Dra. Franzinelli agreed that we should sample the ends of the bottom tubes of each of hers for radiocarbon dating. In this way, both teams should be able to report on the ages of all four lakes.

The Monte Alegre–Prainha lake cores turned out to span the whole of the Holocene, with a string of radiocarbon dates, when recalibrated to calendar years, showing that we had a detailed history of the last 10,800 years from Comprida and Tartaruga, from Geral only 8,300 years. This would give us the best history of the central Amazon lowlands yet obtained. Not an ice-age history. But the drive that had sent me south to the equator from that long ago day at Yale was not to test a "beautiful" hypothesis of ice-age climate, which did not then exist, let alone the paradigm that followed it. The goal was to describe environmental change at the equator that might help explain the extreme diversity of species to be found there. The details of 10,800 calendar years was a coup in its own right: even if it had lacked that extraordinary floating forest.

Begin by asking why these sandy holes in a sandy plain began to fill with water 10,800 years ago. The paradigm can easily offer a pretty, but wrong, answer, "Dry in the arid ice age, collected water when it began to rain in the postglacial"; simple. The data, both sediment chemistry and pollen, say otherwise. The sediment from the water of first ponding was calcium-rich, quite unlike the sediment of later times when the water was evidently rain. This allows the inference that the first water was spring water, very likely the same springs that still contribute cold water to the deepest parts of the lake, which are then diluted by mixing with the great mass of rain and runoff water above it. And it is clear why the springs burst into these basins just 10,800 years ago.

The lakes filled when rising sea level pushed up the local water table high enough for the spring line to reach the bottoms of the basins. This happened as the glaciers melted, pouring their water into the sea. The average difference in sea levels between full glacial time and the present is on the order of a hundred and twenty meters. The glacial melt came in great rushes, with much of the rise completed by ten thousand years ago, though a pulse at about eight thousand years ago had still to come and a little more for another two thousand years after that. Lakes Comprida and Tartaruga lie at most a hundred and thirty meters above present sea level and close to a great river on its last flood down to the sea.

The Amazon River itself must have been, as it were, pushed up to is present height by the hundred-meter-plus rise of the sea into which it flows (incidentally, explaining that "seventy meters deep" utterance of the boat captain). And above the river the water table was pushed up as well. Lake Comprida got its first water when the springs which still provide that strange cold water layer at the bottom were lifted high enough to breach the sandy bottom of the basin 10,800 years ago.

We found ample confirming evidence of this early event in the Comprida cores. The early ponding led to shallow water and swamp, the water deepening as sediment gradually sealed the sandy basin. The chemical print of the early springs is clear in sediment initially rich in calcium, the concentration of which tailed off as the lake deepened and the average source of lake water changed with the continual rising of the water table. Perhaps rain water was held better as the basin sealed, or the calcium rich spring water was diluted by rain, or rainfall increased.

Pollen analysis tells the same tale, for good measure adding its own powerful evidence that the land was not arid before the lakes filled. In the first centuries at Comprida, when water chemistry still told of a pond maintained by spring water, pollen of herbs, including grasses and sedges, ringing the marsh were plentiful. But in those same pollen spectra were the pollen of forest trees including Anacardiaceae, Bombacaceae, Sapotaceae and Meliaceae (mahogany), nicely demonstrating that tropical forest was there before the events that turned sandy depressions into lakes.[1]

Within five hundred years a *Mauritia* swamp had developed, presumably the progenitor of the floating forest to be, and more forest trees were detected. Only some 7,000 years ago does green gyttja sediment show deeper open water, a condition that lasted more than four thousand years. Pollen shows the *Mauritia* and associated pollen of a spreading back swamp. Later, some two or three thousand years ago, a blacker, highly organic gyttja, together with a boost in the chemistry associated with dense vegetation, shows the establishment of the floating forest. We must assume a wet time, well suited to the swamp tree *Mauritia*, a flourishing single-species stand, the trees growing cheek by jowl, the root masses interlocking in the soft, wet mud, until the outer rank could stand without any mud under it at all.

Lake Geral started later, then had a simpler history. Its foundation date of about AD 8300 could be a simple result of geography that held the bottom of its basin above the rising water table longer, perhaps as a result of the uplift that effectively dammed the old stream bed. But it is noteworthy that this was the time when Comprida was swelling into a deep lake, suggesting that both events were due to a local increase in rainfall, if true a nice example of climatic change in the Amazon within the postglacial period.

Pollen histories of both Lakes Comprida and Geral show tropical forest as the prevailing vegetation beyond the margins of the lakes from the very beginning of their histories, but Geral pollen show a remarkable change just after 6,000 BP with a massive increase of weed pollen, strongly suggestive of forest clearing and intensive agriculture. Roughly a thousand years later are the first pollen of maize, and in the pollen preparations a huge increase in charcoal fragments, the two together strongly suggesting intensive agriculture.

The region near Monte Alegre, and on the south bank at Santarem, happen to be the sites of intensive archaeological studies. The earliest dating probably related to human activities is from there,[2] and archaeological data have been interpreted as requiring dense human populations in the centuries before western contact.[3] The histories of Lakes Comprida and Geral apparently record some of these stages of human settlement.

But the lake records reveal no fires or forest clearing before about six thousand years ago, suggesting that humans present earlier were more hunters of the forest than agriculturalists.

Charcoal was always rare in the sediments of Comprida. I have a truly "beautiful" hypothesis to explain this. Comprida, with its steep sandy slopes and floating forest, was a lousy place to live. That be as it may, and with my skepticism of "beautiful hypotheses," this should not be taken too seriously. But the broad flattish land round Lake Geral appealed to people long before it became the preferred place for picnics and fishing by modern inhabitants of Prainha. They had good sense, those ancient settlers. Take dry land over a floating forest any day. was enough; try it.

15

The Adventure of the Customs Shed

The existence of inselbergs had been known for more than a century, but mostly as travelers' tales. Arthur Conan Doyle, author of Sherlock Holmes, based his adventure story *The Lost World*[1] on the early reports, creating an imaginary inselberg with dinosaurs of the Jurassic living on top, cut off from the modern world. When published in 1912, this tale was the first incarnation of the "jungle with dinosaurs" theme so lately taken up by *Jurassic Park*.

Doyle's reconstruction of an inselberg got it wrong. He had the land gradually sloping up over many miles to about three thousand feet (a thousand meters in my Amazon thinking) before reaching the base of an immense and unclimbable cliff, perhaps two hundred meters high.

The real Brazilian inselbergs are at once both less forbidding and more spectacular. Seen from a boat chugging up the Amazon or the Rio Negro, an inselberg looms out of the gray distance like a colossal building on the skyline, the architecture of which mixes steep sides with gentle slopes. They stand squat out of the forest, set down squarely on the flood plain, with the river swirling round their almost impenetrable rocks. Doubtless their rocks lie deep below the river line too, like the roots of a tooth. You can walk over them since they have no impassable cliffs.

It was one of these that was being talked about by geologists in São Paulo, even as Paulo learned of the lakes above the rivers, Comprida and Geral. This was a low, well-forested inselberg, well dotted with lakes of a kind limnologists can dream about. The lakes were all small (in the 200 m–400 m range), and so prominent that the geologist explorers in their helicopter called it "*Morro dos Seis Lagos*," the Hill of the Six Lakes. As far as I know it has no other name.

My old worries of "higher than the rivers" were gone as cleanly as those celebrated dreams that go with breaking day. Now I wanted lakes to be low enough on the inselberg to lie down in the tall lowland forest though still prudently above the rivers, that is. A lake less than 300 m–400 m above sea level should do nicely, still in the richer layers of lowland forest but well "above the rivers." By all accounts the inselberg had several such.

The inselberg lay four days' journey up the Rio Negro, at which point the river should be flowing perhaps 150 to 200 m above sea level. This made it possible that we could find a pseudokarst, inselberg lake little more than 300 m above the forest of the plainsland, in effect a lake actually within the lowland forest itself.

Seis Lagos, rapidly transliterated to "Six Lakes" by Paulo's anglophone colleagues, was our obvious prime target. Moreover, Six Lakes Hill stood close to zero latitude at 0° 10′ N. We were not likely to get nearer to the equator than this. The only possible drawback from our point of view was that "Six Lakes" lay near the western edge of the Brazilian Amazon when Ecuador had already given us our fill of the Amazon version of "the Western lakes." But distance has its own meaning in the Amazon. Six Lakes Hill is actually 700 miles, say 1,200 km, east of Mera. Only another two thousand miles to go before we sunk in the Atlantic. And we dared already to have plans for those last two thousand miles.

We were back in Columbus from Monte Alegre in October 1990. My affairs were complicated by my resigning my professorship on one of those points of principle on which academics allow themselves to be exercised from time to time. These processes sap energy. But I negotiated a new laboratory in the Smithsonian Tropical Research Institute, taught my last undergraduate classes, and arranged the futures of the eight doctoral candidates and two postdoctoral workers in my laboratory.

The National Science Foundation and National Oceanographic and Atmospheric Administration grants, which together supported the Amazon work, were portable. They would travel with me to the Smithsonian, as would the essential equipment bought with those funds. This is one of the wisest of all the provisions of the U.S. government's granting system. I was also a principal investigator on part of a collaborative grant from the National Science Foundation Office of Polar Programs to work in Siberia, where I had conducted an expedition early in 1989 in the Kolyma country. I arranged for administrators at Ohio State to take charge of what little was left of this project, with one of my postdocs.

Add in trivia like packing a home and selling a house, and it is hard to deny that I was distracted from the essential task of organizing to drill lakes in remote Amazon inselbergs. On the plus side of the ledger was that this was a government move on foreign service of sorts, because the Smithsonian laboratory was in the Republic of Panama. My wife and I even had red official passports. The government packed both our laboratories and our household goods into a forty-foot container, and took it away.

I even had a new drill design under contract, for this was to be the expedition when the troublesome business of holding the piston in place during its journey down the old hole was to be solved for ever. In the early chapters of this book, I was not shy about the failings of the standard sliding rod (nowadays always a square rod) with which one nudged the piston into its proper place at the mouth of the sample tube, afterwards sliding the rod up and twisting until it locked into the drill string. When driving, all you could do was to hammer on the lugs that held this sliding bottom rod to the drill string. Every since my summer failure at Alaskan Imuruk Lake, I had used my rigid "cork and wine bottle" principle. Even this lacked perfection.

In my last years at Ohio State I met an engineer of the kind who seemed to work magic, Vincent Vohnout. We had interests in common in crew rowing. Each bought a sculling boat, joining with a nucleus of others determined to start crew rowing from scratch at Ohio State. The year I resigned my professorship Ohio State sent a crew to the Henley Royal Regatta.

But it was the engineering side of Vince I needed now. Vince had just built a walking machine, a real full-sized one, distinctly reminiscent of the walking tanks in *The Empire Strikes Back*. Vince's used the body of a Ford one-ton truck, and had six legs instead of four. I reasoned that anyone who could do that could solve my piston problems. And it was so. The task took him six hours.

Vince designed a piston that would grip inside the mouth of the sample tube with such tenacity that even direct blows from a heavy hammer could not dislodge it. When it was down the old hole the grip could be broken, and the piston freed to do its work, by a sharp jerk on the piston cable. The system works because the piston is locked by pressing it down against a rotating cam with a screw, thus squeezing the piston seals out against the tube walls until immovable. Unlocking is a simple matter of rotating the eccentric cam with the tug on the cable, thus releasing the pressure on the seals. Works like a charm.

Vince wasn't satisfied with just a giant slain. He now redesigned every part of our rig, new rods, new couplings, new casing, new anvils to lock on the rods whenever you need to hammer, an extraction tool, even a specialist tool kit. The newly designed rig is now, as it were, standard issue and widely used in the profession, but this was the prototype: beware of bugs. The bugs duly appeared, but are now all dealt with.

Paulo went to Brazil as a pathfinder. He went to Manaus and made detailed logistic arrangements right down to chartering a river boat, visiting the area to scout out an approach route, and arranging to hire porters for the difficult trek to the hill. He had also set out pollen traps in the forest blocks near Manaus that were under intensive study by the Smithsonian's forest fragmentation project, traps that were to be of immense value for our eventual unraveling of forest history.

Paulo, like many of the advance guards of history, also came back with a warning of "Do it soon." The streams were high enough to let canoes close to the farthest, low side of the Hill, where a trail was known by locals for a relatively easy climb. Also there were perhaps twenty men in the nearest village who needed employment and might be hired as porters. But if the coming dry season was in the dry end of the range, the streams might be impassable. In addition, porters might no longer be available.

A third "Do it soon" was to prove even more ominous. To take part in the field work the team from the *Universitas Amazonas* had to fit the work into their summer break. This meant our getting there in the order of six weeks sooner than we managed. It could not be done. And it was to mean even bigger trouble than what was to come by having a "film crew."

"Soon" was in thrall to my leaving Ohio State and taking service with the Smithsonian. Resigning a professorship (actually two since my wife was moving her tropical marine laboratory to the Smithsonian also) and leaving all the people who have become dependent on you is neither easy nor pleasant. Even squeezing the vital dictate of "make haste slowly" beyond the point of reason, we could not leave for Brazil until October 1991. By then all three of Paulo's warnings had come to be fulfilled. And I had added to our potential troubles by taking on some additional obligations.

I had an unusual influx of recruits. Shana Weber, an undergraduate honors student whom I advised, won a university research prize that paid for her to go with us to the Hill of Six Lakes as the subject of an undergraduate thesis. Her joining brought the expedition up to five: Paulo, Mark, Mike, Shana, and me. Then I added Nick, who needed an assistant, bringing the total to seven.

Nick Carter was a graduate student seeking a doctorate in an unusual cross-disciplinary undertaking, essentially a doctorate in "communications," part of the "thesis" for which would be a documentary film of a new kind. He wanted to use one of my expeditions as his subject, which meant now or never. I was attracted by Nick's enthusiastic ego and had added to the burdens of that year by writing an additional successful proposal to the National Science Foundation for supplementary funds to pay for his equipment, state-of-the-art video cameras, etc. And of course, to hire his assistant. He chose an undergraduate, Melissa Carter (no relation).

I had set up trouble for myself. Filming in the Amazon requires a special government permit, something which I did not know and could (can) scarcely conceive. We were going to record our approved collaborative research on videotape as well as on pencil and paper! But it was set to make bad trouble.

What I had done was to recruit two teams, a coring team and a two-person film crew. The coring team was myself and three veteran colleagues, sure of each other from long experience and friendship, with one new recruit, Shana, who gave every sign of being as good as the best of us. The film team gave clear promise of artistic excellence as well as of competence in the rigors of work in remote places. But the two teams did not know each other, were actually to meet together for the first time at the airport en route to Brazil. And thenceforth, of course, their goals and needs were different, even though they would journey together to the same remote site. In a less distracted time I would have planned for these special requirements of a joint expedition, at least getting the teams together for a "dummy run." I failed to do this. I was to be spared paying the price for this failure only by the human qualities of the seven who came.

Research in cinema had a long history at Ohio State where, twenty years before, I had backed the film-making of another graduate student. She had made a splendid documentary of Dee Boersma's study of Galapagos penguins that was later used as an educational tool by Ecuador's national museum in Quito. Naively, I felt that a similar effort in Brazil would be public-spirited.

That there were risks to smooth working of the expedition in letting it grow like this I was well aware. Four is a good number, four people who can work together or take on tasks in pairs if need be. With four there is never an odd man out. But I had let it grow to seven, and then I committed the folly of allowing one more (making eight). I took the eighth to oblige someone: never done it before, never done it again, don't do it (even if you have resigned a professorship and are otherwise hassled). Everyone added is a chance of friction, everyone added means more supplies, more visibility, more chance of crossing the bureaucracy, or upsetting a countryman you want on your side. "Need" should be the only reason for adding personnel to an expedition, and this requirement had not been met.

Doing it as soon as possible exacted another heavy cost, even though 'twere not done soon enough. There should have been a trial field trip in Ohio, preferably in the pouring rain, when we all eight of us went out to core a lake with the new coring equipment, and in the eye of the new cameras. But there was none. Unthinkable not to have a dress rehearsal, or in military parlance no training. Go, go, go, we must do it now.

With this lack of preparation hanging round my neck, the eight of us met at an airport with our mountain of baggage, and flew to the customs shed in Manaus.

Parkinson missed one obvious variant of his "law": "All baggage expands as a function of the number of travelers with needs." The team was twice as big as usual. I have no record of the number of pieces this time but memory seems to say forty. The baggage included Customs attractants like drilling equipment and video cameras of advanced design (Nick had proudly told me they were the same model used by CNN).

Customs pounced. No one was there to meet us, the first fruits of my too late arrival resented by our local collaborators. I waved the permit with the President's signature, Paulo argued and went for help. The baggage languished in customs for three days.

I had parked the team at a resort hotel run by the airline, Varig, an expensive choice for so large a team, but chosen because the resort had its own dock on the river where our chartered river boat could come alongside to take our frightful cargo. This grew as our foraging parties headed for markets in taxis, pockets dangerously stuffed with the very used bills of a currency in the last stages of inflation. I got the impression that the hotel people were not too pleased with the ever-mounting baggage. But we were eight foreign tourists paying appropriate rates, they gave us tolerance.

Paulo's boat charter then went belly up, one more cost of being too late. Bread and butter for freelance boat owners on the Amazon is cargo. River boats are the eighteen wheelers of the Amazon road; more perhaps the equivalents of twentieth-century tramp steamers trading the oceans. When you charter for a long journey, the skipper feels free to fill the left-over cargo space with commercial cargo. Fine. But the skipper with whom Paulo had contracted stretched this custom to accept a platoon of soldiers as supercargo. We had not contracted to ship "before the mast" on a troop ship. We had work to do on the four-day voyage, so we canceled the contract. Paulo endured threats from the aggrieved skipper, but found another owner ready to charter. In two days the new charter berthed at the Varig dock for us to load.

My crew of eight had its first communal experience of labor in the 95% humidity and hot-as-human-blood Amazon air, working as stevedores in the ancient way, loads perched on shoulders or heads, walking the plank between ship and shore at the hotel dock. I remember wringing out my shirt to rid it of the accumulated weight of sweat. We checked out, and sailed.

We crossed the celebrated "meeting of the waters," where the black mystery of Rio Negro water batters against the turbid brown flood of the Solimões. This "meeting" has the rare quality of living up to the press releases. An abrupt line separates brown from black, first diagonally across the immense spread of water, then downstream parallel to the shores. Eddies, whirlpools and foam bounce along the meeting line. Mixing the waters after the joint river has become the Brazilian "Rio Amazonas" is slow because densities of the two waters are different. The "black" water is colored by the intense sepia dye leaving the water mass "lighter," the milky brown water by suspended clay from the Andes making it "heavier."

We chose the black stream, leaving the milky brown of the turbid Solimões well to port, and headed into the peace and quiet of the Rio Negro. Only the silence of the forest, the cheerful thumping of the diesel engine, and our wake bubbling golden foam out of the black water.

A speck far behind us appeared, grew larger, soon to resolve into the prodigious bow wave of a ship at high speed. It looked unpleasantly like pursuit. Customs Launch, Police, Bandits? None of the above, though we were in fact being pursued.

The pursuing ship resolved into a river boat just like ours, going flat out, belching black diesel smoke, its whole length buried behind its huge bow wave. With its decks lined with the jeering soldiers of its supercargo. It was the cancelled charter giving chase, and as we were to find out later, determined to reach our joint destination first in order to make trouble with the local authorities.

The pursuer came in obliquely on a collision course, offending our skipper's machismo rather than his ship. Neither captain would risk a ship; they owned them and lived off their earnings. But honor must be satisfied. The pulse rate of our diesel rose, our own bow wave climbed until, as it seemed to me, we had a slight advantage in speed. The confrontation resolved into a game of chicken with river boats as the two captains snaked past each other's bows. We crept into the lead. Honor satisfied, our captain throttled back to cruising speed and let them go past, the supercargo of soldiers jeering to the last. They would be there first, and we should learn about it.

Four days' peaceful cruising up the black river followed, the dark green forest sliding by, golden crystal flashing in our bow wave and wake as sunlight shone through the humic dyes. Our binoculars scanned the forests as we passed. We never saw anything of note, except the river birds themselves, and not many of them, learning as many have before us that a boat in midstream is no place for a forest naturalist.

The legendary white sands first appeared as a white band between the black water and the brown mud capping the river banks. Where I first saw the sand it was only a meter or so thick before it plunged out of sight into the dark water. I trained my binoculars and stared, my memory fixed on the many soil sections I had seen in my youthful days playing at soil survey in Nigeria, Portugal, and the

Canadian east. But this was not a soil profile, it was a deposit, more or less undifferentiated to the resolution of binoculars from a boat in midstream. My first sight of the legendary "white sand formation" of the central Amazon.

Further upstream the banks tended to resolve into rolling mounds of glistening white, hurtful to eyes attuned to black water, dark forest, and gray skies. We stopped at one mighty mound the size of a soccer pitch, a smooth parabola of sand, the individual quartz grains bleached whiter than the shirt in an advertisement for detergent. I ran along it like a child at the seaside, impossible not to.

The white sands had been used by proponents of refuge theory as one of their ancillary props. Where had they come from, what did they mean? Certainly they speak of great changes in times past, sand does not form under a rain forest let alone get bleached. A great expanse of sand suggests bare ground where now there is forest. The white sands seem to spread over hundreds of square miles. Nobody claimed that all that sand was formed in the ice age, and I doubt if anybody secretly thought it did. But the presence of a former bare ground formation did set an agreeable tone for those who wanted it. Sand also has the habit of blowing into dunes, whether in the Sahara or the beaches of Cape Cod or the Norfolk coast of England. And sand dunes are a beloved symbol of those who seek evidence for past aridity. From the air the white sands seemed to oblige.

Over most of their extent the sands are covered with savanna merging into various forms of scrub forest, much of it the peculiar South American form known as *caatinga,* a form of open woodland closer to savanna that it is to true forest. Very many of the plant species in these open communities are what we find in the great savannas outside the Amazon. Which allows the weaving of a lovely tale. These caatinga forests and savannas of the modern white sands (usually conveniently referred to as the central savannas) can be cast as the relics of the great paradigm savannas of the ice age. Not a very strong casting, really, and even in the early days of refuge theory readily dropped. The plant cover is clearly edaphically constrained (that is to say dependent on the properties of its substrate, a great thickness of sand). Geology, not climate, sets this vegetation, and would do so, ice age or not.

More telling though were reports of imagery from the skies, where side-scan radar had shown serial ranks of mounds under the caatinga woodlands, surely dunes of sand, a great fossil field of them? The mounds, seen only by radar, were *under* the modern vegetation, so the mounds came first, the plant cover second. But no one had visited a dune on the ground; above all no one had dated a dune (a difficult thing to do unless you are lucky). Without a date the whole story of sand dunes in ice-age savannas was just a "just suppose." The supposition was an example of what evolutionary theorists call a "Just So Story," a phrase they picked up from Rudyard Kipling.[2] Phenomena were interpreted to fit a fable. This is a no-no in science.

Sticking to what I had seen, I was inclined to river action as the source of sandy mounds in this land of rivers until a good "ground truth" survey was made. It would be nice to see those reported dunes for myself. I thought about looking for datable sediment in ponds that might form in swales even between true sand dunes when rains came.

By the second and third days of chugging upstream, the Rio Negro was changing from an Amazon giant to an ordinary big river, ordinary except that it was still black, still silent, still wrapped close with dark forest. A few very small townships slipped by, a colorful church, some houses, a dock, and a few villages or isolated huts with canoes drawn up half out of the river.

But on the fourth day we came to rapids, rocks breaking through the black water under mists of golden foam and golden spray. Journey's end. We slid into the quiet water embayment where the town and local administration of San Gabriel da Cachoeira mark the head of navigation. And went ashore.

Paulo was greeted by officials with a summons. Not an arrest, a civil suit requiring him to appear before a magistrate. The captain of the boatful of soldiers had rushed ahead to lodge a complaint of breach of contract. His reckless chase out of Manaus had been not only machismo, he meant to get there first for another try at the suckers, using a rural magistrate.

Often it is said that "the law is an ass"; whether Surtees's character, Jorrocks, or Dickens's Mr. Pickwick used the phrase first, I don't know, though I tend to Jorrocks. The aggrieved boat captain must have been sure that it would be proved true by an Amazon country magistrate. But he was wrong. This law was not an ass. The magistrate began by deferring a hearing until after we came back from the Hill of Six Lakes, leaving us free to proceed.

The hearing on our return was in a small room, I rather think a veranda, the only places for civilized discourse in that climate. Four or five people sat in a circle, the magistrate, Paulo, plaintiff, and officials. I was there off to one side as Paulo's sponsor and possible witness. The magistrate appeared to be a middle-aged countrywoman with a no-nonsense air about her, a stern but kindly headmistress perhaps. The question and answer was civilized but time-consuming, and apparently thorough. Being ignorant of the Portuguese language, I never knew what was being said. It ended with "case dismissed" and respectful goodbyes.

It is good to leave a far away foreign place with a sense of having received justice. I think we had seen a neat cameo of how life is lived in Brazil's country districts.

All this was very different from the dirigiste laws that controlled our work in Brazil. When we came down from the Six Lakes Hill, the deferred civil case from the aggrieved ship's captain was but one of the problems waiting for us. The other was much louder, being a meeting with the collaborating scientist from the University who had heard that I had a "film crew" with me without a permit. From the noise level, we had done something shocking, indeed something that might bring down adverse consequences for our collaborating institution: two students with video cameras making a record of our research for a university thesis! Hard to see the evil in this. But then I and my friends had not just emerged from years of working under the dirigiste government of a military dictatorship.

I sensed that we might be near the beginning of the end of the University collaboration. Not only the video, there was also anger at our coming at an inopportune time. I grant the justice of that complaint. Fate gave me little choice, but they could not be expected to understand the drive to resign a professorship. The one thing I could do to make reparation was to help a project of the University that should be nicely parallel with ours (and so subsumed under my grants). Whereas my interests were directed at lake sediments, theirs concentrated on riverbanks and the histories they might reveal. This meant boat and canoe charter (there are no cycle paths along the Rio Negro and the Solemões!), so I provided funds adequate for their whole expedition of two or three weeks in the field.

We soon had reason to believe that some damage really had been inflicted upon us, presumably at a higher administrative level when we returned to Manaus and tried to ship out our equipment and cores from the airport. The airline, Varig, was fine, taking efficient charge, but were blocked by customs. As on the way in, we had all the right pieces of paper, including the impressive one signed by the president of the republic, but paper, as before, was not enough. You needed clout, and this had mysteriously withered away.

Mike and I spent three whole days in the customs shed, mostly just sitting on our piles of boxes in the sweltering heat, enlivened by occasional trips to the office of the Director of Customs, or standing in line to get in. You had to be respectably dressed to get into, or near, the great man's office, meaning no shorts; expedition garb specifically forbidden. Documents with the President of the Republic's signature made no impression, nor did anything else. We should not take our equipment and cores out of Manaus.

Paulo spent those three days desperately seeking help. His troubles can best be described by his return to the shed where Mike and I fretted long after the customs closed one night. Ashen faced and haggard, he blurted out, "I have had a terrible time." No one at the university, our official collaborator, would do a thing. Paulo had even worked up to the director of research who had personally welcomed us, only to be brushed off. South American bureaucrats are extremely good at brushing off. I think they learn the process in admin. school.

Late on the third day I was back in the customs office yet again, to protest, plead, or whatever was necessary. Without a word, the Director of Customs, reached for a document, signed it, and thrust it at me. We were free. I have never been sure what had happened, though I know Paulo had been canvassing possible contacts at INPA.

Nobody knew it, including ourselves, but we had in those cases of core tubes finally trucked from the customs shed to the Varig airliner the most complete and longest history from a site in the central Amazon forest known. It was to be four years before the first publication in *Science*[3] and thirteen years before we completed the history of the inselberg.[4]

After we had left, the University of Amazonas decided to host a conference on the history of the Amazon. They had the misfortune of choosing the exact weeks of a major international congress of botany that was bound to attract the world's leading botanists, I was planning to go and so were others whom they would hope to lure to Manaus. But the organizers wrote to me in the hope that from the Smithsonian Institution I should contribute $5,000 to their organization. That of course was out of the question. In my letter saying so, I said I would use laboratory funds to come myself and bring Paulo, denying myself that important international botanical congress in the process. I also suggested that they abandon so ruinous a conflict and organize with more lead time in which to gather funds. They went ahead anyway. I kept my promise and went to Manaus. At my first meeting with Dra. Franzinelli she formally canceled our collaboration.

I needed a new collaborator. Several leads took me to the Federal University of Parana in Curitiba. I taught a full university undergraduate course in ecology there, *pro bono*, and using a simultaneous translator. We talked, we agreed, we did all the paper work, we had signatures from CEOs in Parana and Washington. A committee in Brasilia, on which I am told INPA was heavily represented, denied permission for the collaboration. But I did get to live for a few weeks in the astonishingly beautiful city of Curitiba.

My bureau director at the Smithsonian, Ira Rubinoff, arranged a visit that might help. The Smithsonian Institution had built a traveling exhibit on the tropical rain forest to tour South America, opening in Curitiba. He gave me the honor of opening the exhibit for the Smithsonian, to cut the ribbon, as it were. After this I would go on to Manaus where I could retrieve the last of our pollen traps from the Smithsonian research plots there. And while there, I should hand carry a letter from Dr. Rubinoff to the director of INPA. For this mission of opening the traveling exhibit and calling at INPA I was an official representative of the Smithsonian Institution, traveling on my red official passport as an accredited agent of the United States government.

This was my opportunity to have a friendly talk about future research plans with a man so well placed to help with Amazon research. When I called at the director's office on my mission I was asked to wait a few moments in an adjacent waiting room. Much later I was told that the director had just left for the airport to fly home. I had been the naïve victim of one of the oldest diplomatic tricks in the book, avoiding a meeting by leaving by the other door.

This continuing saga had the curious effect of making me feel that I was not entirely welcome in the Brazilian Amazon.

16

The Adventure of the
Inselberg That Leaked

The only vehicle for hire in the little town of San Gabriel da Cachoeira was a bus, quite a big unwieldy bus. But to get eight people and our appalling baggage to the canoe landing for Six Lakes Hill, it should do nicely. Paulo hired it. For a second time we played at being stevedores; slippery plank from ship to shore, across the dock and hoist your load through a hole made by removing the escape window of the bus, where those within found room for it on one of the seats.

At the last village on the road only six men were available out of the twenty we had hoped to find. Still there were eight of us, together with a strong young post-doc from the university in Manaus and two plant collectors from the university who were not equipped to carry loads. The six villagers looked strong, and they had canoes and an aluminium flat-bottomed boat for the trip up the little rivulet to the foot of the "Hill."

Paulo reported that the men were far from enthusiastic muttering that the water was low, so we would not be able to get through to the trail on the low side of the inselberg but would be stopped miles short. We also had the advantage of a guide lent by the Brazilian Park Service for the day, this being within the boundary of a national park.

The doubters got it right. "Paddling" those overladen boats was often just wade and drag. With some hours of effort the great inselberg loomed above us, the steep side only. The rivulet was hopelessly jammed with debris and it was obvious that we could not drag our heavy boats through miles of that to get to the low side of the inselberg. Unload and climb it then.

The stream bank was a miserable place, gloomy, almost a swamp. I thought of mosquitoes, malaria, and sleeping. There were too many resistant malaria strains about for us to put faith in prophylactic drugs, and we remained healthy by keeping clear of dangerous mosquito country. I did not like this landing. We had several hours of daylight left. We should go up that hill and sleep on top of it. And that is what we did.

There was no beaten trail, sensible countrymen would wait a few months until the rivulet was up and they could canoe round to the other side where a reasonable trail scaled to the heights across a five mile incline. Or so I am told, I have never seen it. That luxurious detour was out for us. The government geologists who had followed *Radambrasil* to find the lakes had used a helicopter. That comfy ride was out for us too. We had to walk, and do it now before mosquito time. We could come back for heavy equipment tomorrow, and climb the hill again.

It took less than four hours, the Park Service guide pointing the way to the first lake. The climb was steep and slippery, the damp environment as hot as human blood by now becoming familiar. We had about 500 m to go up before a gentle slope down to the first lakes. The seven I had taken with me strong and cheerful as they strode up that apology for a trail under heavy packs.

I had attempted the stupid undertaking more suitable to a younger lead-from-in-front type, adding another storage bag of supplies on top of my own load, and a third day-pack on top of that. With that the hill beat me. On one fall I saw a student assistant with a video camera leering out of the bushes. I should not say "leering," but something like "a reporter's earnest expression." The reporter was absent at another fall when another student witnessed the overbalanced tortoise trying to get up, swooped down, and saying "We have had enough of this" grabbed half my load. Shana put it on top of her own, and took off up the hill. So much for the alpha male.

Near the top of the hill we came out on a rocky place where the enclosing vegetation was thinned by the barren ground. We could look back and beyond to the south where the Rio Negro lay invisible in the misty distance. The tall lowland forest was at our feet, spreading as it were to infinity. Nothing to be seen but forest. We knew it to be all around this little inselberg, and right up to the top in the less rocky parts with more soil. Now let the "six lakes" have the sediment we need and we shall make history . . .

The inselberg was rough and lumpy on top, mostly forested but with large spaces where thin rocky soils confined the vegetation to dwarf shrubs or bromeliads, or to a profusion of dry-looking grass and daisy-family weeds reminiscent of a scrap metal yard in Ohio. The going is up and down past rocky outcrops, once or twice skirting deep holes with the superficial look of craters. One pseudo-crater held a lake a hundred yards or so down a steep slope thick with bushes, but we pressed on after our park service guide. And so we came to Dragon Lake (*Lagoa Dragão*), where the Brazilian geologists had plunked down their helicopter on a convenient rocky ridge.

The water's edge lapped at a berm of pock-marked rock, fiercely hard and firm footing for a boot, but bad news on bare feet. The rubber sandals or boat shoes we all had for work on the drilling raft became our house shoes when we pitched our tents to face over the water. The water at the edge was three meters deep (ten feet) and a comfortable temperature. For the two weeks we lived there, we had a private swimming pool at our door; start with a clean dive, no wading through muddy shallows. And the water met all our cooking and drinking needs. Ever cautious, we filtered or boiled the water first, though I doubt there was real need. There are worse places to live.

But this lake, idyllic as a camp site, was not so good for our purpose. Our camp was not actually in the rain forest. Nor was *Dragão*. The place was habitable because the rocky soils and outcrops around the lake gave no footing for tall trees, only scrub forest and shrubs, though these could be dense enough to please a "maquis" fighter. The Dragão surroundings actually looked unreasonably bare because the Brazilian geologists had let a campfire get out of control and set the shorelines ablaze. Like all sensible people they had gone there in the dry season, but this is when the more scrubby vegetation will burn, even in the Amazon. Years later their fire was partly responsible for the excellence of our camp site.

The geologists' wildfire burn apart, the patches of thin soil and scrub, with their tufts of grass and daisies and the rest, should leave a pollen signature quite different from the forest we had seen spreading at our feet during the climb. These small open spaces brought a source of grass pollen to the very shores of Dragão, perhaps to the other lakes of the inselberg as well. Unspoken the thought, but might these inselberg lakes "above the rivers" be compromised as pollen traps, with the weed pollen from small rocky drainage basins drowning out pollen signals from the outer forest?

By this time we had the outline lesson of our pollen traps learned. Amazon pollen inputs are in three parts: pollen rain that comes from weeds, grasses, and a few tree taxa of the understory: those great trees of the forest whose pollen reach lakes by being washed out of the watershed: and the totally silent component. The published pollen of Carajas suggested that they had got only the weeds from a site even more barren than Dragão. We had no desire to suffer the same fate.

We cored Dragão anyway, a coring operation of textbook simplicity that should at least train the new members of the crew. But it was not a happy event. The sediment was miserable, and in miserable supply. Mike's sonar map of the bottom showed the lake to be actually two deep basins with a submerged sill between. We chose the deeper, 14 m, and stationed the raft in the middle of it. The mud was soft, so soft that it was difficult to feel for it on the end of a 14 m string of rods. The core barrel penetrated with the lightest of pushes, then hit something hard. I pulled it out. We had about 70 centimeters of soft, wet mud, an unusual orange yellow in color, like nothing I had seen before. I went down the old hole with a fresh tube, found the "hard" bottom easily enough, and set out to hammer my way through. It could not be done. The deadly clang and shock felt through the drill string told the tale, we had hit rock. 70 cm of sediment was all there was.

Move the raft several meters, well clear of the offending rock, if "just a lone rock" was all the trouble. Core again. This time I made sure of the mud-water interface by deliberately starting to core too soon, when I knew the core barrel was dangling in open water, near to but not touching, the bottom. I cored. Same result. Less than a meter of soft sediment, orange on top, black at the bottom. Then rock, most uncompromising rock.

Gloom over the camp. The vision of sugar plums that drove me to Six Lakes Hill was of a generous length of sediment that would span perhaps thirty thousand years, mud that should be saturated with rain forest pollen. What I had got was less than a meter of soupy, strangely colored gyttja, with a private worry that much of the pollen in that soupy goo would actually be from weeds among the rocks. Had someone tapped me on the shoulder and said, "Paul, don't worry, it is really an ice age history of the forest you have got there," I should have tried to smile with a wan "Some hope!" In fact that imaginary optimist would have been right. Those seventy centimeters actually spanned more than forty thousand years of forest history. But up on the inselberg I did not know. Gloom was the order of the day.

Our logistics failed us too, with the classic problem of the unwieldy big expedition that this had become, porter trouble. Getting a party of eight up that hill, with all their stores and food for an extended stay, plus provision for a party of poorly equipped teachers from Manaus to join us meant relying on paid help: porters from the village. We had six, not the "up to twenty" that once seemed available. And because the stream round the inselberg was closed by low water, we had been forced into a frontal assault of the steep side of the hill. The six porters did not like the work. Paulo reported that they called it "The Hill of the Americans," not, I think, in tones of admiration.

The six did not reappear. Paulo went back down with some of our party to check. The porters did not like the work and would do no more. Paulo doubled their wages. Three of the six consented to try again. Paulo's party and the three came back laden, and the three went back home. We did not see them again.

We eight, plus the tough young Argentinian postdoc working out of Manaus with the Franzinelli team, were on our own. So be it. Progress would be slower, so we must watch the commissary as we must feed a party, now swollen to twelve with the Brazilian botanists, for longer than planned. To make this task more interesting, the porters had pillaged our supply depot on their way out. The choicest preserved food was largely gone and some of the staples too. Worse, some items of coring equipment, one (our Phleger bottom sampler) critical, had also been stolen, presumably for metal or parts.

Nick reported that the climate was getting to the video cameras, only one remaining fully functional, even that with its "moments." Gloom continued.

Press on regardless. We had found two more lakes, each far more promising than Dragão. Both were in forest: real closed, tall forest, perhaps not quite so grand in stature as that on the Amazon lowlands three hundred meters below us, but hugely diverse and with the familiar rain forest species present. The modestly reduced stature could be explained by the poor soils over old ironstone rock, otherwise this was the real thing. We had lakes in watersheds of lowland rain forest. It was an immediate boost to morale.

The drainage basins of the two lakes had rocky outcrops like every part of the inselberg, but these were small intrusions. Most of their drainage was under true lowland, Amazon forest. Surrounded by diverse forest, safely "above the rivers," in a rain forest climate with 3,000 mm of rain, these were the lakes of my dreams. One of them even had a *Mauritia* swamp on one of its banks, a tiny version of the floating colossus of Lake Comprida.

Shana and Mark had reached these lakes with the help of the Park Service guide, while I cored Dragão. First they sought out the most interesting lake reported by the helicopter-borne geologists who had first explored the Six Lakes Hill. The geologists had noted the lake as near the site they had chosen for hard rock drilling, an enterprise that led to a drill core right through the rock of the inselberg to near its base. Their drilling operation was, of course, far different from ours, with heavy power equipment to drive a rotary rock-drill, brought in by helicopter. But they had seen the lake nearby, blazed a trail to it, and reported what they had found.

The blazes had long since vanished, as had the signs of the geologists' passage. But our guide had been there in the early days and was able to puzzle his way back from memory. The geologists had called it *Lagoa Pata* (Duck Lake). It had a satisfying "normal" look after the eccentric *L. Dragão*, with sloping, muddy shores, tree roots of the surrounding forest awash, and a *Mauritia* swamp. Shana made a sketch map while Mark shouted back depths as he swam with a sounding line. I thought "Shana's Survey Lake" a better name than "Duck Lake," but Brazilian geologists got there first; Lake Pata it is.

They found the second, more remote lake too, this one pea green, obviously from blooms of algae. We had no problems, therefore, with the name of *"Verde"* for this one, particularly as a bright green lake in the Amazon was an oddity we had not seen before. Like Pata, Verde was deep in the forest. Otherwise it was different: the green water spoke of productivity and nutrients far above those of the other lakes and it was in a crater-shaped, forested hollow, so that its hydromechanics were sure to be different.

After the "failure" at Dragão, these two forested lakes were our hopes for success. Our park service guide went home, our porters had all deserted, and we packed the coring rig over our own blazed trail to Lake Pata. From then on my having a too large expedition party paid off, including as it did tough, enthusiastic young women and men. The best porters you can have.

We did not move our camp, it was just too good a camp site, with good water and swimming. Rather we improved the trails to Pata and Verde and began life as jungle commuters Each day we had our therapeutic swims in Dragão. But curious things were happening to these evening swims. Every day we found it harder to climb out. No, we were not getting tired. The lake was losing water, its surface receding almost, it seemed, with the rapidity of a tide.

In ten days the lake fell something like two meters, say five to six feet. Not having expected anything so remarkably odd, we had not measured levels, but the scale of the water loss was clear. On our first day the lake had lapped at the rocky sill, giving that swimming pool feeling. Near the tenth day I had dived in from what was now a high cliff, tired muscles relaxing in the cool water after a too strenuous day under the Amazon sun. Then I was faced with a five-foot wall of pock-marked rock to climb back out. It had tiny ledges and finger holds that made the climb possible, but this time I could not do it. Muscles cramped. I fell back into the lake. I called to Mike on the bank for help, strong arms reached down and I was hauled clear. I may not have measured that two meter drop of water level accurately, but I do vouch for it.

I have a vision of myself on my back, writhing on the ground as I desperately tried to uncramp my limbs, and of overhearing a conversation. A woman's voice, "What is the matter with him?" and Mike's reply, "Cramps, hurts like hell."

The falling level of Dragão brought the question, where did the water go? Like Pata and Verde, Dragão has no outlet stream. When level falls in such a lake the stock explanation is that the water evaporated. This is what the Carajas team had claimed for the low-water episode in their sediments. Could be of course. But

this was not what was happening at Dragão. Two meters in ten days was not evaporation. This lake must be leaking, and in a big way.

But why shouldn't it leak? This was a pseudokarst lake, a basin made by the
slow dissolution of underlying rock by percolating water. We know, as the spe-
lunkers of the world know, that in the most highly soluble of rocks, the lime-
stones, this process goes on virtually without stopping. Beautiful lakes are made
but they leak through the channels pierced by percolating water, and it takes a
rainy climate to keep them topped up with water. Pseudokarst lakes slowly dis-
solving silica should be the same, though the process of solution seem infinitesi-
mally slow.

We had good reason to know that the dry season had set in by our inability to
penetrate far by canoe. We had actually welcomed the absence of rain on our
campsite. Before we got there, rainfall must have kept pace with the leak and the
modest evaporation combined. But with the rainfall stopped, Lake Dragão was
seeping away before our eyes.

Dragão as our first lake was luck, because it must have an unusually big leak.
When we went to look for leaks at Lakes Pata and Verde the evidence was clear
enough in water lines on the bushes that surrounded them: they had fallen
something less than a meter since the dry season began. That just might have
been evaporation if left much longer than our sojourn on Six Lakes Hill, but in
that time, no.

As theory suggests, all pseudokarst lakes leak, the water slowly percolating
down through the inselberg to some remote water table. This makes them extremely sensitive to changes in rainfall. For their levels to fall they do not need
the drought of evaporation. Just let the rain slow enough to stop balancing the
leak. Looking around me on Six Lakes Hill, I was sure that this is what the Carajas team had found with their siderite layers and marshland yielding weed and
grass pollen. They were the first to find the mild drops in rainfall characteristic of
specific glacial intervals, but they collectively allowed this to grow in their minds
into the savage, forest-fragmenting drought required by the refuge paradigm.

It was our peculiar good fortune to set our camp on the shore of Lake Dragão.
It could well be the Amazon's champion leaker because two years after our visit
a Brazilian geological survey party again visited the lake and found it entirely
without water. They told of playing soccer on the lake floor.

Dragão showed that the inselberg leaks, but seemed to give us a poor history of
the past in those 70 cms of yellow goo. Perhaps one of the reasons for this failure
is that because of its mighty leak, the basin has been without significant water for
too much of the time. Our swimming pool was a great rarity following an unusu-

ally wet period. Analysis would show. Meanwhile we had two beautifully forested lakes Pata and Verde to core.

Mike rigged his sonar and made his usual bathymetric map. Pata had a flat bottom under seven meters of water for most of its extent, with gently sloping sides. But there was one small deep hole at one side. The sonar suggested a funnel-like depression running down to a small bottom area about twenty-five meters down. To core the flat or the hole? Or both? I did hanker after the hole according to my private mantra of always coring the deepest part of a lake. But I settled for the flat. Partly for expediency, coring down that funnel might be a pain, and partly for worries that the bottom would be covered by slumping from sediment higher up. Only later came the realization that the fissure, for that is what it amounted to, was probably where the lake leaked, a bad place to get an undisturbed record. We did haul up more casing from the bottom of the hill as a preparation, and I apologise to my colleagues who labored in the haulage. But we cored only the flat. And did it twice.

The Pata cores were nearly seven meters long, longer even than those from Carajas, where they had plunged well past that bogey of 30,000 years. It would have been nice to have longer though, I had that certain feeling that this might be the make or break core. Be optimistic. Put on a bold face. But I had cored too many lakes across the Andes and lowlands of Ecuador with disappointment my reward. Alone in my little sleeping tent the scent of another failure would not go away.

We threw ourselves at the last lake, Lake Verde, Mark taking charge of the coring of a lake that came to be, in effect, his. The sediments looked no more promising than Pata's had been, more colorful from what we could see when coring but about the same thickness. I can say "threw ourselves," however, because it was here that we lost the first coring rig of our South American years through our determination to beat our way down to a long core.

After six meters or so of good sediment we were stuck in something very sticky and resistant, not rock because a furious blow from our ten-pound hammer would get us down further, perhaps half a millimeter at a blow. We took turns with the hammer, the echoes of the blows ringing round the crater-shaped depression in which we floated. We got the tube all the way down in the end.

We could not pull it out again. We struggled, in pairs for only two could get their hands on the core rods at once. All that two strong men could do was sink the raft as the two rubber boats were thrust under water without budging the stuck corer. So we tried to hammer up with the sliding hammer, clipping an anvil between two connecting drillrods and striking upwards. This is awkward.

But a ten-pound steel weight wielded by a fit young man can still strike a considerable blow, even when flung upwards with the follow-through of 180 pounds of good condition behind it. The sampler would not yield. Instead one of the rod couplings, deep under water, broke. So we lost our connection to the drill.

But not quite. We still had the end of the piston cable clamped to the raft, our last tenuous connection with the drill now nearly twenty meters below us through water and mud. One last chance. We filled the boats with water, as many of us as could got a grip pulling to shorten the cable as the raft went down. Then we clamped the cable to a strong part of the wooden superstructure, and bailed out the boats. The cable was now under the pull of the maximum lifting power of two three-man rubber boats, tauter than a violin string, a remorseless steady pull on the stuck drill. But it did not budge.

The last act I have seen, because Nick was floating round us in the tender with his video camera recording the whole thing. We threw overboard the last of our ballast, ourselves. At a one-two-three NOW we all jumped over the side. Now the cable was under the maximum strain possible from our raft. It snapped, in the sudden vicious way that steel cable will snap under too much load.

The drill is still there, deep in Verde's heart, a nice relic for some archaeologist of the future.

Our "war council" met. No point in rigging our spare drill and banging again at the bottom of Verde? Promptly agreed. Hunt for more lakes on the inselberg? Every indication was that the others were smaller, probably less good. True the Lord had been sparing in sediment; on the other hand these tiny drainage basins on thin soils or actual rock could hardly be expected to provide much mud. If the lake basins were really ancient, perhaps we already had as much sediment as the Six Lakes Hill could offer. The mood was "quit on the lakes"; we were running out of food anyway.

But one passage in the old geological report still nagged. Their long, hard rock drilling appeared to have started in a swamp, and their drill log remarked deep, compact organic layers on their way down with an age for which the figure "70,000 years" was suggested. Wow, 70 ka was nearly the beginning of the last ice age, beyond the dreams of avarice. We knew where their swamp must be. Have a go at that with our spare corer, then head for the ship.

Mark led this effort as well as Verde, and named the swamp *Esperança*. This name "hope" is a fair indication of groundswell opinion among us on what Six Lakes Hill had yielded so far. The swampy "hope" failed to lift us very high when we hit rock after a very few meters.

And so back to the ship, wanting to hurry, but with no porters our core party had to haul everything down again with the added burden of the cores. Clearing a work camp is never light work. We did our best to follow our own wilderness code, "Leave no trace of your passing." The fire was stirred until the last, miserable relic of survival was gone and the ashes buried. Things we did not want to carry, were carried out all the same. Up and down the "hill of the Americans" went our willing student contingent under the humid halo of the hill and its temperature of human blood.

One horror stared us in the face, a big, rectangular can of kerosene for our cookstove, still unopened; awkward, heavy, as nasty a pack load as could be imagined for tired people. To me the responsibility; burn the kerosene? Our geological predecessors had done something of the sort and set fire to the place. I hauled it into the woods as far out of the immediate watershed as I could get and poured it out. Where even the lakes leak, that kerosene will probably find its way down to distant great rivers where it will join the effluents of unbridled commerce.

Our ship was there waiting when, three days after our "Abandon the Hill" decision, our rented bus wheezed along side that slippery loading plank. We could load, but not yet sail. Paulo and I had to "go to court" on that deferred summons first. To win time, Mark and a student assistant flew on to Manaus from the local bush airstrip, joining the Brazilian contingent who had to get back for classes. Mark would use the head start to check on the pollen traps that Paulo had set out in the Smithsonian's forest blocks at Manaus, then fly back stateside.

Three days by river boat down the Rio Negro were blissful rest. Hammocks gently swaying by night, or by day when the mood required, the silent forest slipping by, someone else to do the cooking; even the pokey, hand-pump toilet a luxury after wandering off among the rocks to dig your personal hole in the dark.

We were content. We had done it. Whether triumph or another failure now rested on news from a distant radiocarbon laboratory. Looking back on the peace of that voyage, now sixteen years ago, I realize that I already expected at least some good news from radiocarbon. Those rows of aluminum tubes in the ship's hold might well yield the first sedimentary history of the Amazon rain forest, but even the budding optimist in me did not dream the truth, that the Lake Pata mud would eventually be shown to span two whole glacial cycles.

At the ship's home dock in Manaus, Paulo found a truck and driver for hire. We checked in at a pensão. I counted pennies. There followed days when great wadges of currency featured in my life. Sums running into thousands of dollars; bills for boat hire, and a large advance to the Brazilian team to finance their proposal to collect river bank sediments.

Prodigious inflation had made Manaus a cash economy, with "high street" banks reduced to mere intermediaries for changing greenbacks or travelers checks into inflated currency. More uncomfortable journeys from the bank, passing out through the cordon of armed guards, pockets bulging with great wads of currency that could pay many a man's salary for a year, counting laboriously through the piles in a ship's cabin down in the docks or a room at the university, demanding receipts, handwritten on scraps of paper, for my NSF paymaster.

The student crew flew back stateside, but Mike and Paulo remained with me for another go. A long record from the central Amazon near Manaus would be a nice touch, whatever we had from the Six Lakes Hill. This was the ultimate heartland in all thinking about the Amazon. We had kept all the equipment with us for a last throw.

This was the Amazon lowlands with a big "L," no chance of finding a lake "above the rivers" here where the Solimões collected the massive influx from the Rio Negro, to swamp the land as much as any place on earth. The combined river, *Amazonas,* carried so much water that it regularly spilled out of its "seventy meters deep" channel to flood the country for miles around. There were lakes aplenty, but I had long since written them off: until Paulo led me to some geological opinion.

A clue was that channel of the Amazon "seventy meters deep." The explanation of this remarkable trench is not that the modern river carved it, rather that it was the channel of the ice age ancestor of the modern river, when sea level was more than a hundred meters lower than now. This had doubled or tripled the "fall" of the river over its last thousand miles. That ice-age River Amazon had cut a deep gorge on its passage to the sea, and the river kept the same channel when melting far-away glaciers pushed the sea back up again. If this was true of the river, it ought to be true of the riverain lakes also, or at least some of them.

Across the river, to the south and east of the city, lies a suite of large, narrow lakes. Obviously chunks of ancient river valley long since dammed and abandoned by the parent river, I had scorned them when I first studied the large-scale maps. But what if it had been the Pleistocene river that had cut the channels leaving lakes behind during the last ice age itself? Perhaps ancient deposits of the ice age lakes were lying there still, undisturbed, under five or ten meters of Holocene mud. A long core (perhaps very long) in these lakes might well span into the ice age.

Such was the geologists' tantalizing argument. I can see holes in this logic, but we wanted to believe. We three had to try. We loaded every last bit of core rod, casing, rafts and supplies onto the rented truck, and took off by ferry and dirt roads to reach the chosen lake, guided by the large scale map, determined to stay there as long as it took.

The truck owner was he who had hauled our gear from the docks. He was strictly free lance, an entrepreneur, ebullient, a "comrade" in any language, whose like you can find with "taxi," truck, ox-cart or boat in any society. He adopted us, with more than a trace of bravado. Fine, he was game to take his truck through the worst dirt roads we could offer. Clearly the person we needed, though we did look doubtfully at the worn-smooth tires on his precious truck. It was not the truck that was our undoing, it was the machismo.

Our *companero* trucker succeeded in nearly totaling his truck in the middle of the Amazon, close to the meeting of the waters. Perhaps not literally "in the middle" but near enough. Crossing the Amazon from Manaus to the south bank means a ferry ride, quite a long one. The Manaus ferry is a simple structure, basically a large steel caisson with the usual flat bed, a diesel engine, minimal superstructure, and with "drawbridge" flaps at each end. The design is ideal for a great river with minimal waves, deep water but hugely fluctuating level, and muddy banks for loading and unloading.

Our driver drove us on, easily nursing his truck down the steep bank at the Manaus side. We settled down once again to admiring the "meeting of the waters," walking the peripheral catwalk to stare down at the mixing flood.

I trace the first move toward disaster to when we approached the south shore and drivers at the head of the line began to move their vehicles while we were still going full speed ahead. They were maneuvering to be first off. We remained firmly in the "standees" place as our driver's machismo won out. He secured first place in line and looked proudly up at us.

The landing was an appalling slope of red mud, rising at a steep angle to a terrace with a road, perhaps twenty to thirty meters above the level of the river, then at its annual lowest. The lower section looked impassable, like one of those absurd advertisements in the glossy papers of an SUV designed for shopping malls going up a slope so steep it looks to be falling over backwards.

The Manaus solution to this problem is to station a very large earth mover with shovel and blade at the landing to squash the red mud into as mild a slope as possible, smoothing it to meet the lowered drawbridge of the ferry where it has rammed the mud. This takes time. But patience and machismo do not mix. Our driver edged forward to the very edge of the ferry, keeping at all costs his "first place" in line. We scrambled up the bank to watch the unloading from on high.

The earth mover was one of those with a massive counterweight built into the rear end. The huge machine worked down to the drawbridge ramp and backed on to the ramp itself for the last smoothing of the boundary between ship and shore. As it backed rapidly onto the ferry, it was the counterweight that crunched into our truck with annihilating momentum. Bumper, fenders, and other parts were smeared over the front wheels.

We watched sadly from on high, then we took a passage on the ferry for a return trip to Manaus, afterwards steadily refusing our *campanero's* entreaties to give him a little time to "fix" his truck. We had had enough.

We also did not have time to try again. For our schedule, the undertaking had been "one bridge too far." We settled for a reliable trip up river in a motorized canoe instead to sample flooded areas, islets, and grassy areas for surface pollen to estimate the pollen signature of river communities. We also cored to the bottom the sediments of Lake Manacapuru, a large embayment of the Rio Solimoẽs sufficiently away from the stream to trap sediment, in the wild hope that this might mask an older history as we had hoped for the old riverain lakes that we had failed to reach.

Canoes were more to our liking than steel ferries, or beat-up trucks with macho drivers. The two days of this venture were fun. The Manacapuru cores proved to be only five thousand years old, and yet they gave us, with splendid certainty, the pollen signature of the river communities for those years. This was to be of real value when we came at last to interpreting ice-age pollen records from the forest.

Within the Amazon forest there can scarcely be a place with more "grassland," broadly defined, than near the river itself. Temporary islands, seasonally flooded areas, floating mats of grass in tributary streams and still-water eddies, might, it seemed, introduce masses of their wind-blown grass pollen to form a real pollen rain, possibly even masking part of the pollen signature of the forest itself. Manacapuru and the surface samples settled this one nicely. The grass signature was there, but much smaller than we had feared, some ten to fifteen percent total pollen. To obliterate the forest pollen signal you would have to obliterate the forest itself, replacing it with something like the savannas of which the refugialists dreamed. Nice to know.

And so home. For Mike, Cincinnati and his duties as a professor. For Paulo, Chicago, and his stint as a Research Associate at the Field Museum of Natural History on a salary from my NSF grant. For me Panama City, to take up my duties as Senior Scientist at the Smithsonian Tropical Research Institute.

Within a few weeks I had the answer I needed. Lakes on the Six Lakes Hill held ice-age sediments. We had triumphed! Samples for radiocarbon dating had been rushed off as soon as we could open the first core tubes. Mud from near the bottom of Pata yielded a "more than" age, way back in glacial time at the very least. So did a sample from the diminutive core from Dragão. Stuck in my mind is a phrase from a note from Mark, who was working on Verde, "It's older than

God." All three lakes spanned the last glacial maximum, and on into the distant past. After a dozen years' vain search of the Amazon lowlands it felt very nice.

Soon it was to be nicer still. Paulo, on Pata, and Mark, on Verde, rushed out a few quick pollen analyses. Forest, forest, nothing but forest. We still had to look carefully at every interval. But this should be strong meat in a science community that expected to find only savanna.

17

ICE AGES FROM AN
AMAZON INSELBERG

Harold Urey won the Nobel Prize discovering the heavy isotope of hydrogen: deuterium. But he then separated isotopes more exciting for we of the ice-age earth, the stable isotopes of oxygen, ^{18}O and ^{16}O, noting that their roles in chemical reactions depend on temperature. He gave us our thermometer. He deserved another Nobel, but this is against the rules. A passage I loved reading to my students was Urey's postmortem on an animal that had been dead a hundred million years, based entirely on the oxygen isotope ratios of its annual growth rings. *This Jurassic Belemnite records three summers and four winters after its youth . . . warmer water in its youth than in its old age, death in the spring, and an age of about four years.*

If you are not moved by that, perhaps you will be by what followed. An Urey student, Cesare Emiliani, used oxygen isotope ratios on the tiny carbonate skeletons of *Foraminifera* and other animals in deep-sea cores. His goal was more than poetic justice on the death of a Jurassic belemnite, nothing less than an accurate chronology of ice ages. He hoped to show a finely dated chronology of ice ages by allotting strata in the deep-sea cores to cold or warm, thus to glacial or interglacial time, from their oxygen isotope ratios. Once done, the full weight of modern dating methods could be applied to yield an accurate and detailed time scale, perhaps for the whole of the ice ages. I was a young scientist when Emiliani wrote the first of these papers. Compare this dream with the methods of those of us doomed to bulk radiocarbon dating of mud. My hopes were with Emiliani.

But like many a pioneer before him, Emiliani had his dream improved on, though it must have seemed like a hijack to him. Nicholas Shackleton at Cambridge saw that although both Emiliani's logic and his sequence of glacial events were correct, an additional step in the logic was needed on the way to the definitive chronology of ice ages. The oxygen isotope ratios, though based on temperature as claimed, were not so much direct measures of the water temperature in which animals of the sea lived in ice-age time as they were of the relative volume of ice held on the land.

Glaciers are made of snow, the making of a snowflake from a water droplet is a physical change of state that will affect the oxygen isotope ratio in the snow, discriminating against the heavy isotope. Both isotopes ^{16}O and ^{18}O are stable. When the snow is imprisoned in its glacier, the isotopes stay with it. In glacial times, therefore, glacial ice is "light" (that is deficient in ^{18}O) and the other great earthly reservoir of water, the sea, is "heavy" (that is has an excess of ^{18}O, usually expressed the Greek letter *delta* thus "$\delta^{18}O$"). This was Shackleton's point. The "heaviness" of the isotope ratio in the sea is a direct measure of how much glacial ice is locked up on land, and hence the state of glaciation.

From then on the game has been played by the paleooceanographic community in part as Emiliani imagined it. The $\delta^{18}O$ in carbonates in deep sea cores were measures of relative glacial state, but they referred not to the temperature of the water but to the volume of ice on earth, an even better and more direct measure of glacial state.[1] It was to this glacial history that every ingenuity was brought to bear to assign ages.

A subplot in this story is that the chronology of ice ages emerging from radiometric dating of the glacial episodes represented by $\delta^{18}O$ were hauntingly similar to those previously calculated from the astronomical theory of ice ages, usually known as Milankovitch theory.

It is a matter of direct observation that the motions of the Earth in its orbit round the sun are subjected to four distinct and measurable periodicities. As a spinning object the Earth both precesses and subtly changes its angle of tilt. In addition its elliptical orbit round the sun is off center so that the earth is closer to the sun for one half of the year than the other half, and finally the shape of the earth's orbit slowly, rhythmically changes.

Each of these motions makes subtle changes in the seasonal reception of solar heat at the surface of the Earth. The periodicity of each can be, and has been, measured with extreme precision. If, therefore, these seasonal changes of heat received at the surface of the earth are enough to drive the coming and going of ice ages, then Milankovitch calculations, based as they are on precisely measured data, should yield a chronology of ice-age time with astronomical precision.

Upon this foundation is built our understanding of ice-age time. In a process known as SPECMAP that compared radiometric dates on marine glacial episodes with Milankovitch calculations, and by cross-correlation with ice-core data as an added support, glacial time has been divided into Marine Isotope Stages (MIS).

For the purpose of our inselberg study we are concerned only with the top of the long sequence, starting with the present and counting backwards. We now live in MIS 1. The last glacial maximum of recent memory was MIS 2. Then comes MIS 3 as a minor letup in glacial stress, before which MIS 4 was the real start of the last glaciation, as severe an episode probably as was MIS 2 that ended it. Before MIS 4 was the last interglacial, quite like the present, MIS 5. And before that another full glacial episode, that we include from start to finish in a single unit, MIS 6. From then on the sequence goes down far into the Pleistocene epoch, the numbers ever mounting as ice ages alternate with interglacials until the crispness of the details clouds into the mists of geological time. In current seminars, deep-sea folk have been arguing the properties of MIS 11.

Yet we must not sell the inselbergs short as messengers of great spans of time. Those almost indestructible columns of rock have been slowly emerging from the Amazon plains since about Miocene times, say something over twenty million years. Admit I am no Amazon geologist and divide by two or three, and that still leaves plenty million years for an inselberg to grow old. Take from this time the long span needed to dig those pseudokarst lake basins by solution of silicate rock in plain water and what is left remains marvelously old by the standards of paleolimnology elsewhere.

When measured on the bench the two cores from Lake Pata were each just short of 7 meters long: 691 centimeters on our final reckoning. But only in the second meter down was the mud still young enough for accurate radiocarbon dating: accelerator mass spectroscopy (AMS) dates 35,000 years old between 1.1 m and 1.2 m in both of the parallel cores. The error quotes were in the range of plus or minus 400 years in both measures, believable dates.

This meant that the last glacial maximum had come and gone in just the top meter of sediment, or putting it differently, all the sediment from MIS 2 was included in the top meter. And we had nearly six more meters below that. Triumph at last! Heady triumph if you played the extrapolation game. Say thirty thousand years per meter all the way down, multiply by seven and round the answer down, answer: *two hundred thousand years*. That calculation let the cores span at least to MIS 6, and perhaps a bit over. But that is a wild extrapolation, best kept to oneself in the watches of the night. What we can actually demonstrate by radiocarbon is quite good enough for the purpose I had resolved on at Yale in 1964: a section right through the last glacial maximum.

The time-line given to the last glacial maximum by climate historians and modelers is the interval represented by the uncorrected radiocarbon age 18,000 years BP (Before Present), now corrected to a calendar age of 22,000 BP. This interval was marked in the first meter of the Lake Pata core by two radiocarbon dates on samples between 70 and 74 cm down. Theses dates were squarely within a distinctive stratigraphic stratum, 20 cm thick, and like no other in the paired cores. It is the stratum that we called, "Yellowish nodular clay," and it is certainly not gyttja, nor a normal variant of lake mud. Superficially it looks as if the lake had dried up.

Lake level was certainly low, but it had not "dried up." The nodular clay contained both pollen and algal cysts, enormous quantities of them. This is not possible when mud really dries up, because these are destroyed by oxidation, even the so-indestructible pollen. Instead, the pollen concentrations were higher, by a factor of three or four, in the nodular clay than in any other part of the cores.

Our knowledge of what happens to pseudokarst lakes on this inselberg gives a very clear idea of what had happened to Lake Pata in MIS 2 time. Local rainfall fell for at least part of the year, during which episodes precipitation was no longer sufficient to offset the leak. If we are correct in thinking that the principal leak at Pata is through the bottom of the 25 m funnel, then the residual water would be likely to collect in puddles over the broad expanse of the flattish bottom where we cored.

The lake was transformed from a body of dark, even blackish, water several meters deep into a system of shallow puddles and mud, sometimes even to little more than yellow wet mud. The yellow color was already familiar to us in the sediment of Lake Dragão, the lake that leaked before our eyes as it progressed from a lake fourteen meters deep to its use as a soccer field two years later. The yellow clay at both lakes is, of course, the color of the weathered mineral component of Six Lakes Hill when not stained by the black humic content of the gyttja that forms under the dark water.

It was here at last that my two Amazon lives could be peeled apart. Long ago I had been elegantly shamed to abandon the Arctic for the equator by Ed Deevey to seek some among the properties of the wet tropics that might help to explain the grotesque diversity of the place, what I have called "The Reason Why" (chapter 1). Then, years after starting, I found myself up against a paradigm purporting to do the explaining for me with what I considered to be misconceived evidence of the past. Lake Pata was to start the drive that fended off the paradigm to leave me with at least the basis of the original answers.

Paulo's pollen analysis told the tales of both my Amazon lives. After the first excited look (and it was exciting) he settled down to extend our learning and reference collection even beyond the level we had reached through the widespread trapping. We knew how Amazon pollen behaved, this was the time to extend the vocabulary so that it should be rare to contend with an unknown pollen grain.

Both Paulo and I had been made adjunct researchers of the Chicago Field Museum, which just happened to have one of the finest Amazon herbaria in the world. With lists of plant species from the Six Lakes Hill region from his Brazilian colleagues, my NSF grants provided for Paulo to spend two years at the Field Museum to make our collection nearly complete. Add visits to the New York Botanical Garden and the Missouri Botanical Garden who made us free of their great collections, accompanied by a Colombian pollen analyst, Enrique Moreno, whom I hired as a curator of pollen at the STRI laboratory in Panama, and we were ready for definitive results.

First, pollen from rainforest elements never faltered throughout MIS 2. The massive concentration of pollen in the yellow nodular clays held all the familiar forest plants of MIS 1. There was no increase in grass or herb pollen. It was thus impossible that the Six Lakes Hill, or the great plainsland which it overlooked, had been covered with savanna in a fragmented forest as required by the refuge paradigm. More, the pollen data made all chance of savanna unlikely for a very great distance because such grasslands would have lofted pollen clouds able to travel far and wide to drown, in a true pollen rain, any local pollen being worked slowly across the watershed from insect pollinated trees.

Our climate records were also striking, even of precipitation, and despite the persistence of the forest. The yellow nodular clays told of changing patterns in the rains throughout MIS 2. This was not the drying meant by those who dreamed of Amazon aridity, but it was still a significant and systematic shift in the rains that lasted the whole of MIS 2 and thus through the last glacial maximum. This shift told us that ice-age climates did perturb the Amazon, though not catastrophically. I can easily imagine perturbations on that scale causing a flutter in the lives of small things like insects, sponsoring migrations, realignments, selection for different breeding patterns. With minor climatic signals like these it is quite possible to imagine chance isolations of small populations in far away patches on that enormous piece of real estate. This is all the vicariance natural selection should need.

But the biggest thing was that the plants of the cooling signals from Mera and San Juan Bosco were there on Six Lakes Hill too: *Podocarpus, Hedyosmum, Humirea, Ilex* (holly), Ericaceae, *Eugenia, Myrsine,* and even *Weinmannia* (the elfin forest shrub from the high Andes). Finding them on Six Lakes Hill was not like finding them at Mera as we first did, with great mountain populations poised high above them, as they were in the Ecuadorian foothills. Ericaceae and *Myr-*

sine are actually sparingly present in the surface pollen of Lake Pata, suggesting that the low-elevation population is not extinct, even now. But *Podocarpus* pollen is not present, quite in accord with our experience that *Podocarpus* pollen never spreads far for all that it looks like the ubiquitous pine pollen of floating pollen clouds in the northern hemisphere.

These cool-adapted plants at Pata in the last glacial maximum were not replacements, not a new forest type replacing the old. The essential vegetation type, the "biome," had not changed, it had merely accommodated a few immigrants and some population gains and losses, but it was still the same forest, the same biome. When one distant day we have a really detailed survey of the trees and bushes of Six Lakes Hill, and of the tall forests at its base, I dare say we shall find rare living specimens of at least some of those plants I use as the cooling signal. Then the MIS 2 data will show that those rarities of modern living merely became more abundant with the modest 6°C cooling that lasted ten or fifteen thousand years.

But our little list just might include some that were real invaders from on high. I am still suspicious of *Weinmannia*, the elfin shrub of the Andean tree line, but my colleagues, better pollen analysts than I, are increasingly sure of the *Weinmannia* identity, and there are reliable reports of it in the lowlands of the Chocó. And Paulo, who has worked particularly hard to learn the species of *Podocarpus* from pollen alone, is certain that some of the Lake Pata fossil *Podocarpus* is from high altitude species. So these plants might genuinely have invaded the lowlands in glacial times, though apparently securing no more than a footing as minor populations within the existing forest.

The cooling signal was nice. We had now found it at the foot of the Andes, on this inselberg deep into the lowland forest, and on the southeast corner of the Amazon basin where the *Araucaria* trees swept their 600 km north to the tropic. Indeed the only published pollen record of MIS 2 in the Amazon with no evidence for cooling was the "partial" pollen diagram from Carajas. As it happened, the pollen analysts of Carajas were about to be joined by fresh blood who discovered in the old notebooks that the original counters had found both *Podocarpus* and *Mysine* (the two most prominent signals for cooling at Lake Pata), but had left them out of their account without mention of the fact. The prose of the new paper is masterly in the way it avoids any mention of the original discoveries at Mera and San Juan Bosco, avoids all mention of the previous Carajas omission, and even lets Pata appear as a mysterious follow-on from the new authors' later finding in another lake (all that appears in the title).[2] A beautiful cover up.

Cooling across the continent had little bearing on claims of the refuge paradigm but had real punch for the diversity issue that had sent me to the equator in those blissful pre-paradigm days. If proved as an expected glacial condition, it must be seen to be as an environmental vector that surely had an impact on the spectacular biodiversity. But the claim for cooling was opposed as strongly as our denial of aridity. The opposition this time came from CLIMAP paleooceanographers who had mapped the sea surface temperature in MIS 2, finding severe cooling at high latitudes but nothing more than the 2–4° C range, in equatorial latitudes.[3] For some years, this conclusion had prepared biogeographers of the land to follow the lead of oceanographers in assuming that equatorial lands scarcely cooled. It was the loud discussion on this point between me and a paleoclimatologist that had driven the pencil of a reporter from the *Economist* a few years before.

The Pata data showed that equatorial cooling was real. It could not be set aside as an effect of nearby mountains, such as the effect of winds (so-called catabatic winds) blowing off mountain ice caps, an argument of which we had been made familiar among attacks on Mera.

Opportunely, a conference came to my new home, the Smithsonian laboratory in Panama. I talked with Richard Fairbanks, paleooceanographer and student of the ice ages, telling him that I was going to present new evidence that the Amazon cooled. I was then prepared to argue that the equatorial oceans must have cooled also, whatever CLIMAP said. His eyes twinkled as he said to me, "Paul, wait till you hear what we have found." His seminar followed mine. He reported the first results of his student Tom Guilderson showing with isotope data from coral reefs near Barbados that the tropical ocean surface had cooled in glacial times. His estimate, a fall of six degrees Celsius, was the same as ours for the Amazon.[4] One remembers days like that.

I wrote up the first Pata results for *Nature*. The journal *Nature* had published my Bering land bridge history in 1963, the Galapagos results in 1968 and 1972, our first abortive attempt on Amazon lakes, and the Mera paper, both in 1985. Now I could send them the biggest thing I had done, the first trans-glacial record of the Amazon forest. I did. *Nature* rejected it.

Comments of the peer reviewers told the story. Reviewers asked where was the evidence for ice-age drought that was so well-known a property of glacial times in the Amazon and East Africa alike. Something must be wrong with our methods, or interpretation, or both that we had no record of the famed aridity. Suspicion obviously centered on the "nodular yellowish clay," with the unspoken admonishment, "There is the evidence of aridity, why don't you admit it?" And one anonymous reviewer leaked the group attitude with the forthright sentence, "I am not ready to abandon the refuge paradigm."

This was the first time the word "paradigm" was used against me. Dispute a fact, if you must, but deny a paradigm at your peril. I edited the manuscript to fit the different style requirements of the journal *Science* and sent it there instead. Not only did the editor of *Science* publish it, he published in the same issue an essay by a science reporter discussing the importance of our finding with the provocative title "Ice-Age Rain Forest Found Moist, Cooler" (1996).[5] The story was out.

Had we stormed the ramparts of received opinion? I knew we hadn't. This was one study from one site. A shrug was all I could expect, a shrug was what we got. Ernst Mayr's prediction was coming true: "Nobody will believe you because the refuge theory is so beautiful."

Of course they did not believe us. Professional lives had been spent constructing scenarios of rain forest evolution based on the notion that the Amazon forest had been fragmented by aridity in ice ages. By 1996, the paradigm had bitten deep, all the way from textbooks to the patter of tour guides. The idea of ice-age refugia was strongly entrenched, and not to be changed. As the wise Nobel Laureate, Peter Medawar, once wrote, "The human mind treats a new idea the way the human body treats a strange protein, it rejects it."

Reasons for rejection were easy to find. An obvious one was to say we must have hit a forest refuge, that was why we found the forest always to be there. Just redraw the maps with a forest refuge round Pata, then business as usual. I knew we had no defense against this one, except ridicule if refuges sprung into being wherever we dug a hole. We joked about this possibility. We shouldn't have.

I saw a manuscript of a letter to the editor by very eminent protagonists of refuge theory playing this very game. In their latest reconstruction of the ice-age Amazon, they said that the borders of the "forest refuge" in the northeast had been moved some hundred or two kilometers; the Hill of Six Lakes was now squarely in the refuge. The letter went on to claim that our work, far from denying their theory, had in fact proven it to be correct: by an actual demonstration that a patch of forest did persist in arid glacial times. That letter was never published, but we were to see the words again later, in lesser venues, in print.[6]

These redrawers of maps failed to notice that they were in head-on collision with more dangerous critics of the kind who had blocked publication in *Nature*. These others were not going to allow our evidence for ice age forest at all. They said that the real arid period in Pata history had only been hinted at by that "nodular yellowish clay" of ours and that the really arid part was missing in our records. Loudly they claimed that there must be a gap in the core: we had missed the critical interval when the lake held no water at all. In this critical time of total dryness, the non-lake left no sediment and no pollen. That was why we had found only forest pollen. The arid land record had been destroyed in the great oxidizing time of aridity and we had passed it by.

This "gap" hypothesis was as "beautiful" in its own way as the refuge hypothesis itself. It predicted that the refuge hypothesis could not be tested by the classic methods of paleoecology. The only trace of the passing of the great drought should be a gap in the record. Finding no data is thus proof of your hypothesis. Even a non-scientist might smell fish in this argument. And yet there were papers on this theme in serious journals.[7]

Those who said we had found a forest refuge in glacial times and those who "found" a gap in the Pata record were of course in direct conflict. I thought it better to look for more data than to engage in the polemic.

There were many more data from Pata, Dragão and Verde on Six Lakes Hill itself, so many more that we were years in the extracting of them, scattered as we were in different institutions. Mike Miller in Cincinnati provided the first clue to a profound discovery when he sent us his chemical analyses of the long Pata core while I was still in the chaos of settling into the Smithsonian laboratory in Panama. An outstanding oddity was in the potassium and sodium analyses. Potassium fluctuated violently throughout the history of the lake, but its close chemical cousin sodium did not. This was without precedent in our thinking. Sodium and potassium usually move in lockstep: they erode together, flow in solution together, deposit together. This oddball behavior of potassium was what thrilled. Out of it was to come the best climatic history of the Amazon lowlands yet.

The cause of this peculiar behavior of potassium must lie in life process within the lake. Potassium is an essential plant nutrient. The potassium "highs" surely were times when pea-green blooms of algae concentrated potassium but not sodium, leaving rich potassium spikes in the mud as testimony to their passing. Nice record, but did the lives of these long dead algae mean anything for us? And there I had let the matter rest for nearly a decade. Inexcusable. Trying to tell myself I was preoccupied with a new laboratory far away, with my people scattered, does not even defend me to myself. I almost forgot about it.

Then, around the turn of the millennium, Mark Bush called from Florida. We had both moved our labs and lives by then. I had settled at the science-cum-fishing village of Woods Hole on Cape Cod. Mark had joined the faculty of the Florida Institute of Technology. He had called to talk about the potassium record in the Lake Pata core, "Did I think it looked like Vostok?"

Search my memory, and a gulp or two later, well, yes I did. What we were agreeing was that the shape of the $\delta^{18}O$ curve of the top part of the glacial sequence in Vostok, the Russian and the longest of the ice cores from Antarctica, matched the shape of the percent potassium curve in the sediments of an equatorial lake in the Amazon. This could sound like a daring folly, Antarctic ice matches Amazon mud! But we both knew at once that this was probably true, and what it might mean. The curves had a common denominator in climate, which almost certainly meant that one or more of the orbital motions of the Earth calculated by Milankovitch was responsible for both.

Mark obtained software that would print out the temperature deviation curves, for all latitudes for each of the astronomical perturbations, but it was obvious that the precession of the Earth in its orbit should be the one, because its period of 22,000 years was on the right order for the distance between the potassium peaks. Precession calculated at zero latitude for the three wettest months of the year, June, July, August, fitted beautifully.[8] The curves did match.

Now was the process made plain. Precession, acting through temperature (as yet we know not how), reduces precipitation in the wet season until this is just low enough to be unable to keep up with the leak in Lake Pata. Lake level falls, but it is not a drying. Instead the lake is made sufficiently shallow for its black dye no longer to block light, photosynthesis by algae is possible, which promptly bloom, drawing on the nutrients that are freely available from the chemically reduced mud in the bottom. The algae concentrate potassium in their bodies, eventually to leave a concentrated layer as testimony to their passing.

Comparing astronomical precession in the Lake Pata potassium with the $\delta^{18}O$ in the Vostok ice core, provides us with an accurate dating tool for the Pata sediments. They reach back to roughly the boundary between MIS 6 and MIS 7, effectively through two complete glacial cycles and roughly 170,000 years. Our "absurd" extrapolation from the 30,000-year radiocarbon date was not far out after all.

Two points about the Lake Pata record from the potassium dating. The lower six potassium spikes were much less visible as physical changes in the sediment, though we could feel the clay nodules in several of them. It was the concentrations of potassium themselves that revealed both the events and their identities. The second finding was major. In the top potassium/precession sediments the pollen signal of lowland tropical rain forest was not disturbed, except by the addition of cool-adapted elements. This is a history of remarkable forest stability in the face of known climatic change.

Paulo, counting Pata pollen, and Enrique counting Dragão, both found the entire series of 400 types of pollen grains that are included with Enrique's photomicrographs in our pollen atlas. Mark Bush had counted the Verde pollen before the pollen atlas was finished when our working total of familiar grains was only 197 pollen types. For statistical consistency when comparing the three lakes we used only the 197 Verde types for all three lakes. The 197 pollen types were fully adequate to show that the forest patches that occupied all three watersheds through all stages of an ice age were statistically similar.[9] In plain language, except for repelling a minority of cool-adapted plants in interglacial times, the forest of Six Lakes Hill remained unaltered through the different stages of a glacial cycle.

The most significant differences among the three lakes lay in the shapes of their watersheds and their depths. Pata was shallow, with a comparatively broad watershed round it, complete with a small grove of *Mauritia* trees (like those of the Comprida floating forest, only these did not float). Dragão and Verde were much deeper, with steep shorelines and smaller catchment areas. These depth and shape factors made their response to the precessional reduction of wet-season rains quite different. In Verde, the potassium peaks were subdued and not markedly different from the sodium curve. We attribute this to the depth and shape of the basin which minimize the effect of lowering the level by leakage. I have already commented on the ability of Dragão to turn into a soccer field. Although we got ancient sediments from it, the column was incomplete, most of the sediment we did have being of the yellow kind left by a shallow lake. And yet we saw it when it was fourteen meters deep. Probably we were specially favored by a quirk of the rains. But for all the differences in slope, size of catchment, and depth, the flora of tropical forest trees were similar throughout this long history in all three watersheds.

Our overwhelming impression from these years of work on three rain forest lakes on the inselberg of Six Lakes Hill is that the lowland forest of at least the central Amazon is a stable system even through the vicissitudes of an ice age. The climate changes that inevitably happen are taken, as it were, in the forest's stride. Furthermore, the forest can accommodate species with different tolerances both when their populations grow and when they come as invaders.

These thoughts come from the data of 170,000 years from three lakes on a single forest inselberg. The lakes are all different, but the forest history they describe is the same. It is one of stability and tolerance. Might not one of the secrets of Amazon diversity lie here in this history of tolerance and stability? Should these properties turn out to be ubiquitous over the vast domain of the lowland Amazon forest, it surely cannot escape being a major cause of diversity particularly, as our few data hint, these conditions might have persisted since the start of the Andean orogeny.

Would that Ed Deevey were still alive to read this.

18

Two Thousand Fathoms
Thy Pollen Lies

The Amazon basin is one giant watershed, the largest on earth. Over this huge watershed, the great diversity of pollen from the Amazon forest is moved by water after reaching the ground near the parent tree, just as it is over the smaller watershed of a lake. These watershed wanderers always account for the long lists of tree names. They do not come as the pollen rain that hay fever sufferers or pollen analysts of the temperate lands know. The contributions of these diverse watershed wanderers to the pollen sum are a blessing given to tropical pollen analysts by the custom of rain forest trees of using animals, rather than wind, to carry the plant equivalent of sperm.

Pollen counts of 300–500 grains from Six Lakes Hill show in the order of a hundred different kinds per sample, perhaps only as single grains per count, but visible evidence that the parent tree grew nearby. Never having been subjected to a large "pollen rain" of grains blown from far and wide its presence is not swamped and hidden from statistical view by the rainstorm of pollen that classical pollen analysis knew. Once a European text claimed that tropical pollen analysis was pointless since the tree pollen you wanted should never enter the pollen rain and so should remain "silent." The reverse is the truth, as the power of the diminished pollen rain to swamp the forest signal, though not "silenced," is very much diminished.

Nevertheless, a restricted pollen rain does exist in the tropical rain forest. It comes from plant families with old fashioned ways (*Ulmaceae*, the elm family, which has tropical genera of pioneer trees), or weeds with habits akin to those of northern weeds which promote the use of wind (*Cecropia*, a small tree of the fig family which can grow like an invasive weed on disturbed ground, or all members of the grass family, from tiny fescues to bamboo).

If ever the great forest were to be struck down or fragmented, it would be these weeds that would benefit. The pollen rain from such as savanna grasses and pioneer trees would soak the landscape, its streams and rivers, drowning out the pollen of surviving trees or forest. Any long core from within the rain forest, almost anywhere in the great Amazon basin could detect forest catastrophe by the falling of this pollen rain, making truly "silent" what pollen record of forest diversity it hid from the pollen analysts' view.

This is what makes the mighty oddity of the whole Amazon basin being but one watershed of such tantalizing interest. Go core in the lake into which this watershed drains to seek the trail of forest catastrophe, if there ever was one. That lake is the Atlantic Ocean. No need to core the whole ocean, just the fan of sediments off the Amazon mouth.

Pollen works its way from where it fell beneath its parent tree into the smallest of Amazon streams becomes part of the sediment load of bigger streams and bigger streams still, until at last it takes its place in the final flood after the "meeting of the waters" that is the Rio Amazonas, seventy meters deep, for the thousand-mile journey to its resting place offshore in the Brazilian sea.

This dumping of mud and pollen has been continuous since the Amazon basin was tilted in that direction late in the Miocene, say fifteen million years ago. What a core this mud might yield. Forget the Miocene, and the Pliocene that followed, with their millions of years to parse, just know in your guts that the ice ages must all be there near the top of the section. That core might answer questions of diversity and stability that were the true reasons for my decades of work along the equator, what were the changes through the epochs, was diversity always higher here? To the arid Amazon of the refuge theory I no longer gave credence, but a core from the fan could settle that too with the simple measure of, "Had there been a pollen rain of grass and weeds obliterating the surface-moving pollen record of the last of the trees or not?"

Face the fact that this core should not be easy. The pollen record was miles out to sea and some 2,000 fathoms down. A bad place to reach with a hand-held drill while standing on a raft of rubber boats. Two thousand fathoms is my poetic way of describing the slope between 3,000 and 4,000 meters, which is the surface of much of the fan. Use the mundane measure of two or three miles, frighteningly deep but not poetic at all. I follow Shakespeare's, "Full fathom five thy father lies, of his bones are coral made." That wonderful pollen lies at 2,000 fathoms.

But oil men and geologists are not fazed by 2,000 fathoms. They have drilling ships that can reach that far and farther, maintaining station not with anchors but with signals from Global Positioning Satellites and a system of propellers to nudge the ship in all directions. They reach for the bottom with the huge rigs oil men know. The National Science Foundation had sponsored such a ship for the use of science, in a program called the international Ocean Drilling Program (ODP). Unknown to me, the ODP had resolved to use their drilling ship, *JOIDES Resolution*, to core the Amazon fan. The project got the name "Leg 155" and was scheduled for March to May 1994. The timing could not have been more opportune, just when I had settled into my new laboratory in Panama and was completing the initial study of the Lake Pata sediments in preparation for the paper in *Science*. But I hadn't known about it.

The ODP announces its proposed projects to the science community well in advance, seeking collaborators. But in the critical year of decision I was moving my work from Ohio State to the Smithsonian, or perhaps the key announcements came when I was actually on the Amazon trails to Six Lakes Hill. I finally learned about it sometime in 1993, too late to get in on the planning.

How I hungered to be on that ship! But I forced myself to accept that I should be able to request samples later. Then a letter from Australia plunked down on my Smithsonian desk.

Many a romance has been written in which our hero gets a lawyer's letter from Australia announcing the death of a long lost uncle who had made a fortune. Our hero is the sole heir. My letter was even better. An Australian postdoc who was already slated to sail on Leg 155 asked if I would support his fellowship to the Smithsonian so he could join my laboratory to do the pollen.

Simon Haberle won his Smithsonian fellowship. He duly sailed on the good ship *JOIDES Resolution*, and arrived at my laboratory bearing samples of mud from the Amazon fan. His samples were already dated to span the last glaciation.

Fan deposits are more tricky to date than lake mud. The deep currents set up as the dense, sediment-laden streams from the great river pour down the continental slope can snake their way, carve new channels, pick up older sediment on the way down. The result can be that blocks of mud need not be in sequence. The answer is an array of parallel cores, and a long, painstaking time of cross correlation.

The dating was in good hands, not our own. The whole team gathered by ODP on the *JOIDES Resolution* was devoted to this problem of dating. Isotope decay dates of many kinds, remanent magnetism, cross correlation of a suite of cores, time-stratigraphic marker fossils in deep-sea deposits, methods learned the hard way through decades of work. These are the group strengths to be marshaled by big projects, more even than the costly equipment and technology available, these are the means of success. From these efforts, Simon was able to retrieve surely dated samples of most of the last glacial period, MIS 2 (the last glacial maximum), and MIS 3 (the period of let up). No core segment had been unequivocally assigned to MIS 4 (the massive start of the glaciation), but MIS 6 (the glaciation before last) was recognized, so Simon sampled that. This stretched his reach to a total of 130,000 years. The Leg 155 team dating also supplied cores identified as from the last interglacial (MIS 5).

Getting MIS 5 sediments from the fan was an unexpected piece of luck. Being an interglacial like the present, the interglacial mud is expected to travel the same route out of the Amazon mouth as the modern mud of our interglacial, and this is not, or not mainly, to the fan. The reason for the difference is sea level. Deep water covers the modern continental shelf in interglacial times, and long-shore currents carry and disperse sediments northwards. Most are deposited on the continental shelf itself, never reaching the deep-sea fan.

In glacial times, however, the fall in sea level makes the shelf so shallow that the dense, sediment-laden stream flows straight on, pushing aside the less dense sea water, eroding its own channel as it goes. The gorge of this glacial channel shows up clearly on modern sonar maps. The Amazon pours into its underwater gorge, across the shelf, and swoops over the edge and down the continental slope, cutting more gorges as it goes, until at last its mud and pollen settle on the fan 2,000 fathoms down.

Finding MIS 5 deposits in the fan tells us that some, at least, of the Amazon's muddy stream can find its way through the old underwater gorge even when sea level is high, perhaps when the longshore current lags. But it is certain that the best way to get a postglacial record (MIS 1) is a piston core from the shelf. This was done and Simon added to his glacial collection calibration samples for the last 5,000 years.

He also brought a still more important contribution. He had found time to travel by commercial river boat down much of the navigable lower Amazon, collecting buckets full of muddy water as he went. Mud filtered out from his buckets should include the pollen now flowing down the river. Taken with my surface samples from mud flats near Manaus, our 5,000-year core from the Manacapuru embayment, and the piston core samples from the shelf, we now had all we needed to reconstruct the pollen load coming down the modern Amazon, the MIS 1 stage of glacial history which should serve to calibrate the pollen record of the past.

Simon had the annoyance of encountering many pollen grains and spores in bad condition, corroded, or colored in ways familiar to petroleum palynolgists but not to us of the Pleistocene. These we could safely dismiss as ancient grains reworked by the rivers from older sediments, perhaps several times oxidized on the long journey to the sea. They were clearly marked and easily recognized for what they were. Simon kept count but did not include them in the pollen sum, correctly, because they did not represent trees living at the time sampled.

The fan samples were also less rich in pollen than those from Amazon lakes. This is easily understood because they were mixed with more and more mud as they journeyed down the muddy rivers and were impelled by turbidity currents into the deep sea. For Simon this was an extra handicap on top of the "reworked grains" as we call them. Nevertheless, he settled on attempting a pollen sum of 300 grains at each interval though not always reaching it.[1]

It is important to keep the scale of this enterprise in mind. We were reducing this vast watershed and its varied forests to a single point sample. Dangerous in itself, this reduction to a point sample held the additional hazard that the method of pollen accumulation was uniquely different from the familiar norm of pollen trapped in the mud of local lakes. Transport in giant rivers tending to erode older pollen with mud from their banks preceding deposition in the deep sea after a four-kilometer plunge through shifting ocean currents, could hardly be compared with the push of surface runoff over a local watershed and sinking through placid waters. Some questions we could ask of the data, many others we could not.

The easiest of the askable questions was, "Has there ever been a time of real pollen rain in the Amazon when the wind-blown pollen of anemophilous plants swamped the pollen of the great host of animal-pollinated trees of the forest?" The answer was a simple and uncompromising, "No." This is a deadly statistic for the hypothesis of an arid Amazon needed to support a savanna-clad Amazon basin or one of the grass-rich open forests like caatinga.

The data for this conclusion are truly striking. The percentages of grass pollen in fan samples from MIS 2 (the last glacial maximum), MIS 3, MIS 5, and MIS 6 are all in the same range as the percentages of grass pollen in the sediments of the embayment at Manacapuru, or suspended in Amazon river water, or deposited in MIS 1 sediments of the continental shelf. In simple words, there has never been more grassland in the Amazon basin at any time in the ice ages than there is now.

In those two short paragraphs we are making the claim that, on the basis of the fan samples alone, were there no other data pointing to the same conclusion, the aridity hypothesis of the refuge paradigm was tested, found wanting, and should be discarded. To defend the paradigm after this, the insignificance of grass and other wind-blown weed pollen from the fan must be explained away. They tried.[2] Only one tactic was possible, and this they used. Simon Haberle's pollen, all of it in their model, came only from the river bank, and of course nobody claimed that savanna grew right to the edge of so massive a source of water. The great rivers were lined with riverain forest. All the fan recorded was pollen of trees that grew at the water's edge.

When European pollen analysts of the stature of T. Van der Hammen and H. Hooghiemstra had published this explanation, the subsequent literature suggests something like a sigh of relief from exponents of the paradigm. Pollen from the fan gets the briefest mention in many a paper, with a reference to Van der Hammen and Hooghiemstra to show there was no need to worry.

Yet this apparent escape from worry was a chimera. In the straightforward sense the model of riverain forest pollen swamping the grass signal could not work. Grass pollen from massive savannas of the interior is presumed by the tactic to have dribbled into the rivers from the banks along with the riverain pollen, which is absurd, as our mathematical friends say of dud calculations. But grass pollen travels on the wind, in great clouds whether from hay field, prairie, or savanna. Its descent, often in real rain made of water drops, gave rise to the term "pollen rain." A rich European literature long ago investigated this process showing that lines of forest along waterways could not impede the passage of the cloud because the tree "barriers" would be overtopped.

Haberle's grass pollen from the fan, therefore, remains a deadly and direct refutation of the refuge hypothesis and of the paradigm that grew around it.

The fan does more because its pollen gives generality to the forest history culled from the Six Lakes Hill. They were both long histories: 170,000 years from the hill and 130,000 years from the fan. Both show the cooling periods with *Podocarpus*, *Hedyosmum*, *Wienmannia* and the rest showing up as significant percentages in the pollen. None of these can be fairly thought of as part of the pollen rain, their radius of spread being local. But there they are in glacial sediments of the fan just as they are at Six Lakes Hill, at Mera and San Juan Bosco, and, as had at last been admitted, in Carajas. The fan even includes *Alnus*, a shocker if the Van der Hammen observation that it is found only above 3,000 m in the Colombian Andes is correct. *Alnus* is a true member of the pollen rain, pollinated by wind and traveling great distances. I have found *Alnus* pollen on the Galapagos Islands though the nearest alder is 2,000 km downwind in the high Andes. I have also seen living alder trees lining a stream at 1,000 m in Panama. The safest explanation of the *Alnus* pollen in the fan is that it fell with the pollen rain in the western Amazon and then was carried by the rivers with the forest pollen down to the sea.

The pollen list that Simon compiled for the fan was not as long or detailed as that from Six Lakes Hill, eighty-plus taxa instead of four hundred, though counting in fern spores brings the number closer to a hundred. This lesser number of taxa is surely an artifact of pollen distribution by surface runoff and stream flow over huge distances, one of the consequences of the absence of a pollen rain for forest pollen. In that vast real estate, many a small watershed becomes the habitat of taxa that are not ubiquitous throughout the forest. It would be glorious if our single point sample caught them all, as a net across the mouth of a stream catches all the fish making their run for the sea. But the fan is not a net, and pollen are not fish.

The two lists do overlap. The few trees that use wind to pollinate, so to make a pollen rain appear, as expected, are in both: *Celtis*, Moraceae/Urticaceae, *Cecropia*, Melastomataceae/Combretaceae, Myrtaceae. So are many of the Amazon stalwarts, what have been called by some Amazon "oligarchs." Many of the Leguminoseae are in this group. Doubtless these get through the filters and pass the net because of their large number of starts to the obstacle course, whether pollen rain of the first group, or from the sheer numerical abundance of parent trees in the second.

Strong in my mind by now was an Amazon vegetation, mostly forest, that shimmered down the years over the vast spread of slightly different watersheds, exposures, aspects, soils, rainfall, response to the reductions in wet-season rain with the precession cycles every 23,000 years, and how forgiving each local community was to invasion during the six-degree cooling that came at intervals of maximum glaciation. How this "shimmering" of the forest might influence, or even drive, the high diversity (the object of my work since the beginning) was far from evident.

At least one thing now seemed clear, there had been no biome change during the last 170,000 years of the kind posited by the refugial paradigm.

19

MAICURU, THE LAST ADVENTURE

Maicuru was the last site: the last promised in my grant proposals, the last remaining from Paulo's talking to the men of *Radambrasil*, the last best chance for me to include the eastern Amazon in my equatorial data set.

It was also the last in the "hurry up and get it done" sense. In the year 1998 I was about to leave the Smithsonian, with its isolation in a small institute in Panama, and my wife and I had already put our condo, perched over the Bay of Panama, on the market. It was then thirty-four years since I had started the transglacial project for the American equator and I was two years away from my own three score and ten. Time to move on. Maicuru would not only bring to a close the project started at Yale so long ago, as much as it could be done in one lifetime, but was also likely to close the net round that pesky side issue of the refuge paradigm, which had popped up while I was already engaged in the larger project.

Maicuru was an inselberg to duplicate Six Lakes Hill, but a thousand kilometers further east. It was, as near as life offers, "on the equator" at 0° 30' S, 54° 14' W. It had the same scabs of pseudokarst lakes on its ironstone surface, the domed plateau rising from 500 m to 600 m.

Maicuru was an explorer's site. We had the *Radambrasil* map sheet on which the inselberg was near the middle. On the whole sheet was not one settlement, not one road, not one airstrip, not one navigable waterway. Beautiful.

Once the great rivers must have torn at its base but now there were none within a hundred kilometers, the gap between being filled with tall lowland forest, probably as virgin as any to be found in the whole Amazon basin. No trails or roads penetrated this forest. Dense and undisturbed, the forest jungle served the inselberg castle as a defensive moat from human works, at its narrowest part 100 km wide from the frontier village of Recreio.

And like other near-impregnable castles of old, it held gems of great price: in Maicuru two such gems. The greater of these was the third or more of the plateau covered with splendid tall forest despite its ironstone foundation, implying the presence of real soil. As far as we could tell from the aerial imagery, the trees were as tall, diverse, and grand as those in the 100 km moat. And perched, half in this forested portion and half in a lesser forest of some sort, was a lake that looked the equal of Pata.

The second gem was a broad area at the opposite end of the plateau where the surface was mostly bare rock, shown by the images to support no more than a fuzz of vegetation. This had a dozen or more conspicuous scabs that we took to be nearly drained, or marshy, basins like the one described for Carajas in its "dry" phases. This rocky landscape, with its specialized, edaphically constrained, and drought-tolerant plants, should serve as a surrogate for the Carajas Plateau flora that we had never seen.

Maicuru was the last best chance to provide the data needed to close the most glaring gap in our coverage of the equatorial Amazon. I felt a compelling need to core that forest lake, and to survey the Carajas-like vegetation, before our Smithsonian departure. But to get to it we had to cross that 100 km moat of virgin forest, carrying all the impedimenta of a drilling crew. The habits and dreams of a life time said, "Walk, this is your last chance for a real walk." This would mean a large expedition: rejected on grounds of expediency, no time to organize, no time to walk, almost certain rejection of an application for another expedition permit with my name on it, and the private knowledge that the arthritic knees of my sixty-eighth year would likely make me a burden to my companions. So I yielded to common sense and thought helicopters.

Use a helicopter, but how? There was no helicopter available for charter nearer than the coast >500 km from the outer edge of the forest moat, and no airstrips near enough to Maicuru to serve a helicopter as base and refueling station. The solution was to build our own heliport outside the 100 km moat of trees. All we needed was a field and a store of fuel. Then charter a helicopter to fly in to our private heliport from the coast.

We already had our equipment stored in Paulo's family property in São Paulo, a decision taken after those grueling experiences in the Manaus customs shed. There was nothing to stop these stores being used by a Brazilian team, headed by Paulo. He invited two U.S. friends, myself and Jason Curtis, the latest postdoctoral fellow in my laboratory, to join him. We three were all agreed that it should be prudent for me to stay out of Brazil until the organization was complete. Paulo and Jason (traveling as Brazilian host with guest), should go out first, set everything up, then call me.

Recreio, as the nearest frontier village to Maicuru, became our base. The distance from village to inselberg neatly defined the width of the forest moat as one hundred kilometers. A dirt road into Recreio is its lifeline, good enough for the passage of a three-ton truck. The really vital physical property of Recreio, however, reflects a national passion. It has a field on which to play football. On this we should build our heliport.

Paulo and Jason drove to Recreio to confer with the village chief and the people. They made friends, and more. Young men of the village were keen to come with us. The chief had no problems with our renting a corner of the football field to establish a dump of aviation spirit and land our helicopter for the two or three weeks of our operation. For the village it should be both fun and profitable.

The helicopter company, Kovaks, advised on the purchase of sufficient aviation spirit, stipulating the type of plastic drums suitable for manual refueling. Paulo and Jason rented a large open truck, loaded the required quantity of aviation fuel in its plastic drums, and drove to Recreio, where many willing hands stacked the fuel on the football field, covering it against the sun. They set out the markers for the helipad as instructed by the helicopter pilots. They had done a "Clausewitz," not only "securing our base," but doing so among very friendly people, even allies. They returned the truck, and called me to come.

I left my official (on the service of the U.S. government) passport behind, preferring the anonymity of a tourist guest of a Brazilian citizen. I used my British passport, which the Queen never lets you give up, citizen though you have become (remember the trouble this caused in 1812). A British passport did not need a visa: I was anonymous. I flew overnight directly from Miami by Varig to the rather small and obscure airport of Belem, landing in the early morning of May 30. No problems. I walked with deliberate slowness, pushing my cart loaded with the latest of Vince Vohnout's improvements to the coring equipment as well as personal effects down the long customs and immigration line in the "nothing to declare" lane. I was not stopped. A man was holding one of those signs you always see at airports with people's names on them. The name was mine. The man was from Kovaks, the helicopter charter company. We were off.

Fixed-wing aircraft late in the afternoon to the airstrip of the small Amazon town of Monte Dourado for a night of catch-up sleep. Early morning start for the long truck haul to our frontier village of Recreio. Pitch my little tent on the football field, look around, plenty of daylight left. A hum in the distance, a speck, and the helicopter was circling the village before cautiously lowering down to the soccer field and the "pad" laid out near our fuel dump. It had flown from Belem in stages, with refueling stops on the way. 'Twas still the last day of May. Immaculate arrangements.

Paulo had already hired six strong men to go with us. Early next morning the advance party of Paulo with two of the villagers took their seats and the helicopter rose slowly from the field. A circle of women chanted prayers as this strange chariot of the air took their men away. None in the village had seen a helicopter before, probably not even on television, something the frontier was without. And now this thing had scooped up two of their men folk to take them over the forest and far away. I could hear worry in the tempo of their chanting as the circle slowly shuffled round.

Three hours later the helicopter was back and it was my turn, with more of the men. Rain forest, rainforest, all the way, canopy of shades of green, straggly stems of emergent giants glinting with silver bark or epiphytes, occasional blotches of red or yellow as trees were in flower. One great tree was alive with monkeys that dropped helter-skelter out of sight as we came near, the response, either innate or learned the hard way, to large flying objects in a forest where monkey-hunting eagles lived.

I had seen this same forest from the air before, in far away Ecuador. But this was different because there were no great rivers, no glint of water beneath the trees, no clearings, no trails, no grass huts, no smoke, no hint that people were anywhere on the earth.

In less than an hour a great rampart of land was before us, the ironstone inselberg of Maicuru. The helicopter climbed and made a pass across the plateau to show me the lakes that Paulo had seen, one in particular that we called "Lake 1" but changed to "Bean Lake," the name given to it by our Recreio volunteers from its shape. To my biased eye it was Pata, or Verde, or Dragão all over again, a lovely black enigma, ringed with forest, the stuff of dreams.

There was to be no landing anywhere near Bean Lake; on half its shores high forest, including some forest giants with trunks nearly two meters thick. The other half had more scrubby forest, from which we deduced, correctly, rocky ground with little soil. Extensive shallows at one end of the lake had emergent water plants visible from the air, but there was very little in the way of beach. Definitely no place to put a helicopter down; unless it had floats, which ours did not.

Our pilot flew back to land at the other end of the plateau in the large area of rocky surface that we had seen on the radar images. The biggest of those scabby depressions that we had thought, correctly, to be pseudokarst basins, held a shallow lake, giving the pilot a clear run in before putting down close to the shore in an area free of bushes. This was some three kilometers from Bean Lake.

There he had dropped Paulo with the advance party, who had spent the last two hours clearing brush and marking out the classic shape of a helipad. The pilot had given precise instructions to the Recreio men on what was needed and these forest men understood completely. And there we landed. Out we jumped, grabbed the baggage, and the helicopter was off for another load.

The pilots got in two more round trips before nightfall, after which all nine of us, Paulo, Jason, myself, and the six Recreio men, bedded down among the rocks for the night. One more flight for the morning with the heavy baggage in a helicopter freed of passengers, and the ferry job should be complete.

Easy really. Or so it seems when baldly stated. Here we were on this impossibly remote inselberg that had haunted my thoughts for years, whisked there in a couple of days. It would have been different had we walked, as the early adventurers had had to walk. It was the little helicopter and the surgeon-like skill of its pilots that made the difference.

The machine was a Bell Jet Ranger II, a model known in that part of Brazil as "the Volkswagen of helicopters" because it just keeps going. The registration letters were PT-HON. A few months after our return from Maicuru Paulo read in a Brazilian newspaper that helicopter PT-HON had crashed in the forest with the deaths of all on board. The pilots' names as given were different.

Now we must walk to Bean Lake, all across the very rough surface of the inselberg. The lake of the landing place was useful only for a baseline study. I had been able to wade all across it, probing at the mud with drill rods. It was a puddle. A waste of time. We had to pack all our gear the rest of the way. And that meant we had to cut our way by dead reckoning to find that small, Pata-like lake three kilometers away through dense forest and scrub.

The men from Recreio had a better plan: start at the lake and cut their way back to make a trail. Then use the trail to pack in. They persuaded the pilots that, if the helicopter took three men, and hovered just above the water of Bean Lake, they could jump out into the shallows. They should need no more than what they carried, including their vital sacãos to chop their way back to us.

The pilots said OK. Paulo thought it should work, trust these men to find their way, it is their forest, OK. I gulped, then affirmed, OK. We all helped the pilots remove the doors from the helicopter to ease the jump, in climbed the three volunteers, the helicopter bore them away, to return empty perhaps twenty minutes later. All was well, the three had waved cheerfully after "landing." The doors went back on the helicopter. The pilots left, promising to return at nine o'clock in the morning exactly fourteen days later. Be packed up and ready.

The men "dropped" at Bean Lake were away for three days. I confess to being a little anxious as the second day passed, and no sign of them, but I should not have been. The forest they had to cut through is better described as "jungle," a dense and unyielding screen, cloaking piles of rocks, or fissures. They were never at loss for direction, but they had to hack their way through. The other three from

Recreio were not bothered. This was their forest. The three would just naturally choose the right direction, and come (had they known the term) "with all deliberate speed." They did, making good time for trail makers, one kilometer per day.

Those three days' stay on the rock and ironstone plain, with its pseudokarst basins in various stages of draining, was a third gem of price offered by Maicuru. We found ourselves in scenery, and with botany fitting closely to the descriptions given for the land round the Carajas lake. This might have been the *campos rupestre* itself, or the high altitude savanna, were it high enough. In fact it was probably a little less than 600 m as opposed to the 700 m at Carajas. We could see exactly what the Carajas pollen actually meant, and it had nothing to do with an arid landscape or pollen blown from imaginary savannas on the Amazon plains.

Begin with the grass and the weeds, the collection of wind-pollinated types that made up what at Carajas had been decreed to be a pollen rain blown up from a savanna below the plateau. Grass was growing healthily everywhere. The lake providing the helicopter a safe run in, everywhere less than a meter deep, was a meadow of rooted aquatic grasses, their fronds waving above the water like a hayfield ready for the cutting. Grasses of more terrestrial habit grew in the numerous mud-filled cracks in the ironstone, on the long slope down which the leaking lake had retreated. Others of the smaller pseudokarst basins within twenty minutes' walk had no standing water, but plenty of standing grass (probably the condition of the Carajas lake during its driest episode), and one was filled with football-sized grass tussocks giving a superficial similarity to an Alaskan tundra north of the Brooks Range. Such were the sources of the grass pollen that the analysts at Carajas had convinced themselves came as a pollen rain from distant savannas. If their data were to support the paradigm, they had no choice but to make that eccentric claim.

The weeds were more interesting still. The commonest of them, both in the Carajas pollen and as living plants at Maicuru, were of the tropical weed genus, *Borreria*. As weeds of dryish places, species of *Borreria* grow in all the sorts of place that weeds do grow, they are the brownish ugly weeds at the edges of paths across fields, at roadsides or railway tracks, and they certainly find space in a savanna. I would hazard a bet that they grow on slag heaps too, though my experience of tropical slag heaps is vanishingly close to none at all. At Maicuru *Borreria* species are as abundant as grass in the cracks in the ironstone slopes round the landing-ground lake, and in "puddles" of dried lake mud above the retreating shoreline. Most beautiful of all, a healthy population of an aquatic species of *Borreria* grew actually in the water. The pollen of all these species looks the same by our technique of the light microscope. Once it was clear from the state of the mud that the Carajas lake had drained, *Borreria* plants were sure to come, their swelling populations recorded in the pollen. They did, of course. But the aquatic species of *Borreria* was probably dense in the last shrinking puddle, too. From a distant savanna? Rubbish!

Compositae, the daisy and dandelion family, the collective pollen of which were claimed among the "savanna pollen indicators" at Carajas, were everywhere in the cracks and mud patches at Maicuru.

But the really beautiful error was the shrub *Byrsonima*, undoubtedly a critical component of many South American savannas. Triumphantly Van der Hammen and Absy found its characteristic pollen in the ice-age mud of their Carajas lake. But *Byrsonima* bushes were common on Maicuru now. The clincher came for me on our trek of the second day to another shallow lake when I was wading through a swamp, grasping at a bush for support. Paulo beside me observed dryly that the bush was *Byrsonima*, what is more, it was in full flower. There it was, one of quite a community of its kind, its feet firmly planted in wet mud under water, its scores of little yellow flowers dropping pollen directly onto the surface of the receding lake. Savanna indicator? Again rubbish.

Cuphea is another plant used by the Carajas team as in the family of savanna and caatinga "indicators." It is a pollen analyst's delight, its pollen both beautiful and distinctive. It is wind-pollinated, producing pollen in some caatingas in such quantities that it can be thought of as a true part of the pollen rain. This plant we did not find at Maicuru, but specimens in the Field Museum collections from the Carajas Plateau describe them as weeds of stream banks. The few pollen grains found at Carajas, surely like the *Byrsonima*, *Borreria*, and grass, came from the shrinking lake margins, and eventual marsh, when leakage exceeded wet season precipitation as at Pata.

So the botany of the stony draining part of the Maicuru plateau was a virtual model of ice-age time at Carajas. None of the herb pollen were "savanna indicators"; the pollen came from local inhabitants of the draining swamps. The idea that the pollen blew up from the bottom lands from mythical vegetation that replaced the forest 600 m below the plateau was absurd. Every excuse would continue to be found to deny permission for my team to go to Carajas and core for ourselves. But Maicuru had already shown the more grotesque errors of Carajas interpretation for what they were.

Towards the end of the third day of anxious waiting for our three men to reach us, I saw a great bird glide up from the forest edge at the far side of the rocky expanse. The flat vee of wings that could be a vulture. But it was not a vulture. The wingtips were square, more like a goshawk than anything I had seen. And huge, a rectangular flying machine it looked. A couple of circles and it vanished back into the canopy. No time to run for my binoculars, but there could be little doubt what it was. My first harpy eagle.

A shout from the forest on the other side. The Bean Lake party had arrived. They had a plan for us all to move camp to Bean Lake. We would take advantage of a spring-fed stream near Bean Lake, in real forest, like the forests near their home village rather than this rocky mess where we were.

In the morning all nine of us would use their trail to take the coring rig and minimal supplies. We should start our lake work the day after, while the Recreio men would ferry our remaining stuff across the inselberg, say goodbye to the rocky base, and build the work camp near Bean Lake.

But what about the helicopter, it still can't land in the forest? No problem, the six woodsmen had more than a week to make a clearing and the pilots had told them what the helipad should need. Evidently they had already colluded with the pilots. I could only rejoice at the grown-up boy scouts we had with us. And accept their offer.

Why had they taken three days to cover three kilometers? True they had added a detour to blaze the way to another of the lakes near the line of march. But the real reason was only clear in the morning when, coring rig and stores spread into nine pack-loads, we followed their trail to Bean Lake.

They had cut away enough for a man with a loaded pack to squeeze through, always avoiding the densest growth, the highest rocks. We ducked, and edged, and sidled, wading for a time up a rushing stream in a narrow gorge, climbing a small rapid, almost a waterfall, on the way. We wended through wild pineapples on a rocky flat, tearing at our legs with their savage spines, and a grove of *Cereus* (organ pipe) cactus, also among rocks without soil. The route was anything but straight, chosen solely as terrain dictated. Pretty familiar stuff to those who wander in wild places.

But it was the forested part of the first kilometer of our march to Bean Lake that was the hardest going, and on which our friendly natives of the forest marveled as they cursed. This was something outside even their experience. A forest of large trees grew without soil, solely on a mat made of their own roots. As one woodsman remarked, it was like a carpet a meter thick, looking as if it could be peeled off the rocks and rolled up. Not a carpet of luxury though, for the weave was coarse. The threads were tree roots, many of them ten and twenty centimeters in diameter. Brobdingnagian, the whole thing, with Brobdingnagian-sized spaces between weft and warp. These spaces were often ample enough to swallow a boot, together with the leg that followed. Larger roots on which we had to pick our way were slippery with rain.

I vote that forest on its carpet of roots the hardest going I have known in a long life of field work. But perhaps my vote is influenced by that sixty-eighth year. I locomoted on knees that had lost their old zest, to the point that the joints have since been totally replaced with concoctions of metal and plastic. Anyway, the truth is that I fell repeatedly until being virtually trapped like a fly in a web. A man from Recreio hauled me out, took off my pack, went into the forest. Chopping sounds, he reappeared and, with a broad grin on his face, handed me a stout stick to use as a crutch. He added my packload to his own and I hobbled after.

An hour or two later we were in normal rain forest on real soil, with giant trees soaring around us. A trail that looked like a trail, and our pace quickened until we reached the spring feeding the little brook, our home for the next ten days.

We had taken a full half day to walk the three kilometers to Bean Lake, even with a marked trail. Do it with neither trail nor horizon, blazing and clearing a path as you go, and three days is pretty good.

We three "lost not an hour," as Nelson once put it, in starting the last ten-minute walk to Bean Lake, clutching rubber boat and skeleton limnological equipment. I treated myself to my customary first row down the lake. Glorious to hear the water bubble under the forefoot as long ago I had heard it bubble from my position as bow oar in an eight-man shell. The sheer sense of freedom of being alone in a buoyant little boat, in so secret and remote a place, would need a poet to describe.

But a naturalist can describe what he saw. On the bank from which I had come, tall lowland forest, fair enough to be called rain forest with its giant emergent trees protuding above the canopy of merely tall trees. Unlike the great rivers of the Solemões and Negro this secret little lake was faced with little that could be called riverain bush. Instead, the whole forest just stopped at water's edge.

Yet more than half the banks had no such forest. Trees yes, even woodland, but a scrubby sort of woodland quite evidently growing in very rocky ground. At one place a fine grove of organ-pipe cactus, at another a steep slope of scree. At the platform where scree met the water a largish caiman lay sunning itself.

Jason and Paulo took over the Avon for a sonar survey of the bottom. They quartered the lake, at one point reporting a tapir on the bank that rushed off into the forest. I have still to see a wild tapir, though I have seen quite a few piles of fresh tapir dung.

The sonar reported two basins forming a dumbbell shape, though much shallower than dumbbell-shaped Dragão had been. A quarter of the lake surface had floats of the trailing water weed *Ericaulon*, a new plant to me. It snagged oars, a trial when rowing. But the deeper basin had five meters of water, too deep for the rooted weed to reach the surface.

Knobby eyebrows of caimans, with a third knob for their nostrils, floated in the *Ericaulon* beds. In the heat of the day caimans sunned themselves on a rocky part of the bank. Idyllic. But we were to see less and less of the caimans. Our woodsmen companions killed them with their sacãos, for our dinners. They imitated the caiman calls at dusk, bringing them within reach of a swift blow.

We cored Bean Lake in several places, a task of pure routine. The only novelty was the resistance of matted *Ericaulon* weeds to the passage of the heavily laden raft on its two hulls. Our only tug-force was the single pair of small oars in the Avon yacht tender. Rowing the tug on the end of a tow rope felt like moving an immovable object. My notes record trying the alternative of towing from the banks at the end of a long rope, but this was worse as the tow rope was grabbed by the weed over its whole length. Minor irritant. That my notes record it shows how free from real difficulty the whole project was. We came, we saw, we cored; just like that.

The best core, some five meters long, seemed quite comparable in the taking to the cores at Lake Pata. Bean Lake was a bit like Pata in general appearance too. We had little doubt that we had got what we wanted. And we were right. As at Pata, radiocarbon was to show that the top meter of sediment alone spanned more than 40,000 years.

I lived in my little sleeping tent in the forest, just twenty feet from a buttressed forest giant more than three meters across at the base and two meters at the top of the buttress, the canopy reaching up to about 40 m: in plain English a monster tree, six feet thick and a hundred and thirty feet high. The camp was a ten-minute hike from the lake, chosen for its flat ground and a tiny spring-fed stream for drinking and washing. I slept to the music of the stream's gurgle as it sounded ten feet from my tent.

The only hostiles I saw in these two weeks of forest life were army ants, who revealed themselves but once when I carelessly stepped on a column. A furry tarantula on my little path by the tent was harmless, but I stepped round it all the same. And I kept back from a savage buzzing as a "blue-tail fly" (blue hunting hornet), with a body rather larger than my thumb, rummaged at the base of the great tree. A naturalist's peace.

The men from Recreio built what I can only call a house, and they did it in little more than a day. Their only tools were their sacãos. Start with two living trees, six or ten inches thick and as far apart as the length of the house. Clear the space between for the floor. Cut a sapling for a ridge pole, and four more saplings to make long poles ending in crotches of stout side branches. Lift the ridge pole, hold beside the living tree trunks at the desired height, and use two of the crotched poles, one at each end, to prop the ridge pole against the living trees. Then add the remaining two poles to the opposite sides so that the crotches again support the ridge pole. You now have completed the framework for a strong house. The foundations are the living roots of the sentinel trees at each end. The "A" frame at each end is held by the weight of the ridgepole lying on the crossed crotches of the end poles that form the sides of the "A"s. You can lash the poles to the ridge pole and the living post trees for additional peace of mind, but this is not necessary: the structure will stand on its own.

All you need now is the vital roof to keep out the rain. It did rain on Maicuru, every day, sometimes in heavy, drenching showers. Our roof was a large plastic tarpaulin, the kind now mass-produced somewhere in Asia and ubiquitous the world over. Without it you would have had to thatch with palm fronds; lots of work, prettier, but not as good as the tarp. Elegant housekeepers, the Recreio men added cross members on which to tie hammocks, and made seats from short "Y" posts and cross poles. A neat "A-frame" home for nine men in the jungle, the only thing carried in being the tarp.

A cooking fire smoldered most of the time, used for boiling manioc porridge (our staple food), making coffee or tea, and broiling caiman steaks (good eating if bad for an ecological conscience). It is good having countrymen to look after you.

While we three biological passengers went about our ecological business, the men from Recreio built a heliport in the jungle. This was beside Bean Lake, in the scrubbier end of the real forest where it met the dry woodland. Placed thus, the helicopter could approach low over the water and never be in danger of tangling with trees. The actual clearing was little more than twenty meters across, though of course open on the lake side, but they had topped trees over a larger area. The ground was cleared of rocks and stumps. Poles were cut and arranged flat on the ground in the approved hieroglyphics of the helicopter world to direct the pilots in. All was ready by the tenth day.

So were we ready, the coveted long cores from Maicuru safely in the stack of core tubes awaiting the helicopter. We had followed our guides for days as they hacked a way to other lakes, learning botany and the facts of inselberg life, but finding nothing to compare with Bean Lake. Even the other "prime" lake of the advance party's detour on their trek to the rocky landing ground was a shallow disappointment with a rocky bottom. Bean Lake is probably the only ancient lake on Maicuru. But one is enough.

Bean Lake level fell in the ten days we lived by it, but not by as much as had Dragão in the same period, about half a meter compared with two or more meters. And my notes record that we had rain nearly every day. Clearly Bean Lake has a significant leak, making a low-water episode spanning the last glacial maximum probable.

As the fourteenth day dawned we had packed all our belongings and were hauling everything to the "heliport" by the lake. We had a ten-kilogram sack of manioc flour to take out, our remaining life support (other than caimans). Now we trusted to the human contract to ferry us over that hundred plus kilometers of forest. If Maicuru was any guide to what the "going" would be like in that forest (which it wasn't) it would be quite a walk if we were forgotten.

Just before nine o'clock that morning we heard a hum in the sky, saw a speck coming straight at us, and then the helicopter did one circuit before lowering over Bean Lake to plunk down between the poles marking its station. Astute men of the forest who had mastered the landing needs of a flying machine. Elegant pilots of the "Kovacs" company.

So ended half a life-time of field work. With those twenty-odd tubes of mud we had found the last of the samples needed to meet a great professor's challenge, made in an ancient Yale laboratory in the year that Kennedy was shot. Our transect was from sea to sea; our way stations followed the equator from the Galapagos ocean site, to the line where Andes met Amazon at Mera, at the Brazilian equator of the West at the Six Lakes inselberg, and a thousand kilometers east on Maicuru Inselberg, then the last eastern station from deep in the Atlantic Ocean two thousand fathoms deep in the Amazon fan.

From the rain gauge lakes, in the sea on the Galapagos islands, the western anchor, came a glittering story, outside Professor Deevey's challenge, the oldest history of the Eastern Pacific Ocean yet found. This was more than thirty years before its time, when the theory of the El Niño/Southern Oscillation was still the subject of eccentric speculations. Those old papers of mine were found thirty years later by a young generation who knew more of what my discoveries might mean than I did. Such are the joys of a life in science.

Jason and Paulo had meanwhile "proved" that Maicuru and Bean Lake were what we had dreamed. Paulo's pollen analysis showed that there was forest, forest all the way, periodically enhanced by the plants of colder periods, which we had come to expect. Jason prepared the tiny samples, all that were needed for modern radiocarbon dating, 21 samples, all from the top meter of mud, and sent them to the AMS at Lawrence Livermore National Laboratory, where our colleague and friend Tom Guilderson presided. 21 samples from 10 cms to 85 cms depth were dated (a prodigy to we of the old brigade), and the calibrated radiocarbon age from the deepest sample (85 cms) was forty six thousand, two hundred and eighty seven years BP. We have four more meter-sections of sediment below, five meters in all. The sediment appeared roughly the same all the way down. Take a gamble that is was all deposited at about the same rate, and use the dating of the first meter to give us a conservative estimate of 40,000 years per meter and we get 200,000 years as the age of the complete record. Shades of Lake Pata where a similar rough extrapolation in the same range proved to be correct. This gem in Castle Maicuru is still a vision like the cup of the Holy Grail that we have not yet grasped. As yet we have no curious history of potassium to trigger thoughts of the Vostok ice core, but we shall find something.

Yet we do have a shock. No amount of visible searching can find evidence of a break in the monotony of the sediment; at first sight it is as if the lake level never fell, despite its leaking, and despite the 25,000 year cycle of cooling from the Earth's precession. Nothing to see. But there was; if you had an AMS to look with.

Jason and Tom Guilderson found the episode of water draining away by dating every centimeter of core across the most likely interval, their reward coming between 76 cm and 77 cm. There had been no sediment, thus no water in the lake, for seventeen thousand calendar years, yet leaving no sign for the detective with a hand lens to notice.

A detective's "Aha!" all the same. The actual dates, roughly 18,000 and 35,000 years ago spanned both the last interval of equatorial cooling that we attributed to the precession cycle at L. Pata and taken by the ice-age community to be Last Glacial Maximum, Paulo's pollen from those adjacent samples at 75 and 76 cms are closely similar, essentially the same 136 taxa (types) before and after, roughly 70% of both lots being from forest trees. Not a sign of a savanna interval. A nice support to the conclusion that the forest persisted right through the time of lake draining is that on each side of the hiatus in our record are some of the cool adapted trees and shrubs that we found at L. Pata and the Amazon fan, after our first discoveries at Mera.

When we have brought to bear every method now available, I expect not only to confirm these findings of perpetual forest while rainfall fluctuates through an ice age only within the tolerances of Amazon Tall Forest. This is the essential truth against which our understanding of high Amazon diversity must be forged, although I suspect we shall have more to say about the climate dynamics of the Amazon basin as a whole. Meanwhile those rain gauges in the Eastern Pacific Ocean, the lakes of the Galapagos, the first fruits of my "Amazon Expeditions" will be delivering their gems to a generation who know better than I how to look.

An outline history of the Amazon rain forest from side to side is traced by this life-time's collection of ancient records. The technical advance that let this record be read was the breaking of the pollen code for a vegetation without the signals of pollen rain, because few trees in the forest were pollinated by wind. This breaking of the code was largely the work of Mark Bush and Paulo de Oliveira with their inspired pollen trapping. We learned that the concept of a pollen rain was virtually without validity in such vegetation as this, that the arrival of copious pollen traveling long distances on the wind was relatively rare and confined to very few Amazon plants other than grasses.

The pollen data show that there has been no biome change in the Amazon throughout the last glacial cycle. By extension, this conclusion implies the forest persisted through the whole of the long succession of ice ages spanning some two million years. Quite possibly the forest has changed little since its first gradual establishment in the Miocene, ten million years ago. The explanation of the Amazonian high diversity must be sought in evolutionary mechanisms that act in stable systems of immense size and great antiquity rather than from vicariance brought about by fragmentation on climatic time scales.

From my decision to go south to the equator at the urging of Ed Deevey, to seek evidence for the influence of ice ages on diversity, the task had taken me thirty-five years. Neither Deevey nor I had had any idea that, after the first few years of work, our goals would be subsumed in the world of evolutionary biology by a paradigm so plausible and entertaining though wrong, that many an Amazon study of potential worth should be subverted to conform with the paradigm. Showing that data and paradigm did not fit became necessary as a secondary task of the enterprise, adding the spice of scholarly struggle, but at a cost measured in years.

20

PARADIGM COUP DE GRACE

The title of this chapter is taken from a comment by Stephen P. Hubbell upon the *Amazoniana* article described below, "I really enjoyed your coup de grace of the Haffer refuge model. It was really fun to read."

Scholars regularly get letters from editors asking them to review manuscripts. Anonymous, unpaid, unrewarded reviewing is one of the dubious privileges of academic life, better for the intellect and self-regard than grading term papers but even more time-consuming. This invitation was different. The manuscript was due to be published whatever reviewers said. It was written by two of the biggest names in Amazon research for a special edition of *Amazoniana* to be published as a "festschrift" to the pioneer scholar of the Amazon river, and founder of the journal, Harald Sioli. The authors were Jürgen Haffer and Gillian Prance. What did I think of it?

I thought it was meretricious, not to use a more earthy adjective. The guru and the prophet of refuge theory had come together to reassert all their claims in the teeth of the mounting evidence to the contrary. Their paper was clearly planned to rebut a recent review of mine commissioned by a subsidiary of the international office of the Intergovernmental Panel on Climate Change, a subsidiary describing the measurement of past global changes known as PAGES.[1] The authors denied not only my conclusions of years of my work but virtually all the work of my colleagues too. For Haffer and Prance, Lake Pata merely demonstrated that the Hill of Six Lakes was in a forest refuge, Haberle's pollen from the fan was from riverside forest only, and therefore without value. It was the Van der Hammen school pollen work from Rondonia and Carajas that mattered, true demonstrations that the lowland savannas of the Amazon did exist in ice ages. And they took their stand on the geological data by repetitive citing of the original authors of the claims for sand dunes and stone lines as arid soils, while carefully ignoring the provocative phrase "stone line" itself. There was nothing new, except a sliding away from the term "savanna" in favor of "dry forest." I thought of the whole enterprise as stout denial. As P. G. Wodehouse put it in the words of Bertie Wooster, "You can't beat stout denial."

My recommendation to the editor was that he invite a review in rebuttal to be published alongside the Prance and Haffer paper in the festschrift volume. Not by me, I said, as certainly to be thought biased. The critical evaluations ought to be by those knowledgeable of the land surface: geomorphology, sand dunes, and soils: those were more concrete, and less liable to subjective interpretation than Amazon pollen. Two men stood out as obvious choices, Georg Irion of the Senckenberg Institute in Wilhelmshaven, and Matti Räsänen of the University of Turku in Finland.

Both these men had spent years studying the physical properties of the Amazon surface, soils and surficial geology. Each was preeminent in his chosen Amazonian specialty. Irion had first been sent to the Amazon as a young man by Harald Sioli himself, the honoree of the volume, to assess the properties of soils in the central Amazon. He was by far the best scholar to evaluate claims that phenomena such as stone lines and deep dissection of central Amazon soils were evidence for glacial drought. Räsänen led a team from Turku investigating the geomorphology of the western Amazon basin. So active was the Turku group that Arctic Finland was becoming a major center for Amazon research with papers appearing in the journal *Science*. Irion or Räsänen had the rare geological qualifications for the job, but if the editor really wanted a paleoecologist, I suggested Mark Bush.

For once in my life I seemed to have written something an editor wanted to hear. This editor, Joachim Adis, himself an Amazon scholar of renown, promptly invited all of us to do the job together: Irion, Räsänen, Bush and, against my advice, me too. I ended up as collator and rewrite-man as well as contributor. I was the only one of us no longer with the duties of a regular academic appointment and so able to devote full time to the enterprise. The review effort occupied a year, by which time we felt we had evaluated every relevant Amazon claim.[2]

There could be no better team with which to work, strongest in expertise on land forms and soils, the subjects that I knew least. In a flurry of e-mails we agreed that we should each draft a quarter of the text, in essence four separate essays that I would mold into a continuous whole, all drafts and rewrites to circulate until we were all satisfied. This should be a true group effort in which we, and our several reputations, were equally involved.

Georg Irion led off with a daring adventure that was probably critical to the success of our enterprise. He settled the matter of the sand dunes.

The weak spot in my PAGES review was my dismissal of claims for fossil dune fields shown in remote sensing images. I was vividly conscious of this vulnerability. Fossil sand dunes where forest now presides had been apparently irrefutable evidence for former aridity. But no one had actually seen the critical dunes on the ground. They had been "seen" only as images from remote sensing. In PAGES I had denied their existence, suggesting that interpretation of the images as dune fields was mistaken. I was out on a shaking limb. Georg Irion picked it up at once. And did something about it.

The first e-mail came from Wilhelmshaven in January, 2001. I have lost it from my files, but it was short. The key sentence still rings through my head all these years later, "Paul, I am going to look." Irion was about to leave for Brazil in early February, at a month's notice, to look for the dunes on the ground.

The next e-mail was dated February 12, 2001; sent from Cuiayá in Mato Grosso.

> Dear Paul
> In the meantime I traveled to the Fazenda Rio Negro in southern Pantanal and to the area west of Coxim where KLAMMER 1982 showed examples of well-developed aeolian dunes. I could not find any dune — definitely the structures he described are no dunes.
>> In ten days I hope to be at Rio Aracá
>> Best regards, Georg

Laconic and to the point. Irion had walked the ground where the fossil fields of dunes of the Pantanal had been "seen" on remote sensing.[3] He found no dunes, only low sand structures, behind which were many small ponds less than a meter deep. Apart from these, the land was nearly flat, with local relief no more than three meters, a far cry from rolling fields of dunes. A simple journey was all it took.

The sand bars damming the ponds were evidently the source of the remote sensing error. Irion identified them as old river levees. This terrain covered a huge area. A similar landscape was known from across the continent on the plains of the Rio Paraguay and has been shown to be the product of meandering drainage channels of rivers crossing a flat plain. Sand grains are moved along in even sluggish streams by "bouncing" along the bottom. Sand piles up at meanders as dams, and the stream then finds a new channel. The process in the Pantanal had been the same. No fossil dune fields there, just piles of sand. I was vindicated on that one.

Irion's promise shortly to be at Rio Aracá was to have a different ending. This area north and east of Manaus is on the white sand formation, rough country and little known. Sandy ground determines the landscape of this place, barren compared to lush Amazon forest. Rainfall is in fact sufficient for tall forest, as demonstrated by patches of it growing in places where richer soils are imposed on the sands. But mostly the plant cover is caatinga, other variants of open woodland, or actual savanna, all constrained not by rainfall but by ability to grow on sand. Difficult terrain to cross, and not much incentive to do so.

Irion saved his report until he could get back to Wilhelmshaven to write in peace. I quote part of his long letter, dated March 22, 2001:

> The access to Rio Aracá to the place of the open savannah (cerrato) vegetation is very hard. I lost in 10 days 5 kg of my weight. One local person from Barcelos and I had to walk many, many hours, at temperatures up to 50° C, just to reach the area of the described dunes. There we found one dune, but for sure there are many more. . . . about 20 m high. The vegetation is open, some grass, few bushes and very little trees. All are adapted to the yearly natural fires. . . . There are as well sole places with higher forest.

> There are large areas with savannah. On those savannah there are in large distances small about 1 m high dunes. The dunes are obviously active. E.g. some of the vegetation is covered by sand.

There, in expressive German-English of a hurriedly written e-mail, was the missing and critical "ground truth" for dune fields on the white sands. There really were dunes in this part of the Amazon basin. I had been wrong to bracket them with reports from remote sensing in the Pantanal as misidentified mounds of sand. Georg Irion knows well the difference between a "mound of sand" and a dune.

Dunes on the white sands north of Manaus seen by airborne radar, therefore, are genuine, wind-blown sand dunes, contradicting what I wrote in PAGES. But they offer no succor to the arid Amazon hypothesis. Many, if not all, are still active, they are not just relics of an imagined aridity of long ago. Dunes have probably been present through all the several million years since the origin of the sand formation in Cretaceous or early Tertiary times. For good measure, these dunes are not under forest but on terrain where we would expect dunes the world over.

Irion's observations tied in nicely with Paulo's work on the "Little Sahara" dunes of Bahia that we had included in the PAGES article. The Rio Aracá dunes and "Little Sahara" are at the western and eastern ends of a long arc of unusual landscapes running all across the north central Amazon basin on a traverse of more than a thousand miles: caatinga and savannas on the white sand formation at one end and the dry coastal belt of northeastern Brazil at the other: edaphic vegetation constrained by sand at one end, drought-restricted vegetation of similar aspect at the other. At the Bahia, "Little Sahara" end Paulo had shown that the dunes were active throughout the Holocene as well as before, by radiocarbon dates on river peat sandwiched between dune sands. At the Amazon (western) end Irion had now demonstrated that dunes were still active and almost certainly always were active.

Thick sand deposits are always likely to be worked by wind. It happens along the coasts of Europe and North America, or along the shores of Lake Michigan at Chicago. Vegetation on sand is always sparse and unstable. Whenever something happens to the plants, the sand blows. And it is the same in the Amazon, where the "something" that happens to the plants is often dry-season fire.

Georg contributed a photograph of a dune on the white sands engulfing living plants of the savanna, effectively ending the claim that the central Amazon dune field was evidence of ice-age aridity.

To sum up: the southern "dune fields" of the Pantanal were a case of mistaken identity, as I had guessed. The Eastern coastal region dunes of "Little Sahara" near Bahia had been active throughout the Holocene, as Paulo's radiocarbon-dated cores had demonstrated. Now Irion had shown, at the cost of 5 kg of body weight, that dunes on the white sands were active in the Holocene too. Thus were all claims for uniquely Pleistocene dunes within the Amazon basin overthrown.

For completeness consider the great dunes in the Llanos of Venezuela, far outside the Amazon basin but provocative because they help carry the aura of a tropical continent driven to aridity in an ice age, just like East Africa. These dunes are huge, some of them more than a kilometer long, variously described as covered with grass or scrub. Tradition has it that they were first described by von Humboldt in his travels between 1799 and 1804, a nice, early piece of ground truth. I have looked to find his description in his book of travels but so far without luck.

My copy of von Humboldt's "Travels" runs to some 1,500 pages in three volumes with a meagre index, suggesting that the cause of my failure to find the dune account is my own lack of diligence. But von Humboldt does expand on the Llanos country in an illuminating way, calling most of it steppe with great grass plains so flat that a horseman senses the curvature of the earth to the horizon the way a sailor does at sea. He also likens a desert area to the Sahara, perhaps that is the source of the legend that he was the first to see the dunes. And he speculates that the reason the Llanos had never nurtured nomadic cultures, complete with warfare as had the steppes of Eurasia, was only because early Americans had no milk-yielding livestock.

The Llanos was a likely place for dunes indeed. The only question is when were they active? The dunes were mostly well to the north of the Amazon basin,

but this brought them closer to the Caribbean where there was sound evidence
for glacial drying. Lake Valencia beside the Caribbean coast of Venezuela, is
now more than 40 m deep but was shrunk to a marsh before 11,000 years ago.[4]
And the Lake Valencia marsh episode, in turn, is correlated by radiocarbon with
other low-water evidence from lakes in Guatemala. Put together, this is evidence
of lowered precipitation in at least parts of the surround to the Caribbean basin.

The sole published effort to date a Llanos dune described digging under a
dune, finding buried organic matter and dating that by radiocarbon. The age of
the buried soil was between 11,000 and 12,000 years.[5] The dune that did the burying must, of course, be younger than the soil. Unfortunately, organic matter under a sand dune is a prime candidate for contamination by modern carbon
brought down by percolating water. The dunes could be any age.

Giant dunes seem to argue for something drastic. This area, well removed
from the Amazon basin, was likely influenced by local climate change induced
by the closure of the isthmus of Panama some three million years ago. The closure disrupted the exchange of ocean water between the Atlantic and Pacific, inevitably with large effects on regional climate. This could well be the time of
those huge fossil dunes, a suggestion consistent with the recent finding of dust in
cores from the adjacent Pacific marine deposits in the Pliocene, but none in the
Pleistocene.[6]

The true age of the giant dunes of the Llanos merits serious study. It must be at
least as likely that they date to the closing of the isthmus near the Pliocene/Pleis-
tocene boundary as to a Pleistocene glaciation. Whatever their true age might
be, the former movements of these dunes in their modern steppe environment
have little relevance to the history of far-away Amazon forests.

Thus no compelling evidence of excessive dune activity in or around the
Amazon basin in glacial times exists. Verifiable dunes were either still active in
modern climates or too removed from the Amazon basin, or both. The dune
fields of the Pantanal that had been used to reconstruct paleowind directions
over the Amazon basin were chimeras of mistaken identity. The next time that
someone throws at me the question, "But Paul, what about the fossil dunes?" I
have an answer ready, "There were none."

Yet the stronger prop of refuge theory had always been the supposed evidence
of soils, including the almost legendary business of stone lines said to be old
desert pavement, and the seemingly utterly convincing evidence of steeply dissected ground hidden under the trees.

The misidentification of stone lines, accepted by virtually all who worked on
refuge theory, should stand as the prime example of how a paradigm can lead the
unwary astray. Stone lines can be seen throughout the red soils of the Amazon

basin; outside the basin too. No need to dig, they are in road cuts and river banks. A particularly fine exposure can be seen along the road from Manaus airport into town. A meter or two down in the red clay, there they are, a band of small stones buried in the red clay. Very striking.

How did they get there? The explanation beloved of those of the paradigm is that they are relics of the arid past. The dry land surface collected stones, looking perhaps like the red stony ground of Arizona between the cacti. Came the rains and the stones were buried in clay moved by the waters, leaving the stone lines as a fossil land surface of the great dry time of the ice age predicted by the paradigm. Papers written by senior Brazilian soil scientists, particularly A. N. Ab'Sáber, who contributed his assessment of stone lines to Prance's symposium on *Biological Diversification in the Tropics,* the volume that marked the elevation of the refuge hypothesis into a paradigm.

Georg Irion, trained as a tropical soil scientist, knew of stone lines as standard features of red lateritic tropical soils the world over, the learning of literally decades of research, particularly by the Dutch in what is now Indonesia. The stones are concretions found in all lateritic soils. They grow with exquisite slowness as percolating water deposits minerals on mineral cores. Irion had actually published his findings in 1978, four years before the Prance volume.[7]

When no one paid attention, and the arid land claims persisted for stone lines, Irion had taken a team to the classic Amazon exposures of stone lines, undertaking a thorough analysis. The result was an immaculate demonstration that the stones were concretions, what Irion called "pisoliths."[8] Still no refugialist paid attention.

In their *Amazoniana* paper Haffer and Prance still cited Ab'Sáber's work as critical evidence for dry conditions in the ice-age Amazon. In our reply Irion not only elaborated on stone lines but also showed that soil dissection under forest was an expected property of weathered tropical soils under both endless heavy rain and permanent forest cover.

Both these phenomena were properties of the Belterra clays that underlie the tall closed forest of the central Amazon for great distances east and west of Manaus. The clay unit itself is usually one to several meters thick. This, however, is just the top portion of a weathering profile that can be 50 m thick before unweathered surface of the underlying Cretaceous deposits of the Barreira formation are reached. This colossal depth of surface weathering is what can happen under heavy tropical rains.

A good exposure can easily be divided into three horizons, the Belterra clay itself, then an horizon with concretions (the "stone lines") a few decimeters thick, and below that the mass of partially weathered regolith itself. In the deeper part of the regolith the original structures of mineral grains are retained, but the feldspar within has been replaced with kaolinite. To penetrate so deeply, weathering has been gentle and prolonged with sufficient moisture always available. This is the key to understanding that tremendous profile. What we see are not layered deposits but the prints of weathering working down from above. In the technical usage of a professional soil scientist, the whole thing is a "soil," and what we see in the central Amazon is a "soil profile," prodigiously inflated over what we of the temperate north are wont to see, but a soil profile none the less.

Visible layers in soils are not separate deposits, like the stratified layering of geology. Rather they are marks in the ground of soil water percolating down from above. To be clear about what they describe, soils people call their layers "horizons." We, the population of the temperate zone, have all seen profiles of soil horizons at the tops of road cuts, even that great majority of people who have never dug a soil trench to have a look, and who do not care anyway. Profiles may be quite spectacular like those in the boreal belt, the band of Christmas trees ringing northern Europe and America.

These soils of the boreal forest are called "podsols," a practical name for a beautiful soil. Black rotting litter on top over a stark white band, often sandy. Hence the name "podsol," which is Russian for "ash earth." Plough it and the field looks to be covered with ashes. The white band that looks like ash has been bleached by percolating water, both cold and acid so that it dissolves iron and aluminum leaving only the silica behind. Soils people call these surface horizons that have been leached "A horizons." After six inches or so (two decimeters) of this, the earth retains the brown color of iron and aluminum, but with dark red or even black bands in it, some of them hard, like an iron pan. This is the "B horizon" where the iron dissolved from above has been dumped from the percolating water by the curious chemistry of soils (the pisoliths of stone lines are the tropical equivalents of this process). And at the bottom is the color of the underlying mineral mass (regolith) out of which the soil profile was forged, the "C horizon." Nothing of interest below that but mere geology. The whole drama is over in a meter, two meters at the most; no need to dig very deep. Such the normal experience of European and North American soil scientists.

Irion recounted how surveyors of the 1960s went out from Manaus to look at soils exposed on river banks and road cuts, the usual first stops of soil surveyors.[9] But what they saw were profiles tens of meters thick. On top one meter, to as much as ten meters, of what they were to call the "Belterra clay." Beneath this was a layer with stone lines, typically only a few decimeters thick. And below that, tens of meters of regolith weathered red or yellow, before the basement Cretaceous deposits were reached. The whole profile could top fifty meters, a prodigious profile for a soil. The first impression of some surveyors was that they were looking at geological layering, not a soil as they knew a soil to be. Not too unreasonable an opinion for the scale was gargantuan.

Perhaps the Belterra clay of the surface "layer" was lake mud deposited on top of the "stones" of an ancient landscape by a giant lake that once covered the Amazon lowlands? Internal evidence soon let this hypothesis be abandoned. The high concentrations of chromium and titanium in the clay require lateritic weathering, and the fine structure of kaolinites is of a kind found in only the topmost layers of tropical humid weathering. The Belterra clay had to be a product of *in situ* weathering of the surface. It was the extra thick "A" horizon of a deep soil that was long in the making, not an overlying deposit.

The underlying layer with stone lines was another horizon, also formed *in situ*. In this horizon the stones were analogous to the layers of redeposition of iron in the "B" horizons of northern podsols. But in the wet tropics percolating water can carry its load of iron and aluminum down ten meters through the clay before letting go. And at the temperature and pH (acidity) of tropical soils the deposits collect as the iron minerals goethite and hematite, coated onto residual grains of quartz, like the pearls of an oyster building round grains of sand, not in the amorphous layers of a podsol.

The *pisoliths*, as they are called, are painfully slow to grow. Whereas a decent podsol in the Canadian woods can develop in little more than a human lifetime, pisoliths require near geological lengths of time to form. An age for the surrounding matrix can be calculated from the difference in quartz content between the pisolith-rich lateritic horizon and the basement Barreiras formation, assuming likely weathering rates for quartz. This was later done both by Irion's team and a French team from ORSTOM independently, both teams arriving at Tertiary ages in the ten-million-year range.[10] Stone lines do not date to glacial times any more than they are records of aridity. They are geologically old.

Possibly the gross misinterpretation of stone lines was in part a consequence of the misinterpretation of the Belterra clay as a lake deposit. If the clay was lake mud, how come the stone lines beneath? Answer: they were on the old land surface, the lake merely covered them with sediment. So how did they get on the old land surface? Answer: they were the pebbles that collect on surface soil in arid lands. Therefore the land before the flood was dry. I hazard a guess that this was how the mistaken belief of stone lines as aridity markers first planted itself in the minds of Brazilian researchers.

But the desert pavement idea did not die with the lake hypothesis as it should have done. When the lake hypothesis collapsed, the supposed burial of stone lines was still the claim, though needing a hodgepodge of *ad hoc* mechanisms like soil creep, flooding, even wind to account for it. The claim was still there in the Haffer and Prance paper for *Amazoniana* to which we were to reply, but now it was hidden. Haffer and Prance refrained from using the phrase "stone lines" in their latest review, instead citing the papers by the Brazilian soil scientist Ab'Sáber and others for their authoritative statements that the Amazon was once arid, statements that were themselves largely based on stone lines.[11] In the very stubbornness of the desert-pavement idea, once planted, was perhaps the greatest coup of the refuge hypothesis, leading to much of its mischief.

But perhaps for some who hear about it from far away the dissection of the land surface under the forest was an even more powerful jolt towards belief in an arid Amazon. We are used to thinking, quite correctly, that dissection by gullies is a result of deforestation. Local slopes under Amazon forest on Belterra clays can be steep, up to forty-five degrees in places. Ha! Something removed the forest. The arid ice age serves nicely as the required deforester.

Georg Irion slew this dragon also. The steeper slopes remain covered with a horizon of clay up to ten meters thick, just like horizontal surfaces. Not being on a horizontal surface, however, they tend to be better drained. An odd consequence of this is that the soil profile developing on slopes of Belterra clay under forest has properties of a podsol, complete with the white (or whitish) sandy layer at the top. This, true to tropical expectations, can be uncommonly thick, running into meters of it, before the ochre, iron-stained colors of the Belterra clays are reestablished.

These podsols on the well-drained slopes should be vulnerable to deforestation in the classic manner as their surface mixtures of bleached clay and quartz grains would be swept away by the unimpeded force of torrential rain. They have been able to form only under the many layers of the protecting canopy of a rain forest, where rain loses its power and soil movements are restricted to creep. Far from showing that the forest was once fragmented, the steep slopes with their podsols are evidence that the forest canopy has not been broken for far longer than the span of an ice age.

Our *Amazoniana* review concluded that all claims for past aridity based on soil properties in the Amazon basin were wrong. We had now examined them all and found them wanting. After publication of the *Amazoniana* paper I had a letter from a geography professor in England (actually an e-mail, all interesting letters seem to be by e-mail now). Beginning by saying a colleague had lent him our paper on *"the chimera of the arid Amazon"* as he called it. He continued:

> I have not enjoyed a paper so much for a long time. Partly because I remember briefly mentioning that elegant (*refuge*) hypothesis to my students some 20 years ago; partly because having seen so many lateritic soils, stone lines and

concretion lines in Africa (I worked in Uganda for 14 years as a forester before I became an academic geographer), I find it almost incredible that similar soil horizons could have been so persistently misinterpreted in S. America.[12]

"Almost incredible"; yes indeed. The paradigm's compelling power did induce "almost incredible" misinterpretations.

That verdict on the vast Belterra clays tears the guts out of the arid Amazon thesis. The whole central area of the lowlands covered by the classic rain forest is planted on Belterra clay: on the flatter bits, beds of well-leached clays several meters thick over lateritc horizons with lines of stony concretions, and on steep slopes thick with podsols. Under both soils, wet weathering is visibly evident for tens of meters down before the unaltered deposits of the Barreiras formation are reached. Building this huge edifice required prolonged humid weathering starting well before the Pleistocene ice ages, and continuing to the present. There cannot have been interruptions by epochs dry enough to remove forest trees essential to the stability and growth of deeply weathered soil.

The hunt for geological evidence for glacial aridity had not shot its bolt with stone lines, dissected Belterra clays and fossil dune fields. The latest shot was aimed at the western marches of the basin, the front line of the Andes from Ecuador through Peru to Bolivia. This frontier region has the highest forest, the richest forest, and the highest rainfall of any part of the Amazon basin. Up to nine meters of rain have been known to fall in a year at one site in Peru. Bits of Oriente Province in Ecuador are not far behind, rain of the kind that had set my teeth chattering in that canoe on the Napo.

Nearly all speculators have assigned rain forest to this region of soggy foothills throughout glacial times, whatever fantasies of savanna they held for the central lowlands. Haffer and Prance had done so in numerous publications, listing the Andean forelands as the largest of their refugia. Our pollen histories from Mera and San Juan Bosco had confirmed the persistence of wet forest through at least part of the last glaciation. Despite these odds, however, the aridity paradigm was reaching for this region too.

One research team claimed Pleistocene fossils of grazing mammals as evidence that the forelands were clad with savanna, whatever pollen analysts might say to the contrary.[13] Another team pointed to concretions of aragonite and gypsum in river banks, classic indicators of evaporating water, as evidence that a former giant lake, held by a dam somewhere downstream and piled against the Andes, had dried up in the ice age. For the proponents of glacial aridity these were two more delicious hints, both cited in the Haffer and Prance review that we were charged to answer.

I was present when the evaporating lake "discovery" was announced in Manaus. It was at the Amazon conference organized at the University of Amazonas after our Six Lakes expedition, the conference at which I was told that our collaboration agreement was canceled.

The finds were between 9° and 10° south latitude in Acre, the Brazilian frontier province butting onto southern Peru and northern Bolivia, a huge landscape of the Andean forelands known to geology as the Acre sub-basin. Old riverain deposits had been cut through and exposed by the modern Acre and Purus rivers, major tributaries of the Amazon system.

B. I. Kronberg and her colleagues had found great chunks of crystalline aragonite embedded in the mud of river banks, along with and below the usual tree trunks and wooden debris of tropical rivers. Some aragonite chunks were a foot long (30 cm). In the same mud were crystalline deposits of gypsum, scattered in fissures or as encrusting coatings on old bones (said to be crocodile).[14]

The researchers concentrated on the aragonite. Crystalline aragonite, a form of calcium carbonate, has its origin in water. Various marine organisms deposit aragonite in their skeletons, coral reefs being a large source as not only the scleractinian corals themselves but also the massive populations of calcareous green algae *Halimeda* responsible for much of the submerged mass of reefs have skeletons made of aragonite. But calcium carbonate will also crystallize out of evaporating water as aragonite, by far the most likely origin of these large crystalline chunks. So when did this happen?

Re-enter the giant lake hypothesis, "re-enter" because Amazon lowlands flooded by a giant lake is a notion with a long history of several births followed by as many deaths. One of them, perhaps the most serious, was the lake required by those who imagined the Belterra clays to be lacustrine deposits, an idea long in doubt and finally put to rest by Georg Irion's demonstration that the clays were parts of a soil profile weathered *in situ*. Another was the more recent idea of a vast flood at the end of the last ice age as a glacial dam burst high in the Andes sending catastrophic cascades onto the lowlands. The rushing water was supposed to build its own dams downstream out of the old alluvium scooped up on the way.[15]

This "Lake Amazonas" hypothesis was killed by glacial geologists working in the high Andes who denied the crucial glacial dam an existence.[16] There were other versions, a background noise of muttering about giant ancient lakes in the soggy Amazon lowlands. I had once been snared by a local version in Ecuador, an embarrassing memory that sent me blundering about in aircraft, but nonetheless resulted in my meeting the pilots who carried me to the unknown lakes of Kumpak[a] and Ayauch[i].

That the aragonite researchers had the ice age in mind for their postulated lake is shown by their decision to date the deposits, including the aragonite, by radiocarbon. You do not use radiocarbon if your expectations are of ancient deposits: 30,000 years OK, with AMS a few thousand years more. The limits of believability of even AMS radiocarbon dates is rapidly approached in the high "forties." Clearly the researchers expected a late Pleistocene age suitable to the aridity paradigm.

The AMS dates they offered were a hodge-podge. Of nine dates on wood and bone, two were close to 12,000 BP, four were between 24,000 and 32,000 BP, and three were between 42,000 and 45,000 BP. None were in stratigraphic sequence, being just a jumble of transported objects dumped by the rivers. The dates implied nothing of interest beyond the obvious, that the deposits were a jumble of transported debris. The story as offered relied entirely on the remaining two AMS dates, these made directly on the matrix of large aragonite concretions at the bottom of the section. These two came in at 49,000 and 52,000 BP.

This was good enough for the researchers. For them the aragonite crystallized during the last ice age. True, 50,000 BP required a decidedly earlier date than the last glacial maximum, a key time for refuge defenders' thinking. But what the heck, any lake should do as long as it was big enough with plenty of carbonate in the water and an ice-age date. The point was aridity.

The lake they proposed was drying up to deposit aragonite crystals 50,000 years ago. Drying up meant arid times, and they had a "date" that was indubitably in the middle of the ice age. Stone lines or fossil dunes could not be dated (or had not been), now they had a date for the aridity. Triumph.

But for us sceptics the aragonite dates stretched the limits of radiocarbon believability, certainly too much for me. 50,000 years was vanishingly close to "infinity" for any radiocarbon date, the minutest contamination with modern carbon to a "dead" sample should give this result. The lumps of aragonite had recently been jumbled about in wet mud as younger debris was dumped on top, letting replacement of some of the carbon by modern carbon be more likely than not. Something wrong in that dating. If so, where did the aragonite come from? When was there a pan of evaporating, carbonate-rich water in the western Amazon basin?

I drifted from the lecture room confused. Matti Räsänen came past me, heading somewhere. I had met him in Turku a few years before, when I was external examiner for one of their doctoral students. I clutched, "Matti, what is the age of that aragonite?" He stared me in the eye a moment and said one word, "Miocene," before hurrying on.

This was the richest jest yet from the aridity paradigm. It had brought events from ten million years ago into the ice age, a bit like those early days before Lyell and Darwin when bones of extinct animals were ascribed to the great biblical flood. This was a "triumph" of a sort. Haffer and Prance duly cited it in building their "case."

This incident was in my mind when I suggested Matti Räsänen to Joachim Adis to write the *Amazoniana* rebuttal. Räsänen's geological history for our *Amazoniana* review tells the history of the western Amazon basin since the Miocene "like it really was," picking up both the aragonite deposits and the bones of grazing animals on the way.

The Andes are young by mountain standards. They are still actively growing as the Pacific plates continue to crunch into the continent. The land is restless. I had had my own lesson of this with my failure to find ancient lakes in Ecuador. But "young" by mountain standards can still be "ancient" by the measure of an ice-age clock.

Before the plates collided and the long orogeny took hold in the Miocene, what is now the Andean foreland was flooded by a marine transgression. The rising Andes first flanked this inland sea; then, as the thrust line moved eastward, the transgression was pushed aside. Great ridges formed before the thrust, breaking the Andean forelands into a series of sub-basins, of which the modern Acre sub-basin is one.

By the end of the Miocene what were once marine systems increasingly became river flood plains as the rising mountains to the west drained heavy rains into the sub-basins. The middle and late Miocene spanned times when both coastal marine and lacustrine ecosystems were displaced, leaving characteristic deposits, including aragonite and gypsum, as records of their passing.[17] This was the origin of the aragonite to be so misdated to hypothetical ice-age aridity. Aragonite and gypsum deposits were a familiar sight to Räsänen in the Miocene/Pliocene strata.

The mountain building continued full spate through the Pliocene and Pleistocene, its thrusts spreading steadily eastward, its sub-basins trapping huge sediment loads as the cordillera rose on their west, behind them as it were. Upward still, and upward, until the sub-basins became not so much sediment traps as sediment sources for rivers draining them to the east.

As the Pliocene gave way to the Pleistocene the great tributary rivers with romantic names, the Putumayo, the Napo, the Marañón, the Ucayali, the Madre de Dios, carved their ways down through the thick sediments collected in the foreland sub-basins. Uplift continued apace, the rivers continued to cut their valleys until now they are as much as 50 m below their old terraces. These were the high banks we had marveled at along the Napo. They were cut through the

Pliocene to Miocene sediments that had filled the old sub-basins formed earlier in the orogeny.

The abandoned terraces, often high above the modern rivers, held old river gravels, the so-called "lag deposits" thrown up beside rivers. It was in these river gravels that A. Rancy, a Brazilian graduate student of David Webb at the University of Florida, found bones of grazing animals, both of living and extinct species. They included still living *Vicuña* and an extinct llama, *Paleolama*. Both vicuñas and llamas are animals of open spaces, most notably of the paramo, the grasslands of the high Andes. Their teeth shout to a paleontologist that they were accustomed to eat grass. But these bones were at low elevations, not in the bottom lands exactly but at elevations where we now still expect forest.

Rancy argued that the elevations where he found the bones must have been grasslands in the old days when the animals lived there. This thesis then developed in the familiar ways of refuge theory, the "grassland" the animals needed was likely to be the westward extension of the great savanna of the ice age.

Rancy's work had the credibility of coming from the laboratory of the foremost expert on fossil mammals of the great faunal exchange between South and North America, David Webb.[18] You could be sure "they got their fossils right."

Haffer and Prance had added Rancy's discovery to Kronberg's aragonite in their *Amazoniana* manuscript to declare their ice-age aridity to have reached well into the western Amazon, tacitly accepting the diminution of their cis-Andean forest refugium that this implied.

Rancy's claim, however, was in head-on collision with our pollen and tree stump evidence from Mera and San Juan Bosco for the very lands drained by the Napo River. Paleobotany said, "forest"; inference from mammal fossils said, "grassland." Hard for both lines of evidence to be correct. Again Räsänen's history showed the way out.

The bones were in lag deposits, scattered with the other flotsam of river banks. They had certainly been rolled about; perhaps whole bloated carcasses had floated downriver from the paramo homeland of such animals, before the rotting remnants were cast ashore. Letting the river move the bones is far more parsimonious than requiring the forest to go away. But perhaps more parsimonious still is letting the animals move themselves. River flood plains are great places for grazing. Even when cutting through forests, river valleys are prone to patches of succulent grass and herbs on islets or areas of seasonal or storm flooding. Grazing animals wander to such excellent feeding places in many a modern ecosystem. Perhaps these Andean animals of long ago did the same.

The animals of Rancy's bones did not have to move far. Walking a few kilometers down hill was all it took. Some of the bones were probably washed down, but many probably walked. The bones do have tales to tell, but not that the forest was abolished in the ice age. This had been one more instance of fitting data to the arid Amazon paradigm, despite more parsimonious explanations.

Few geological claims remained to rebut. We noted for the record that "arkosic" (feldspar-rich) sands in deep-sea cores off the mouth of the Amazon River were brought from the Andes, not from the lowlands. This had already been elaborated in the PAGES article where I had quoted the lovely remark of a noted hydrologist of the Amazon, John Milliman, that it was "silly" to think anything else. Despite that eminently reasonable opinion, the "arkosic sands" still sit solidly in citation lists, notably to account for dust in Lonnie Thompson's Andean ice cores,[19] latched onto because they have been the only "evidence" for Amazonian "aridity" which carried real radiocarbon dates. An irritating afterlife for a "silly" misinterpretation is part of the mischief worked by the paradigm. But this hoary chestnut seems already to have fallen into disuse among refugialists themselves. Thompson's dust probably came from the outwash plains of the montane glaciers themselves.

This brings us to the position where there are no soils, geological, or geographical claims for Pleistocene aridity in the Amazon lowlands that have survived rigorous testing, literally none. Every claim, from fossil fields of sand dunes in glacial times, through properties of soils near Manaus, to evaporites from millions of years ago highjacked for the ice ages, rested on gross misinterpretation of data.

These wrong conclusions had just one thing in common, they fitted into the aridity paradigm. They were made with the paradigm and for the paradigm. How should it have been possible for soil phenomena like stone lines familiar to pedologists in tropical soils worldwide to be so misconstrued for so long without the protection of the paradigm — and the protection given by the shadows of celebrated scientists who believed in it? The power of the paradigm had distorted science, a power derived from the beauty of a brilliant hypothesis.

One vestige of the grounds for thinking aridity remained, that it was real in Africa. The first link of the chain that had let refuge theory loose in the world was the discovery that East Africa had indeed been far drier in glacial times than it is now. This was where I came in, as a doctoral candidate in Daniel Livingstone's laboratory at Duke, assigned to the Bering Sea as Dan set off to make his discoveries in what were then still the British colonies of Kenya, Uganda, and Tanganyika.

The Haffer and Prance defense in *Amazoniana* was still citing East African aridity as reason for thinking that the Amazon should have been dry. But the

Amazon basin is not in Africa, and the local geography is utterly different. In matters important to climate the two geographies are the direct antitheses of each other.

Apart from spreading over the equator, the two continents have only one thing in common: they sit on the same Earth spinning in the same direction where Coriolis force drives winds, and ocean currents at the land from the same direction, from the east. This is the limit to similarity. But whereas Africa presents its most mountainous region directly to the weather on its eastern flank, South America hides its mountains on the western, Pacific Ocean side. South America's Atlantic coast mountains are modest, doing little to impede wind coming in from the southeast via the south Atlantic high pressure systems. The flow of moist air into the Amazon, its rain constantly recycled by the transpiration of the forest, can blow on some three thousand miles before hitting high mountains.

There lies the vital difference which lets one continent remain moister than the other in the boundary conditions of glacial times. Rain from moisture imported into the Amazon lowlands is attenuated every 22,000 years or so with the precession cycle, but the water imports are still sufficient to prime the water-to-air recycling system of the great forest. The result: wetter lowlands in the ice-age Amazon than were possible on the raised savanna plateaus of East Africa.

No doubt this account is too simplistic for a climate model, but it serves to illustrate the vital geographic difference between the continents so conveniently forgotten when using one continent as a model for the other.

After our paper appeared in *Amazoniana* a heart-warming letter from Dan Livingstone himself approved particularly our review of the geological evidence. He spoke about the long history of the Haffer hypothesis *"rising from the ashes every time you slay it"* and, *"I feel a certain responsibility for the hydra-headed monster, because it was our work in Africa that got Jürgen Haffer thinking along those lines."*[20]

Dan Livingstone really did have a "certain responsibility"; it was for demonstrating glacial aridity in East Africa. Data confirming this from many laboratories are now overwhelming, the results validating fully a remark by G. E. Hutchinson while Livingstone was still in Africa, *"Dan's expedition to Africa will tell us more about the dark continent than all the other expeditions that have gone before."*[21] An enviable level of "certain responsibility," this.

But the "responsibility" that Dan claimed so wryly was not for the triumph of demonstrating ice-age aridity in Africa, but for having given new life to tropical biogeography with the novel idea that equatorial regions had climate change too. Another enviable "responsibility." Haffer had seized the moment and grafted an African climate history onto the Amazon. A curious consequence of this

seizure was that it dated speciation of his endemic bird populations to the last
glaciation. This is extremely youthful by evolutionary standards, making the hypothesis vulnerable to independent dating of the species themselves. When Haffer first wrote, we had no direct way to measure the age of species. Now we do,
through the "molecular clock."

Dan's letter went on to say, "*Even in Africa, however, the taxa seem to be much
older than the late-Quaternary events for which we have information, at least if
one places any credence at all in the molecular evidence. I don't see any reason to
doubt it myself.*"

Similar results are coming from molecular studies in the Amazon. These are
early days for so huge an undertaking, but in the Amazon as well as Africa the
taxa so far looked at seem to be older than the late Pleistocene.[22] More of this
kind of evidence should kill off the Pleistocene refuge idea by itself before long.
Molecular evidence, however, might be particularly vulnerable to the process
that Ernst Mayr had warned me about in my Panama office, "No one will believe
you because the refuge theory is so beautiful." Better to concentrate on the gross
environmental changes predicted by the hypothesis.

Thanks to Georg Irion and Matti Räsänen, we had at last shown that ice-age
aridity in the Amazon was, in the words of the Bristol geographer, "a chimera."
But the primary gross environmental change claimed by refugialists was the collapse of the forest biome in glacial times, and its replacement by savanna. In the
Haffer and Prance *Amazoniana* paper the savanna was getting moderated to seasonal tropical dry forest, but the basic claim remained the same: arid Amazon,
fragmenting a forest, destroying a biome. This claim had been arrived at by inference from biological data, bird distributions initially. The claim of biome replacement should be tested by direct biological data for biologists to accept the
loss of their "beautiful" hypothesis. Call pollen analysis as the next witness.

I have already described the long or glacial pollen histories my team extracted
from the Amazon in my professional lifetime. No need to recapitulate or sum-
marise here. The first from the Andean foothills, Mera, and San Juan Bosco;
next from the Hill of Six Lakes, Dragão, Verde, and Pata (all spanning a glaciation and Pata spanning two), the Maicuru inselberg, Bean Lake, and the record
of two glacials from the Amazon fan.

These were only four sites to describe by pollen the vegetation of an ecosystem
three thousand or more kilometers across, depending on where you measured it.
But those were the only sites we could find, and one of them, the fan, tested the
conclusions of the other three by sampling the whole of the Amazon basin.

We were, and we are, quite certain of our main conclusion, that evergreen lowland tropical forest was present, throughout entire glacial cycles, in all the places were it grows now. It was never fragmented nor replaced by different kinds of vegetation. Being so diverse, the potential home of so many species, it constantly reshuffles local arrays of species with the slow tempo of tree life-times. Not every hectare has the same array of species, though some are so abundant in contemporary forests as recently to be called "oligarchs" by Amazon botanists.

This accommodating habit of diverse Amazon forest shows up in times of glacial cooling when trees like *Podocarpus*, *Hedyosmum*, Ericaceae, and others we have identified as the cool-adapted suite, are able to maintain viable populations within the tropical lowland forest, providing pollen analysts with an unexpected signal for modest cooling in glacial times, a cooling that has no destructive effect on the forest.

We were able to make these claims only after realizing that the classic European methods of pollen analysis, invented for forests of wind-pollinated trees, were not appropriate for the animal-pollinated forests of the tropics. Pollen trapping in forest plots with known species lists, let us divide tropical forest pollen into the three classes, no pollen to reach sediments (silent trees or plants), pollen washed into lakes across watersheds (the majority of species), and the few wind-pollinated types capable of producing a pollen rain. Nothing was hidden, we published to scale specimen photographs of all the claimed pollen types in our pollen atlas.

Our conclusion of unfragmented forests through complete glacial cycles was nicely in accord with Irion's analysis of the deeply dissected Belterra clays, and indeed even of the stone lines. The deep podsols on the dissected clays in particular required that the multilayer evergreen canopy had never been fragmented for spans of time at least as long as the Pleistocene. Stone lines likewise required wet climates for durations measured on geological time scales. The pollen data required continuously moist forest, the soils data require epoch-long wet regimes. The lines of evidence concurred.

I regret that these conclusions were deeply disturbing to a Brazilian faction with deep scholarly knowledge of Amazon biology. After our *Amazoniana* rebuttal was published, a paper in the *Proceedings of the Brazilian Academy of Sciences* dismisses *"Colinvaux's . . . limelight-seeking line of thought [that] needs no further attention."*[23]

The Brazilian who wrote that dismissal, P. E. Vanzolini, is an outstanding biogeographer whose knowledge of the Amazon animals in which he specializes is probably unmatched. Moreover he has a solid claim to having called attention to some peculiar patterns of endemism in the Amazon even before Haffer's seminal paper on birds. He has lived a professional lifetime with the refuge theory and not found it wanting. He describes his long association and friendship with the soil scientist Ab'Sáber, who conceived the aridity hypothesis for stone lines. His dismissal of me as the author of persistently contrary conclusions is both human and entirely rational.

It is illuminating, though, that Vanzolini's summary dismissal of the *"limelight-seeking line of thought"* appealed to the views of the pollen analysts of Carajas and Rondonia who continue to decree evidence for savanna and drought in glacial times. This school of palynology not only insists that pollen from Rondonia boreholes and Carajas lake sediments do really describe a savanna-clad Amazon but they even debunked the pollen evidence from the Amazon fan, claiming it to be meaningless.

Vanzolini and others who think like him are not using pollen data as such to support their arguments, rather they are using the opinions of palynologists. Pollen evidence, particularly from so outlandish a place as the Amazon forest, has some of the properties of a medical X-ray or CAT scan image. To know what the image implies you have to know the rules. The physician who requests the image does not trust his own interpretation of what is shown, he asks a radiologist to interpret the image for him. The radiologist does "know the rules," and backs that knowledge with experience. A pollen diagram is our "CAT scan image" of the past. Anyone can learn to read it, but there are subtleties. At this point the opinion of the expert palynologist replaces actual data as "evidence" and the clash of scientific opinion can resemble squabbles in the law or among rival "authorities" in historical scholarship.

Set a palynologist to catch a palynologist. In law or politics this tactic would be called, "Your expert against my expert." The response has to be to show that "their" experts are wrong. Concentrate on the data, taking the fan history first.

Marine mud off the mouth of the Amazon has collected pollen for as long as there has been an Amazon River. The ship *JOIDES Resolution* had cored these sediments, and Simon Haberle in our laboratory analyzed the pollen in cores spanning the whole of the last two glacial cycles. He identified the influence of river bank communities, and could separate them out. The remaining pollen, the great majority, was rain forest pollen. There was not one iota of evidence that the rain forest had ever been fragmented or replaced by savanna, no surge in the telltale grass pollen at any time. That the record was, nevertheless, sensitive to environmental changes across the Amazon was shown by the clear print of pollen indicators of glacial cooling, *Podocarpus* and the rest. Thus did this, the third long pollen record of the Amazon forest, give generality to the detailed findings from Six Lakes Hill and Maicuru.

When Simon presented his results in our Panama laboratory we joked that "the others" would simply claim, "so what, all you got was riverain forest, our savanna was not on the wet river bank, stupid." Joking was the correct response to this imagined response, because sediment outside other river mouths around the world had already demonstrated that estuarine mud did yield regional pollen signals for the vegetation of the interior. The profession had dealt with this issue long ago. So we joked, no doubt wryly influenced by the way our forest history at Lake Pata had been sidestepped by the simple expedient of erecting a "forest refuge" to take care of it.

Joke or not, this is what "their" pollen analysts did. The fan pollen was merely "river bank pollen," river banks do tend to be wet and have trees, so what? I first heard of the assault through mutterings on the grapevine and at conferences, just as I had first heard of the attempt to debunk the Mera record by ridicule of the dating (how long ago that attempt seems now). The eventual political handling of the anti-fan message was masterly. Print the statement buried in a long article for a specialist review journal, with little elaboration or argument as to why the Amazon should be the exception in not carrying a regional signal.[24] The reputation of the authors was good enough. Henceforth ignore it. Future writers confronted with fan data merely had to cite the gurus and the problem went away.

Haffer and Prance used this tactic in their *Amazoniana* paper, quoting the review paper by Van der Hammen and Hooghiemstra[25] as saying that lots of grass pollen on river banks confuses the savanna grass signal and that changes of sea level might have mixed old pollen, simplistic ideas that had many times been addressed, most recently by Haberle himself. But what really raised the pressure of blood was their inane claim that (and I quote), *"Large gallery forests along the river courses in Amazonia probably prevented pollen grains from savanna vegetation to reach the system."*

This inanity of the gallery forests blocking pollen nicely raised Mark's dander, and we used his prose in our rebuttal.

"Could tall trees flanking the river screen Gramineae pollen? Of course not. Pollen obeys Sutton's model of ground level release. Pollen is quickly carried aloft by convective activity, and then either falls out of the air or is knocked out of the air by rainstorms. Very little pollen arriving to any water body enters through the trunk space of trees. A screen of trees, even one several kilometers wide, would have negligible effect on the proportion of long-distance pollen transported to the site. It is the great convective storms of Amazonia that would

carry pollen from all over the basin and pour it into the river. If there were siz-
able savannas they would be visible in pollen samples from Amazon river sediments, but clearly there were no such savannas."[26]

Mark was doing no more than stating, with references, what had been well established through decades of study by European palynologists, meteorologists,
and students of hay fever. Tempting to paraphrase the Bristol geographer's remarks on stone lines by saying, "I find it almost incredible that well-known pollen
phenomena should have been ignored in this way by trained palynologists."

Mark's passage is also a neat summary of the perils of relying on the opinions
of authority figures to support your arguments. In science this tactic is inherently
vulnerable to data and experiment., the dismissal of the pollen record from the
fan being a glaring example. In essence the attempt to debunk came down to saying that clouds of grass pollen, potentially rising to more than 50% of the pollen
rain, were blocked from the rivers by lines of trees, a notion that had long been
discounted as possible by good science: through observation, experiment, and
physical modeling.

With the fan record released from calumny, the known long pollen histories
from the Amazon forest are reestablished as three, the fan, Six Lakes, and Maicuru, to which can be added the not continuous but still glacial samples from
Ecuador, Mera and San Juan Bosco. Together they embrace an enormous area.
In all essentials their records are the same; the great forest persisted through all
stages of a glacial cycle and was never fragmented by savanna or any kind of
grass-rich dry woodland. This is a direct, empirical refutation of the refuge hypothesis. Even in face of this, however, refuge theory is taking an uncon-
scionably long time dying.

This slow death is at heart because of the very beauty that made the theory
both popular and fashionable. It seemed to pull that nasty thorn of so very many
species coexisting where it was hard to imagine physical, geographic isolation.
Without refuge theory, what are we to do? The reluctance to quit is entirely un-
derstandable, "No one will believe you because the refuge theory is so beautiful." But even last-ditch defenses need a nugget of support. Some of the necessary nugget comes still from "palynologists."

Pollen analysts of the pro-savanna school of thought were trained in continental Europe where pollen analysis had its beginnings and where it had developed
as a trusty craft. Half a century of experience was behind the interpretation of
pollen diagrams of simple, species poor European vegetation in which the important forest trees were pollinated by wind. The vegetation histories they had
mastered were the successive colonization of glacier-devastated land these last

12,000 years. Moreover their ecological training came from a tradition that used "indicator species" to recognize plant "associations." Indicator species worked, more or less, in the simple vegetation of Europe. We know the tradition by the name of the scholar who codified its rules, Braun-Blanquet.

At its heart the Braun-Blanquet system was a way of classifying plant communities. It stipulated that plants lived in recognizable communities; if you have seen one oak forest you have seen the lot. The next one you see will not only have the oak trees, but in addition most of the other kinds of plants that lived with the oaks in the first oak forest. I have dubbed this concept, "The Nation States of Trees."[27] One of its offshoots was that there might well be plants so confined to a particular community that you could use them as "indicators" of the whole collection. If true, this could be a godsend to pollen analysts: find pollen of the "indicator" and you have found the whole lot.

"If true" is the critical phrase. Some among us, of whom I am one, doubt that it ever is really true. It must be admitted, however, that in the simple vegetation of Europe "indicator species" can serve as a useful guide. The indicators might well be doing their own thing, regardless of the community in which they happen to live, which is our theoretical objection. But if their optimum habitat is the same as the optimum for many other species, the whole lot will probably be found together. It was possible to turn a hunt for indicator pollen into a useful practice in European palynology. When the practice was brought to Amazon pollen analysis it was pregnant with the possibility of error.

I have related in the last chapter how pollen types from Carajas claimed to be "indicators of savanna" were nothing of the sort. *Byrsonima*, *Borreria*, and Compositae were all abundant on marshy or rocky flats at Maicuru, their habitat no more than clearings in the rain forest caused by outcropping of ironstone rocks with little soil. *Byrsonima* bushes were actually flourishing in wet mud beside puddles and one species of the weed genus *Borreria* flourished as a true aquatic plant in one of shallow lakes while sister species were abundant in weed communities growing out of cracks in the rocks. So many weeds in the daisy-chrysanthemum family Compositae grow in the tropics that it is little less than outrageous to claim most Compositae pollen as indicators of anything.

The modest surge in grass pollen when the Carajas lake drained was not an indicator of savanna. It meant first a puddle with emergent grasses, then a marsh with more of the same, both draining systems surrounded by weeds. Call this, if you like, a local community of weeds "indicated" by a list of weed pollen. Outside the weedy area of the drained lake bed, the meagre Carajas pollen list of 13 types (including all the weeds but without the two cool-tolerant pollen types, or any others as yet undisclosed that were omitted from the original diagram) is entirely consistent with the persistence of the strange *campos rupestre* vegetation that still occupies the site on the high elevation ironstone massif of Carajas.

And yet pollen from Carajas and the point samples from boreholes at Katira remain the argument of choice for pollen analysts of the "yes there was savanna" school. Katira lies on the southwest ecotone between forest and savanna in Rondonia, where excess grass pollen is to be expected as the ecotone shifts with the fluctuating boundary between air masses. The "savanna indicators" at Carajas were a simple case of mistaking the grass and the weeds of the draining lake basin for a signal of savanna; and a true savanna signal at Katira was no more than to be expected near a fluctuating ecotone between forest and savanna. Ignore these realities, believe in savanna, and the logic goes thus: savanna was predicted by the paradigm, the pollen included possible "indicators" of savanna, therefore there was savanna. A prime example of fitting data to hypothesis.

From there believers in an arid Amazon need only look up climate records for modern savannas and calculate the reduction of precipitation from modern levels to apply this climate to the whole of the lowlands defined by the paradigm to have been savanna, and "eureka." On this attempt at logic is based the claim of precipitation falls of 30%–50% in glacial times by "palynologists." I called this thought process "a guess," there might be more accurate, less kind, descriptions.

Looking at all the pollen data with a modern ecological understanding leads inescapably to the conclusion that none of the pollen evidence so far available from the Amazon basin requires, or even allows, widespread savannas to replace forest in glacial times. The three transglacial pollen diagrams from the forested lowlands, particularly the falsely calumnied fan pollen, demonstrated the continuity of the forest biome. Carajas demonstrates the stubborn stability of its edaphically constrained *campos rupestre* community over the same time interval, and the Katira borehole record shows local fluctuations of the forest savanna ecotone of the kind shown by all ecotones that are set as airmass boundaries.

The pollen evidence thus joins the geological and soils evidence in declaring that the Amazon basin has never been arid throughout at least the 170,000 years of the last two glacial cycles. There are no exceptions. All known evidence from geology, pollen, and soils from the ice-age Amazon demonstrates the persistence of the rain forest biome and a moist environment throughout the lowlands. No wide-spread aridity, and no incursions of savanna, despite modest glacial cooling, and the 22,000-year cyclic reduction of the copious rainfall. This is a direct, empirical refutation of the refuge hypothesis.

After publication of our *Amazoniana* rebuttal my mail included a letter from Peter Moore, senior author of a major textbook on pollen analysis, professor of biology at King's College, London, and for many years plant ecology correspondent for *Nature*. I quote the paragraphs referring to the rebuttal paper in full.

> You have indeed left your mark on the Amazon. I have a copy of your pollen manual and I am staggered by the diversity of pollen types contained. I take the point that you make in this paper, that the use of indicator types, or even the search for meaningful "communities," is a quite impossible task, and indeed a wrong approach in such a situation. Your arguments relating to the misinterpretation and over-simplification of pollen diagram interpretation are well based. It comes of the application of Old World minds to New World problems.
>
> It is, as ever, a joy to read your vigorous style and to ponder your iconoclastic ideas. As the Roman Festus once said to your namesake, the Apostle Paul, "you almost have me convinced"! The paleoecological approach seems to be limited, due mainly to lack of sites. Perhaps the way forward is the evolutionary angle. More in the way of molecular studies may provide better data on evolution processes and biogeography in the area, don't you think? Meanwhile I agree with you that the doubts now cast upon the fragmentation/refugium hypothesis are sufficiently strong to view that model with severe scepticism.

That is nicer to read than the "limelight-seeking" passage from the Brazilian Academy of Science. Also more comforting than Ernst Mayr's "No one will believe you because the refuge hypothesis is so beautiful."

Peter Moore offered the comfort of endorsing our pollen research, but he goes further. Effectively he questions the very beauty of the refuge hypothesis itself. The "beauty" lay in providing periodic physical separations of species populations by uncrossable boundaries. Throwing out refuge theory leaves us with the original conundrum that the richest diversity on earth is in the Amazon, a vast green expanse of ancient forest. Hard to keep populations separate for generations within a perpetual forest.

Perhaps it were better to rethink the rules. As Moore wrote, "Perhaps the way forward is the evolutionary angle. . . . Molecular studies may provide better data on evolutionary process and biogeography in the area, don't you think?"

Yes, I do think. Already the "molecular clock" is demonstrating that the timing of speciation in the tropics is not the timing of ice ages, as Livingstone noted in his letter. More powerful still are genetic relationships between species as indicators of where they used to live. Say we really have run out of pollen records of the ice-age Amazon (possibly true), the genetic signals in living tissues are alternative records of history; no shortage of these records.

Is complete geographic isolation as necessary for speciation as claimed? Genetic isolation of populations, yes indeed. But does "isolation" necessarily mean being marooned on an island in an hostile sea? I doubt it. Perhaps "the way forward" is to use the richness of Amazon species as a laboratory to study species formation when the starkest forms of geographical isolation are absent.

Start with the sheer size of that vast forest, an immensity like no other ecosystem on Earth outside the oceans. Graft to this immensity the properties found by ecologists seeking to explain high diversity in the niche arrays in a tropical forest (chapter 1). Among these properties are elaborate physical structure, year-round productivity, intermediate disturbance cycles of tree fall and succession, the way that high diversity itself lessens exclusion by dominance, and the absence of ice-age catastrophe proclaimed by Alfred Russel Wallace but reinstated by refugialists. Add to this mix the separations by distance inherent in the size of the forest, together with the opportunities offered by its unique age, take all this into account and isolates seem more likely than not.

Take into account also the vibrancy of environmental change on all time scales, from intermediate disturbance to glacial cooling and precessional fluctuations of the rains, and then overlay the mixture with climate shifts within the Amazon as the bordering airmasses clash on Pleistocene time-scales. We now know that none of these local events within the margins of the forest were catastrophic, they did not replace one biome with another, rather they set rhythms to ecosystem dynamics that were not necessarily in phase across the great forest.

We now have allies in the oceans. The highest diversity now known on Earth is not in the Amazon but in the microorganisms living in the deep sea. It never changes much down there. There are no physical places to hinder, nor physical boundaries. How selection works to produce this cornucopia of minute organisms in the deep sea is a subject of our times. We might well compare notes.

As a hypothesis, the refuge model is dead, even if it won't lie down. Popular hypotheses die slowly as those who have proclaimed them most strongly are reluctant to let go, and those who were taught to respect them flounder. But this hypothesis has failed the starkest of empirical tests. Environment, vegetation, and climate of the ice-age Amazon were not as stipulated. To kill a paradigm, however, is not so easy. This amounts to killing a majority belief. The necessary change in thinking encounters the process so elegantly put by the wise physiologist Peter Medawar, "The human mind treats a new idea the way the body treats a new protein, it rejects it."[28]

In *The Structure of Scientific Revolutions*, the work that injected the concept of "paradigm" into science, Thomas Kuhn's great contribution was to show that mass rejection of old beliefs always went along with crucial advances in knowledge. Old paradigms must die as stubbornly held beliefs collapse. Kuhn's most celebrated example was the death of the paradigm of fixed continents and its replacement by plate tectonics, an event that he witnessed. That issue was long in doubt.

The death of the old paradigm, the *coup de grace* as it were, finally came with the crucial evidence for sea-floor spreading, those fateful magnetic stripes on the Atlantic floor, marching in parallel array outwards from the mid-ocean ridge. So it is true. The old paradigm is dead, long live the new paradigm.

To kill the refuge hypothesis we have nothing to offer as dramatic as magnetic stripes on the sea floor. Demolition of wrong geological conclusions, together with the evidence of a few pollen diagrams, will have to do instead. Perhaps the astonishment that people could still believe in the arid past of Brazilian soils, when experts from the rest of the world knew this to be untrue; or the stubborn application to the great forests of old-fashioned methods of pollen analysis that had made sense in very different European vegetation long ago, will shake some minds and start the rout. Perhaps, even better, if Amazonian evolutionists find themselves scooped by molecular biologists from the deep sea.

NOTES

PREFACE

1. Amsterdam: Harwood, 1999.

PROLOGUE: TRIAL IN ALASKA

1. D. M. Hopkins, "History of Imuruk Lake, Seward Peninsula, Alaska," *Bull. Geol. Soc. Am.* (1959).
2. This refers to the Maurice Ewing and William Donn theory of Ice Ages, which held that continental glaciers could not form without the precipitation that was postulated to come from an open arctic ocean. Once ice sheets had so lowered ocean levels that Bering Strait was blocked, no warm water could flow north, the ocean would freeze over, precipitation of snow onto the ice sheets would cease, and glaciers would wither. Strange though it seems now, forty years ago the theory had a wide following.
3. P. A. Colinvaux, "A Pollen Record from Arctic Alaska Reaching Glacial and Bering Land Bridge Times," *Nature* 198 (1963): 609–610; Colinvaux, "The Environment of the Bering Land Bridge," *Ecological Monographs* 34 (1964): 297–329.
4. P. A. Colinvaux, "Origin of Ice Ages: Pollen Evidence from Arctic Alaska," *Science* 145 (1964): 707–708.

1. THE REASON WHY

1. M. Shermar, *In Darwin's Shadow: The Life and Science of Alfred Russel Wallace* (Oxford: Oxford University Press, 2002).
2. F. E. Clements, *Plant Succession: An Analysis of the Development of Vegetation* (Washington: Carnegie Institution, 1916), 242. See P. A. Colinvaux, "The Nation States of Trees," ch. 5 of *Why Big Fierce Animals Are Rare* (Princeton: Princeton University Press, 1978).
3. R. H. Whittaker: "Classification of Natural Communities," *Botanical Review* 28 (1962):

1–239, and "Evolution of Diversity in Plant Communities," *Diversity and Stability in Ecological Systems*, Brookhaven Symposia 22 (1969), 178–260.

4. A. R. Wallace, *The Geographic Distribution of Animals*, 2 vols. (London: Macmillan, 1876).

5. R. H. MacArthur: "On the Relative Abundance of Bird Species," *Proceedings of the National Academy of Sciences* 43 (1957): 293–295, "Patterns of Species Diversity," *Biological Reviews* 40 (1965): 510–533.

6. R. M. May, *Stability and Complexity in Model Ecosystems* (Princeton: Princeton University Press, 1972).

7. G. E. Hutchinson, "The Paradox of the Plankton," *American Naturalist* 95 (1961): 137–145.

8. G. E. Hutchinson, "Homage to Santa Rosalia or Why Are There So Many Kinds of Animals?" *American Naturalist* 93 (1959): 145–159.

9. E. Mayr, *Animal Species and Evolution* (Cambridge: Harvard University Press, 1963).

2. THE GALAPAGOS THAT DARWIN KNEW

1. This work, first published in 1839, has been through numerous editions with different sizes, different pagination, and even different titles (e.g., *A Naturalist's Voyage round the World*). It is thus useless to offer the general reader page numbers for the quotations, but they can easily be found because all are in Chapter 17.

2. Until the 1940s the charts of the Galapagos Islands used by the world's shipping were still headed "Surveyed by Captn. Robt. FitzRoy R.N. and the Officers of H.M.S. *Beagle* 1836." Before FitzRoy, Galapagos charts were so vague as to look like works of the "here-be-dragons" period of cartography. FitzRoy's map, made under sail in difficult waters for sailing, with sextant, compass and sounding line in six weeks is almost indistinguishable from the supervening survey made in 1942 by U.S.S. *Bowditch* with the help of a base on the islands, air photography and World War II technology. It was a most remarkable seaman with whom Darwin sailed and whose quarters he shared.

3. C. S. Hickman and J. H. Lipps, "Geological Youth of the Galapagos Islands Confirmed by Marine Stratigraphy and Paleontology," *Science* 227 (1985): 1578–1580.

4. A good source for early visitors to the Galapagos, including the Essex, is J. R. Slevin, *The Galapagos Islands: A History of Their Exploration*, no. 25 in the Occasional Papers of the California Academy of Sciences (1959).

5. R. I. Bowman, ed., *The Galapagos: Proceedings of the Symposia of the Galapagos International Science Project* (Berkeley: University of California Press, 1966). This volume was published after my return from the islands but I had the good fortune to be able to correspond with Bob Bowman and others who guided the nitty-gritty of my plans. My published review at the time reminds me of the epoch-changing effect that volume was to have for Galapagos research; it achieved what other symposia merely dream of by recruiting a generation of researchers to do its bidding. P. A. Colinvaux, "A Galapagos Symposium," *Ecology* 48 (1967): 701–702.

6. W. Beebe, *Galapagos: World's End* (New York: Putnam, 1924).

7. In Slevin, *The Galapagos Islands*.

8. J. D. Hooker, "An Enumeration of the Plants of the Galapagos Archipelago with Descriptions of New Species," *Linnean Society Transactions* 20 (1847): 163–233.

9. I. L. Wiggins and D. M. Porter, *Flora of the Galapagos Islands* (Palo Alto: Stanford University Press, 1971).

10. The myth and recovery of El Junco. It surely took political courage or popular will to end farming on that fertile rangeland round El Junco. I do not know the story, but evidence of popular will abounds. The lake is beautiful, pure water on the very top of a mountain, hidden by clouds over a desert island, a unique place. Tour guides took their groups to see it. Formerly unknown to geography, El Junco's image reached the illustrated papers, carrying with it always that name I had borrowed for it "El Junco." In the fine new church at Wreck Bay, built to serve that inflated population of the last two decades, is a great mural of El Junco, with an inscription of the lake as sacred to God. The land around El Junco was declared a national park.

11. The tree ferns have been saved along with the *Miconia* (see note 10).

12. Because of Beebe's discovery of this lake, some later accounts have named it "Lake Arcturus" after one of the ships used by Beebe on a later expedition (see *The Arcturus Adventure* [New York: Putnam, 1926]). But Beebe did not visit Genovesa on that expedition, and the Arcturus never went there. We rejected that name in favor of "Tower Crater Lake" or, as now "Genovesa Crater Lake."

13. R. L. Stevenson, *Travels with a Donkey* (New York: Scribners, 1907), 36.

14. P. A. Colinvaux, "Reconnaissance and Chemistry of the Lakes and Bogs of the Galapagos Islands," *Nature* 219 (1968): 590–594. A first test of the tsunami hypothesis was to see if the crystalline deposits were in fact gypsum as claimed. Culpably, I had brought back no samples. Meanwhile publication had its expected result in that someone else did the job for us in the person of Richard Howmiller who visited Tagus Cove later, collected samples of the white deposit, and demonstrated that they were indeed gypsum. R. Howmiller and K. Dahnke, "Chemical Analysis of Salt from Tagus Crater Lake, Isabela, Galapagos," *Limnology and Oceanography* 14 (1969): 602–604.

15. R. T. Peterson, *The Birds*, Life Nature Library (New York: Time, 1963).

16. P. Dee Boersma: "The Galapagos Penguin: A Study of Adaptations for Life in an Unpredictable Environment" (Ph.D. thesis, Ohio State University, 1974), and "An Ecological and Behavioral Study of the Galapagos Penguin," *The Living Bird* 15 (1976): 44–93.

17. The standard work on diatom taxonomy is R. Patrick and C. W. Reimer, *The Diatoms of the United States* (Philadelphia, Monograph 13 of the Academy of Natural Sciences, 1966).

18. P. A. Colinvaux, E. K. Schofield, and I. L. Wiggin, "Galapagos Flora: Fernandina (Narborough) Caldera before Recent Volcanic Event," *Science* 162 (1968): 1144–1145.

3. GALAPAGOS CLIMATE HISTORY: THE EASTERN PACIFIC AND THE ICE-AGE AMAZON

1. Radiocarbon dating as practiced in the 1960s. The method depends on the fortunate fact that the heavy isotope of carbon, ^{14}C, is continuously formed in the atmosphere by

bombardment of Nitrogen (^{14}N) by solar radiation. The rate of ^{14}C production is not quite constant, depending on events in the sun, but, by the standards of 1960s measurements, could be treated as nearly constant. The ^{14}C decays back to ^{14}N with a half-life of 5560 years, emitting beta particles in the process at a truly constant rate. The result is that some tiny, but roughly constant, proportion of the carbon dioxide in the atmosphere is based on ^{14}C; is $^{14}CO_2$ instead of $^{12}CO_2$. Plants take up carbon dioxide of both kinds in photosynthesis, giving the surprising revelation that all living plants, without exception, are, ever-so-slightly, radioactive. So are the animals that eat them. When plants or animals die, their corpses are cut off from replenishing supplies of $^{14}CO_2$- or ^{14}C-rich carbohydrate and what is in their bodies steadily decays away. The insight that won Willard Libby the Nobel Prize was to show that the rate of beta decay from carbon samples of a long-dead organism could be calibrated to show how long it had been dead. This was radiocarbon dating as we knew it in the 1960s. The laboratories extracted carbon from dead organic matter, essentially by burning it in a closed system and collecting the resulting CO_2, then counted the decay rate (flux of beta particles) with a set of geiger counters. Young samples (up to 10,000 years old) yielded a good stream of beta particles so that the rate of flux could be measured quickly and accurately, yielding dates generally within a few hundred years of certainty. But once organic matter had been dead for 30,000 years or more, the decay rate was so slow that you could no longer rely on the accuracy of measurement. The "age" would be given as ">30,000 years BP," meaning anything from 30,000 years to the dawn of time. Even the more accurate "younger" dates depended on rather large samples to yield measurable beta flux. Libby wanted 5–10 grams of carbon, so that his test samples tended to be chunks of old wood. By 1966 some labs had got this down to 2 grams, but that meant 2 grams of actual carbon, not 2 grams of sediment. To make sure of 2 grams of carbon in a sample we had to submit sections of the El Junco core that were ten cm long.

The change that transformed radiocarbon dating in the decades to come followed the ability to measure the actual mass of ^{14}C in the sample instead of the tiny proportion of it represented by atoms emitting beta particles at any one time. This became possible by using a large accelerator as a mass spectrometer (AMS). Tiny samples can be used, down to micrograms (the principal advantage), though the range can also be extended beyond the 30,000 yr effective limit of beta-decay dating, though not to the extent that we once hoped.

2. Nero Wolfe is the detective in the Rex Stout novels whose highest praise is, "Satisfactory."

3. Prof. Dr. Walter L. Kubiena, *The Soils of Europe: Illustrated Diagnosis and Systematics* (Madrid: Consejo Superior de Investigaciones Científicas, 1950, and London: Thomas Murby, 1953).

4. This process of "competitive exclusion" is described in chapters 2 and 13 of Colinvaux, *Why Big Fierce Animals Are Rare.*

5. E. K. Schofield and P. A. Colinvaux, "Fossil *Azolla* from the Galapagos Islands," *Bulletin of the Torrey Botanical Club* 96 (1969): 623–628.

6. P. A. Colinvaux, "Climate and the Galapagos Islands," *Nature* 240 (1972): 17–20.

7. R. E. Newell, "Climate and the Galapagos Islands," *Nature* 245 (1973): 91–92.

8. Now at The University of Washington, Seattle.

9. A less objectionable variant is "ENSO," an acronym for "El Niño Southern Oscillation." This at least separates the crash from the real thing.

10. D. Goodman, *The Paleoecology of the Tower Island Bird Colony: A Critical Examination of Complexity-Stability Theory* (Ph. D. thesis, Ohio State University, 1972). The history of the bird colony has not been published because of the absence of a chronology accurate enough for the numbers to be believable (as described in this chapter), but the mathematical assault on complexity-stability theory was published separately as Goodman, "The Theory of Diversity-Stability Relationships in Ecology," *Quarterly Review of Biology* 50 (1975): 237–266. It remains our contention that solution of the dating problem will let the history of the Genovesa breeding population of *Sula sula* (red-footed boobies) serve as a practical demonstration of a vertebrate system for which diversity-stability models are inappropriate.

4. PLEISTOCENE REFUGE AND THE ARID AMAZON HYPOTHESIS

1. For an account of the discoveries of African aridity see D. A. Livingstone, "Late Quaternary Climatic Change in Africa," *Annual Review of Ecology and Systematics* 6 (1975): 249–280.

2. J. Haffer, "Speciation in Amazonian Forest Birds," *Science* 165 (1969): 131–137.

3. Endemism. "The existence of forms exclusive to a place," which forms are called *endemic*.

4. J. Haffer, *Avian Speciation in Tropical South America, with a Systematic Survey of the Toucans (Ramphastidae) and Jacamars (Galbulidae)* (Cambridge, MA: Nuttall Ornithological Club, 1974).

5. E. Mayr, *Systematics and the Origin of Species* (New York: Columbia University Press, 1942). Evolutionary biologists call geographical separation sufficient to prevent cross-breeding *"vicariance."* The word derives from the Latin *vicarius*, meaning "substitute," implying that in local populations cut off from the rest, selection can substitute slightly different versions of the old thing without the genetic control being flooded by cross-breeding. The term was introduced to evolutionary theory by Mayr, who emphasized the importance of physical isolation to speciation. Mayr's writings are essential reading, his name familiar to every student of genetics, natural selection, and evolution. And Mayr was one of the first to declare the beauty of Haffer's hypothesis; and to declare that some Amazonian bird distributions could not be explained in any other way.

6. An accessible summary of this Brazilian work was given by J. J. Bigarella and G. O. de Andrade, "Contribution to the Study of the Brazilian Quaternary," in *International Studies of the Quaternary*, ed. H. E. Wright and D. G. Frey, (New York: Geological Society of America, 1965), Special Paper 84 433–451.

5. THE REPUBLIC OF THE EQUATOR

1. The original home of the potato plant was the high Andes of Ecuador and Peru, where the plants thrive up to at least 4,000 m (13,000 feet), even on steep hillsides. They are still grown in small plots, tilled by hand with hoes.

2. This was before the rise of the Inca empire but parts of the inter-Andean plateau had already been settled for centuries when a lava flow created Laguna de Colta. In the local Riobamba region, the Tunchahuan culture had by the time of Greegor's work already been dated to 500 BC, well before the damming of the lake basin; see B. J. Meggers, *Ecuador* (New York: Praeger, 1966).

3. D. H. Greegor, "The Environmental History of Laguna de Colta, a Lake in the Ecuadorian Andes," (M.Sc. thesis, Ohio State University, 1967).

4. Parkinson's Law, given by C. Northcote Parkinson in a celebrated essay, states that "All work expands to fill the time available for its completion."

5. P. D. Boersma: "The Galapagos Penguin: A Study of Adaptations for Life in an Unpredictable Environment," (Ph.D. dissertation, Ohio State University, 1974); in *The Living Bird* 15 (1977): 43–93; and in *The Biology of Penguins*, ed. B. Stonehouse (New York: Macmillan, 1974), 101–114. D. Goodman, "The Paleoecology of the Tower Island Bird Colony: A Critical Examination of Complexity-Stability Theory" (Ph.D. dissertation, Ohio State University, 1972), and in *Quarterly Review of Biology* 50 (1975): 237–266. D. C. Maxwell, "Marine Primary Productivity of the Galapagos Archipelago" (Ph.D. dissertation, Ohio State University, 1974). H. R. Racine and J. F. Downhower: "Vegetative and Reproductive Strategies of *Opuntia* (Cactaceae) in the Galapagos Islands," *Biotropica* 6 (1974): 175–186, and in *Biotropica* 8 (1976): 66–70. J. F. Downhower, "Darwin's Finches and the Evolution of Sexual Dimorphism in Body Size," *Nature* 263 (1978): 558–563.

6. *Introduction to Ecology* (New York: John Wiley, 1973).

7. P. A. Colinvaux: "Reconnaissance and Chemistry of the Lakes and Bogs of the Galapagos Islands," *Nature* 219 (1968): 590–594; "Climate and the Galapagos Islands," *Nature* 240 (1972): 17–20; "Paleolimnological Investigations in the Galapagos Archipelago," *Mitt. Internat. Verein. Limnol.* 17 (1969): 126–130; "Vegetation of a Galapagos Island before and after an Ice Age," *Ann. Missouri Bot. Gard.* 56 (1969): 419; E. K. Schofield and P. A. Colinvaux, "Fossil *Azolla* from the Galapagos Islands," *Bull. Torrey Bot. Club.* 96 (1969): 623–628.

8. P. A. Colinvaux, *The Fates of Nations* (New York: Simon and Schuster, 1982).

9. In *Journal of Researches*.

10. Ruth Patrick of the Academy of Natural Sciences in Philadelphia has been the doyenne of diatomologists since before the start of my professional lifetime. Every aspiring young diatomologist made the pilgrimage to her laboratory to learn, if they possibly could.

11. C. H. Dodson and A. H. Gentry, "Flora of the Rio Palenque Science Center," *Selbyana* 4 (1978): 1–628.

12. Corley Smith, once British Ambassador to Ecuador, told me that when presenting his credentials in the Presidential Palace, he remarked on the beauty of the ancient building. The President's reply was, "It's awfully damp and cold in winter."

13. G. E. Hutchinson, "Ianula: An Account of the History and Development of the Lago di Monterosi, Latium, Italy," *Transactions of the American Philosophical Society* 60 (1970): 1–170.

14. "18 K" was the then current shorthand for the time of the last glacial maximum (an abbreviation for eighteen thousand years before present). The radiocarbon timescale has

since been more precisely calibrated to show that this event had a true chronological age of 22,000 yr, with the result that "18 K" is now archaic.

15. P. A. Colinvaux, K. Olson, and K-b. Liu, "Late-Glacial and Holocene Pollen Diagrams from Two Endorheic Lakes of the Inter-Andean Plateau of Ecuador," *Review of Paleobotany and Palynology* 55 (1988): 83–100.

16. P. B. Baker et al., "The History of South American Tropical Precipitation for the past 25,000 Years," *Science* 291 (2001): 640–643; M. Basgall, "Modern Science Probes an Ancient Lake," *Duke Magazine*, September – October 2001: 49–52.

17. J. Usher, *Annalis Veteris et Novi Testamenti*, 1658.

18. Ann Anton, a graduate student from Malaysia, wrote a diatom history of Yaguarcocha for her Ph.D. thesis. She and Melanie Riedinger went back to Ecuador for their own thorough survey of the stormy lake by rubber boat, with adventures that should be told somewhere. We also published a pollen history.

19. D. T. Rodbell et al., "An ~15,000-Year Record of El Niño – Driven Alluviation in Southwestern Ecuador," *Science* 283 (1999): 516–520.

20. P. A. Colinvaux, "Glacial and Postglacial Records from the Ecuadorian Andes and Amazon," *Quaternary Research* 48 (1997): 69–78.

21. Colonel Fawcett was a British adventurer in the Amazon who vanished when looking for a (nonexistent) lost city in the Amazon forest. P. H. Fawcett, *Exploration Fawcett* (London: Hutchinson, 1953).

6. REFUGE THEORY EXPANDS IN BRAZIL

1. P. E. Vanzolini: "Zoologia sistématica, geografia e a origem das espécies," *Instituto Geografico São Paulo; Serie téses e monografias* 3 (1970): 1–56, and in *Anais da Academia Brasileira de Ciências* 74 (2002).

2. J. J. Bigarella, "Contribution to the Study of the Brazilian Quaternary," *Geol. Soc. Am. Special Paper* 84 (1965): 433–451.

3. A. N. Ab'Sáber: "Espaços ocipados pela expansão dos climas secos na América do Sul, por ocasião dos períodos glaciais quaternários," *Paleoclimas* (Inst. Geogr. Univ' São Paul) 3 (1977): 1–19, and "The Paleoclimate and Paleoecology of Brazilian Amazonia," in *Biological Diversification in the Tropics*, ed. G. Prance (New York: Columbia University Press, 1982).

4. J. J. Bigarella et al., in *Anais da Academia Brasileira de ciências* 47 (1975): 357–472.

5. P. A. Colinvaux, "The Ice-Age Amazon," *Nature* 278 (1979): 209–210.

7. THE PARADIGM AND THE PROPHET

1. T. S. Kuhn, *The Structure of Scientific Revolutions* (Chicago: University of Chicago Press, 1962).

2. See J. Giorgi, *Mid-Ice: The Story of the Wegener Expedition to Greenland*, trans. F. H. Lyon (London: Kegan Paul, 1934).

3. A. Wegener, *The Origin of Continents and Oceans* (London: Methuen, 1924). First published in German in 1915.

4. P. J. Darlington, *Zoogeography: The Geographic Distribution of Animals* (New York: Wiley, 1957).

5. The reference is to Karl Popper, who codified logic thinking for my generation, particularly emphasizing that success depends on using hypotheses as ideas to refute if possible rather than ideas for which to seek support. See: K. S. Popper, *The Logic of Scientific Discovery* (New York: Basic Books, 1959).

6. This Sagan quote is taken from the Oxford English Dictionary. The OED is also the source for my analysis of the different meanings of "paradigm."

7. See P. A. Colinvaux, "The Nation States of Trees," ch. 5 of *Why Big Fierce Animals Are Rare* (Princeton: Princeton University Press, 1978).

8. W. A. Watts: "The Full-glacial Vegetation of Northwestern Georgia," *Ecology* 51 (1970): 17–33, and "A Late-Quaternary Record of Vegetation from Lake Annie, South-central Florida," *Geology* 3 (1975): 344–346.

9. M. B. Davis, ed., "Vegetation-Climate Equilibrium" *Vegetatio* 65 (1986): 1–141; includes both Davis and Webb papers.

10. G. T. Prance, ed., *Biological Diversification in the Tropics* (New York: Columbia University Press, 1982).

11. J. E. Damuth and R. W. Fairbridge, "Equatorial Atlantic Deep-Sea Arkosic Sands and Ice-age Turbidity in Tropical South America," *Bulletin of the Geological Society of America* 81 (1970): 189–206.

12. M. L. Absy and T. van der Hammen, "Some Paleoecological Data from Rondonia, Southern Part of the Amazon Basin," *Acta Amazonica* 6 (1976): 293–299.

13. D. A. Livingstone, in Prance, *Biological Diversification*, page 524.

14. E. Mayr, *Animal Species and Evolution* (Cambridge: Harvard University Press, 1963).

15. J. R. G. Turner (ch. 17); W. W. Benson (ch. 34); T .L. Erwin and J. Adis (ch. 19); L. A. Endler (ch 35). In Prance, *Biological Diversification*.

16. E. Mayr and R. J. O'Hara, "The Biogeographic Evidence Supporting the Pleistocene Refuge Hypothesis," *Evolution* 40 (1986): 55–67.

17. J. A. Endler, *Geographic Variation, Speciation, and Clines* (New Jersey: Princeton University Press, 1977). E. Mayr and R. J. O'Hara, "The Biogeographic Evidence Supporting the Pleistocene Refuge Hypothesis," *Evolution* 40 (1986): 55–67. The Mayr title is a little misleading. The paper does review favorably the long history of refugial thinking as ways of surviving ice sheets in the north and the suggestions for parallel fragmentation of forests in the tropics. But the references to Amazonia are entirely to Haffer and his allies. The bulk of the Mayr paper uses African data to illustrate his refutation of Endler's analysis. And Mayr wisely cautions that many more biogeographic data are needed, especially from South America.

8. AMAZON'S BITTER LAKES

1. Secchi disk. Named for a monk who in the nineteenth century lowered a white disk over the side of the papal yacht in the Mediterranean Sea to measure light penetration into the oceans. If memory serves, his first Secchi disk could just be made out at 20+

meters in that blue water ocean, then vanished. We still report "secchi disk depth," the depth at which the white disk vanishes, as a routine descriptive measure of water properties in lakes.

2. Limnology, the study of lakes, from the Greek *Limnos*, a lake.

9. ON THE TRAIL OF FRANCISCO DE ORELLANA

1. G. R. Enock, *Ecuador* (New York: Scribners, 1916).
2. F. W. Up de Graff, *Head Hunters of the Amazon* (New York: Garden City, 1923).
3. Ibid., 47.
4. Apart from Enock's book, *Ecuador*, this account is based on Prescott's classic work *History of the Conquest of Peru*, which is itself based on contemporary Spanish accounts.
5. R. T. Rivington, *Punting: Its History and Techniques* (Oxford: Rivington, 1983). A rivalry exists between England's two ancient universities about how to punt. At Cambridge they stand on the closed-over stern, putting the punter high out of the water and forcing him to bend his body at the end of the stroke. At Oxford they stand in the bottom ahead of the closed-over portion, lowering the position of the punter. This means that there is no need to bend at the waist, making not only for comfort and elegance, but also for more power. This is the way a racing punt is propelled (or a canoe); stand in the bottom, keep the body erect, and use the greater power of the upper body to drive the boat. I learned to punt at Cambridge, only coming to my senses, kicking and screaming, in later life as a visiting professor at Oxford and with the personal guidance of Rivington. Napo river punters, like river people the world over, punt their canoes Oxford style.
6. I. G. Frost, "A Paleolimnological and Palynological Investigation in the Ecuadorian Rainforest: Evidence of Regional Flooding and Paleohydrological Disturbance in the Amazon Rainforest" (master's thesis, Ohio State University, 1984).
7. P. A. Colinvaux, M. C. Miller, K-b. Liu, M. Steinitz-Kannan, and I. Frost, "Discovery of Permanent Amazon Lakes and an History of Hydraulic Disturbance in the Upper Amazon Basin," *Nature* 313 (1985): 42–45.
8. Lakes are so important to Amazon people that you can usually find a trail leading to a lake, however remote and nameless. It is even likely that you will find a small canoe stored there, if you know where to look. The dugout, typically small, for one or two persons, has been sunk close to the bank where the trail ends. Once Mike Miller and I completed a reconnaisance of a lowland lake by "borrowing" a sunken canoe, finding it a fast and able craft. Afterwards, of course, we put it back, sinking it where we had found it, for such is the self-evident protocol of Amazon life. Sunken canoes can, of course, be left so long that even their resistant wood rots. On a different lake I once tried the same trick of "borrowing," only to find the waterlogged hulk sinking below me in mid-lake, forcing me to swim back towing the wreck. I suspect it had been left unattended since before the rubber barons and slavers had worked their frightful mischief on the people who had built the canoe more than a century before.
9. Colinvaux et al., "Discovery of Permanent Amazon Lakes."

10. ICE-AGE FOREST FOUND

1. The blue of deep alpine lakes, like the blue of the tropical Pacific Ocean, is the color of light remaining after all wavelengths of the red end of the spectrum have been absorbed during a long passage through clear water. The color is named after John Tyndall, the physicist who explained the effect. See P. A. Colinvaux, "Why the Sea Is Blue," in *Why Big Fierce Animals Are Rare* (Princeton: Princeton University Press, 1978) and in *Yale Review* 64 (1974): 135–146. See P. A. Colinvaux, "Why the Sea Is Blue," in *Why Big Fierce Animals Are Rare* (Princeton: Princeton University Press, 1978) and in *Yale Review* 64 (1974): 135–146.
2. C. Clapperton, *Quaternary Geology and Geomorphology of South America* (Amsterdam: Elsevier, 1993).
3. K-b. Liu and P. A. Colinvaux, "A 5200-year History of Amazon Rain Forests," *Journal of Biogeography* 15 (1988): 231–248; M. Bush and P. A. Colinvaux, "A 7000 Year Vegetational History from Lowland Amazon, Ecuador," *Vegetatio* 76 (1988):141–154.
4. K-b. Liu and P. A. Colinvaux, "Forest Changes in the Amazon Basin during the Last Glacial Maximum," *Nature* 318 (1985): 556–557.
5. CLIMAP project members, "The Surface of the Ice-age Earth," *Science* 91 (1976): 1131–1137.

11. POLLEN, ANACONDAS, AND THE
COOL, DAMP BREEZE OF DOUBT

1. P. A. Colinvaux, P. E. de Oliveira, and J. E. Moreno, *Amazon Pollen Manual and Atlas: Manual e atlas palinológico da Amazônia* (Amsterdam: Harwood, 1999).
2. Quadrat. A marked-out area on the ground, usually square, in which every species present is identified and plotted.
3. J. Lyons-Weiler, "Palynological Evidence for Regional, Synchronous, Holocene Climate Changes in Amazonia and Their Influence on Evolution in the Tropics" (master's thesis, Ohio State University, 1993).
4. D. R. Piperno, *Phytolith Analysis: An Archaeological and Geological Perspective* (New York: Academic Press, 1988). Apart from the work on the Mera and San Juan Bosco sections, Piperno demonstrated the presence of maize (*Zea mais*) phytoliths in 6,000-year-old sediments from the Ayauch[i] cores from which we had previously suspected maize pollen. This was then the earliest date known from the Amazon lowlands for agriculture. M. B. Bush, D. R. Piperno, and P. A. Colinvaux, "A 6000 Year History of Amazonian Maize Cultivation," *Nature* 340 (1989): 303–305.
5. *Economist*, September 12–18, 1992, 119–120.

12. THE PARADIGM STRIKES BACK

1. T. Van der Hammen, "Pleistocene Changes of Vegetation and Climate in Tropical South America," *Journal of Biogeography* 1 (1974): 3–26.

2. K. Heine, "The Mera Site Revisited: Ice-Age Amazon in the Light of New Evidence," *Quaternary International* 21 (1994): 113–119.
3. The abbreviation "ka" is used in writing of ice-age time to mean "thousands of years."
4. Inselbergs (literally *island mountains*) are a classical feature of the Amazon region. Outcrops of mineral-rich, hard resistant rocks, they have been left as broad, flat-topped pinnacles after the huge river system had washed away the surrounding land. With steep, even sheer, sides inselbergs loom out of the distance like great teeth. Typically the inselbergs of the Amazon basin itself rise no more than 300 to 500 m above the lowland plain, but north of the basin boundary in Venezuela where erosion is from the Orinoco river system, inselbergs can be several times that height. It was on early descriptions of these that Conan Doyle based his "Lost World" adventure yarn.
5. ORSTOM is a French government program that maintains scientific missions attached to foreign institutions; in Brazil to the University of São Paulo.

13. THE ADVENTURE OF THE DARIEN GAP

1. M. B. Bush and P. A. Colinvaux, "Tropical Forest Disturbance: Paleoecological Records from Darien, Panama," *Ecology* 75, no. 6 (1994): 1761–1768.
2. Our archaeological colleague, Dolores Piperno, has a more hopeful view based on surviving accounts of conquistadors who describe an open landscape over which they rode their horses, a landscape apparently without woodlands that would require more dramatic recolonization by the forest. Mark and I prefer to rely on the evidence of pollen that the forest was always there, rather than old eyewitness testimony. This is an example of a struggle between disciplines in science. We prefer data to eyewitness accounts, probably written to impress.
3. M. B. Bush and P. A. Colinvaux, "A Pollen Record of a Complete Glacial Cycle from Lowland Panama," *Journal of Vegetation Science* 1 (1990): 105–118; D. R. Piperno, M. B. Bush, and P. A. Colinvaux, "Paleoenvironments and Human Occupation in Lateglacial Panama," *Quaternary Research* 33 (1990): 108–116.

14. THE ADVENTURE OF THE FLOATING FOREST

1. M. B. Bush, M. C. Miller, P. E. de Oliveira, and P. A. Colinvaux, "Two Histories of Environmental Change and Human Disturbance in Eastern Lowland Amazonia," *The Holocene* 10 (2000): 543–554.
2. A. C. Roosevelt et al., "Paleoindian Cave Dwellers in the Amazon: The Peopling of the Americas," *Science* 272 (1996): 373–384.
3. B. J. Meggers, "The Early History of Man in Amazonia," in *Biogeography and Quaternary History in Tropical America*, ed. T. C. Whitmore and G. T. Prance (Oxford: Oxford University Press, 1987), 151–174.

15. THE ADVENTURE OF THE CUSTOMS SHED

1. A. Conan Doyle, *The Lost World* (London: Hodder and Staughton, 1912).
2. Just so story. The allusion is to Kipling's (1902) book *Just So Stories for Little Children.* The stories have titles like "How the Leopard Got His Spots" and "How the Camel Got His Hump" — beautiful fairy tales all. Among evolutionary ecologists the phrase "just so story" has become a stock put-down.
3. P. A. Colinvaux, P. E. de Oliveira, J. E. Moreno, M. C. Miller, and M. B. Bush, "A Long Pollen Record from Lowland Amazonia: Forest and Cooling in Glacial Times," *Science* 274 (1996): 85–88; R. A. Kerr, "Ice-Age Rain-Forest Found Moist, Cooler," *Science* 274 (1996): 35–36.
4. M. B. Bush, P. E. de Oliveira, P. A. Colinvaux, M. C. Miller, and J. E. Moreno, "Amazonian Paleoecological Histories: One Hill, Three Watersheds," *Paleogeography, Paleoclimatology, Paleoecology* 214 (2004): 359–393.

17. ICE AGES FROM AN AMAZON INSELBERG

1. The Greek symbol delta is a convenient shorthand for "excess amount."
2. M. P. Ledru et al., "Late-Glacial Cooling in Amazonia Inferred from Pollen at Lagoa do Caçó, Northern Brazil," *Quaternary Research* 55 (2001): 47–56.
3. CLIMAP, "The Surface of the Ice-age Earth," *Science* 191 (1976): 1131–1144.
4. Tom Guilderson is now (2004) director of the Accelerator Mass Spectrometer (AMS) facility at the Lawrence Livermore National Laboratory. We are collaborating on a study using tree rings from Panama to reconstruct detailed climate histories, a technique which, if it works, we hope to expand to the whole of the New World tropics. His paper on the Barbados cooling is: T. P. Guilderson, R. G. Fairbanks, and J. L. Rubenstone, "Tropical Temperature Variations since 20,000 Years Ago: Modulating Interhemispheric Climate Change," *Science* 263 (1994): 663–665.
5. P. A. Colinvaux, P. E. de Oliveira, J. E. Moreno, M. C. Miller, and M. B. Bush, "A Long Pollen Record from Lowland Amazonia: Forest and Cooling in Glacial Times," *Science* 274 (1996): 85–88; R. A. Kerr, "Ice-Age Rain Forest Found Moist, Cooler," *Science* 274 (1996): 35–36.
6. J. Haffer and G. T. Prance, "Climatic Forcing of Evolution in Amazonia during the Cenozoic: On the Theory of Biotic Differentiation," *Amazoniana* 16 (2001): 579–608.
7. M. P. Ledru et al., "Absence of Last Glacial Maximum Records in Lowland Tropical Forests," *Quaternary Research* 49 (1998): 233–237.
8. M. B. Bush, M. C. Miller, P. E. de Oliveira, and P. A. Colinvaux, "Orbital Forcing Signal in Sediments of Two Amazonian Lakes," *Journal of Paleolimnology* 27 (2002): 341–352.
9. M. B. Bush, P. E. de Oliveira, P. A. Colinvaux, M. C. Miller, and J. E. Moreno, "Amazonian Paleoecological Histories: One Hill, Three Watersheds," *Paleogeography, Paleoclimatology, Paleoecology* 214 (2004): 359–393.

18. TWO THOUSAND FATHOMS THY POLLEN LIES

1. S. G. Haberle, "Upper Quaternary Vegetation and Climate History of the Amazon Basin: Correlating Marine and Terrestrial Pollen Records," in *Proceedings of the Ocean Drilling Program: Scientific Results*, vol. 155, ed. R. D. Flood, D. J. W. Piper, A. Klaus, and L. C. Peterson (College Station, TX, 1997), 381–396; S. G. Haberle and M. A. Maslin, "Late Quaternary Vegetation and Climate Change in the Amazon Basin Based in a 50,000 Year Pollen Record from the Amazon Fan, PDP Site 932," *Quaternary Research* 51 (1999): 27–38.
2. T. Van der Hammen and H. Hooghiemstra, "Neogene and Quaternary History of Vegetation, Climate, and Plant Diversity in Amazonia," *Quaternary Science Reviews* 19 (2000): 725–742.

20. PARADIGM COUP DE GRACE

1. P. A. Colinvaux, P. E. de Oliveira, and M. B. Bush, "Amazonian and Neotropical Plant Communities on Glacial Time-Scales: The Failure of the Aridity and Refuge Hypotheses," *Quaternary Science Reviews* 19 (2000): 141–170.
2. P. A. Colinvaux, G. Irion, M. E. Räsänen, M. B. Bush, and J. De Mello, "A Paradigm to Be Discarded: Geological and Paleoecological Data Falsify the Haffer & Prance Refuge Hypothesis of Amazonian Speciation," *Amazoniana* 16 (2001): 609–646.
3. G. Klammer, "Die Plaeosti des Pantanal von Mato Grosso und die Pleistozene Klimageschichte der brasilianischen Randtropen," *Zeitschrift für Geomorphologie N. F.* 26 (1982): 393–416. This was the paper that described widespead fossil dunes in the Pantanal from remote sensing data and used the resulting maps of dunes to plot paleowind directions, and the paper on which Clapperton (*op. cit.*) based much of his discussion.
4. P. C. Bradbury et al., "Late Quaternary Environmental History of Lake Valencia," *Science* 214 (1981): 1299–1305.
5. P. R. Roa, "Algunos aspectos de la evolucion sedimentológica y geomorfológica de la Llanura alluvial de desborde en el Bajo Llano," *Venezolana Ciencias Nat.* 35 (1980): 31–47.
6. P. Molnar and M. A. Cane, "El Niño's Tropical Climate and Teleconnections as a Blue-Print for Pre-Ice-Age Climates," *Paleooceanography* (2001).
7. G. Irion, "Soil Fertility in the Amazon Rain Forest," *Naturwissenschaften* 65 (1978): 515–519.
8. G. Irion et al. in *Geomarine Letters* 15 (1995): 172–178.
9. Confession: my first job in the New World was on the Soil Survey of New Brunswick, Canada, and I had previously spent Cambridge long vacations doing soil surveys in Portugal and Nigeria.
10. A. Chauvel et al., *Experientia* 43 (1987): 234–241.
11. For example, K. S. Brown and A. N. Ab'Sáber, "Ice-Age Forest Refuges and Evolution in the Neotropics: Correlation of Paleoclimatological, Geomorphological and Pedological Data with Modern Biological Endemism," *Universidade de São Paulo. Paleoclimas* 5 (1979).

12. E-mail from Henry Osmaston, Bristol University. Letters about the *Amazoniana* paper were addressed to me because I was listed as the corresponding author.

13. A. Rancy, "Pleistocene Mammals and Paleoecology of the Western Amazon" (Ph.D. dissertation, University of Florida, 1991).

14. B. I. Kronberg, R. E. Benchimol, and M. J. Bird, "Geochemistry of Acre Subbasin Sediments: Window on Ice-Age Amazonia," *Intersciencia* 16 (1991): 138–141.

15. K. E. Campbell and C. D. Fraily, "Holocene Flooding and Species Diversity in Southwestern Amazonia," *Quaternary Research* 21 (1984): 369–375.

16. C. Clapperton, *Quaternary Geology and Geomorphology of South America* (Amsterdam: Elsevier, 1993), p. 169.

17. M. E. Räsänen, A. M. Linna, J. C. R. Santos, and F. R. Negri, "Later Miocene Tidal Deposits in the Amazonian Foreland Basin," *Science* 269 (1995): 1398–1401.

18. A. Rancy, "Pleistocene Mammals and Paleoecology of the Western Amazon."

19. L. G. Thompson et al.: "Climatic Ice Core Records from the Tropical Quelccaya Ice Cap," *Science* 203 (1979): 1240–1243, and "Late Glacial Stage and Holocene Tropical Ice Core Records from Huascarán, Peru," *Science* 269 (1995): 46–50.

20. Letter from D. A. Livingstone to the author, March 18, 2002.

21. G. E. Hutchinson was visiting Duke University from Yale. He made the remark in the author's graduate student office, around January or February 1961.

22. E. Bermingham, C. W. Dick, and C. Moritz, eds., *Tropical Rain Forests: Past, Present and Future* (Chicago: University of Chicago Press, 2005).

23. P. E. Vanzolini, in *Anais da Academia Brasileira de Ciências* 74 (2002): 633.

24. H. Hooghiemstra and T. Van der Hammen, "Neogene and Quaternary Development of the Neotropical Rain Forest: The Refugia Hypothesis and a Literature Review," *Earth Sciences Reviews* 44 (1998): 147–183.

25. T. Van der Hammen and H. Hooghiemstra, "Neogene and Quaternary History of Vegetation, Climate, and Plant Diversity in Amazonia," *Quaternary Science Reviews* 19 (2000): 725–762.

26. O. G. Sutton, "The Problem of Diffusion in the Lower Atmosphere," *Quarterly Journal Royal Meteorological Society* 73 (1947): 257–276, and *Micrometeorology* (New York: MacGraw-Hill, 1953). H. Tauber, "Investigations of Pollen Transfer in Forested Areas," *Review of Paleobotany and Palynology* 3 (1967): 277–286; I. C. Prentice, "Pollen Representation, Source Area, and Basin Size: Towards a Unified Theory of Pollen Analysis," *Quaternary Research* 23 (1985): 76–86; S. Sugita, "A Model of Pollen Source Area for an Entire Lake Surface," *Quaternary Research* 39 (1993): 239–244.

27. My essay "The Nation States of Trees" is chapter 5 in my *Why Big Fierce Animals Are Rare*. Major schools of thought in ecology, particularly in America, have disputed the tenets of the Braun-Blanquet system for years, concluding that plant communities are much more the product of individual plant properties and chance than of any natural grouping. The concept of indicator species seems particularly ill chosen in complex vegetation.

28. Sir Peter Medawar won the Nobel Prize for Physiology or Medicine in 1960.

INDEX

above-the-rivers hypothesis: at Carajas lake (Brazil), 177; at Manaus (Brazil), 230; at Napo River (Ecuador), 136, 140, 141–42, 144, 151; at Six Lakes (Brazil), 193, 195, 222, 224; at Zancudococha (Ecuador), 161, 163

Ab'Sáber, Aziz, 101, 103–4, 115, 118, 119, 273, 276

Absy, Maria Lucia, 197, 258

Acacia, 36

accelerator mass spectrometry (AMS), 69, 152, 236, 279

Acre sub-basin, 277–81

Adis, Joachim, 268, 280

Africa, 73–74, 103, 119–20, 169–70, 282–84

Agassiz, Alexander, 95, 139

age: of Añangucocha (Ecuador), 139; for aragonite, 279; of Atlantic Ocean core samples, 247–48; of Ayauchⁱ lake (Ecuador), 151; of Bean Lake (Brazil), 261, 265; of Carajas lake (Brazil), 177–78; of Chicopan lake (Lago San Pablo; Ecuador), 94; of Cunro lake (Ecuador), 91; in El Junco (Galapagos Islands), 59; of El Valle lake (Panama), 190; for Genovesa Crater Lake (Galapagos Islands), 69–70; of Kumpakᵃ lake (Ecuador), 151; of Lagoa Comprida (Brazil), 205; of Lagoa Dragão (Brazil), 223, 232–33, 250;

of Lagoa Geral (Brazil), 205, 206; of Lago Agrio (Ecuador), 129; of Lagoa Pata (Brazil), 229, 232–33, 236–37, 243, 250; of Lagoa Santa (Brazil), 182; of Lagoa Tartaruga (Brazil), 205; of Lagoa Verde (Brazil), 232–33, 250; of Laguna de Colta (Ecuador), 81, 93; of La Yeguada lake (Panama), 188; of Maicuru (Brazil), 261, 265; of Manacapuru lake (Brazil), 232; of Mera section (Ecuador), 151–53, 173–76, 189, 191; oxygen and, 234–36; of sand dunes, 272; of San Juan Bosco section (Ecuador), 165–66, 189, 191; of San Marcos lake (Ecuador), 95, 96; of Serra Negra lake (Brazil), 182; of Six Lakes (Brazil), 232–33, 236–37, 243, 250; of Surucucho (Ecuador), 97; of Lake Wodehouse (Panama), 187–88; of Yaguarcocha (Ecuador), 93; of Yambo lake (Ecuador), 92, 93; of Zancudococha (Ecuador), 164. *See also* radiocarbon dating

agriculture, 187–88, 207

Aguarico River (Ecuador), 123–24, 129, 139

aircraft, 140, 145–46, 253, 254–56, 262, 264

Alaska, 1–2, 4–10, 12, 77–78, 104, 136

Alnus, 9–10, 166, 172, 250

Amazon forest: Aguarico River (Ecuador), 123–24, 129, 139; Añangucocha

Brazil (*continued*)
232, 238, 257–58; ice ages in, 100–104,
114–15, 178, 179–80, 182–83, 205–6, 215–
16, 226, 236–37; insects in, 200–201,
220–21, 261; Lagoa Comprida, 201–4,
205–7; Lagoa Dragão, 222–23, 225–27,
232–33, 243–44, 245, 250–51; Lagoa
Geral, 204–7; Lagoa Pata, 224–25, 226,
227, 229, 232–33, 236–44, 245, 250–51;
Lagoa Santa, 182; Lagoa Tartaruga, 203,
205; Lagoa Verde, 225, 226, 227–28, 232–
33, 243–44, 245, 250–51; lakes in generally, 123, 193–95; Maicuru, 252–53, 260–
66, 289; Manacapuru lake, 232, 248, 249;
Manaus, 102, 197, 198–99, 210–11, 229–
32, 248; Pantanal, 269, 271, 272; plants in,
178–80, 182–83, 215–16, 238–39, 253,
257–58, 263–64; pollen analysis in, 178–
80, 182–83, 197, 206–7, 226, 232, 233,
237–40, 243–44, 245, 248, 249–51, 257–
58, 265, 267, 286, 289–90; pollen in generally, 222, 223; radar mapping in, 101–2,
103, 114–15, 116, 193, 194, 197, 215–16, 252;
radiocarbon dating in, 177–78, 182, 205,
232–33, 236–37, 261, 265; rain in, 100–
104, 114–15, 178, 179–80, 183, 206–7, 224,
226, 237, 238, 264, 269–70; refuge hypothesis in, 78–79, 100–104, 114–15, 116,
120, 192–93, 197–98; regulations in, 181–
82, 192–93, 194, 195–98, 200, 211–12, 217–
18, 253; Rio Aracá, 269–70; Rio Negro,
213–16, 229; Serra Negra lake, 182–83;
Six Lakes, 194–95, 208–9, 216, 218, 221–
28, 232–33, 236–44, 245, 250–51, 267;
temperature in, 180, 182–83, 202–3, 238–
40, 250–51; travel in, 198–201, 210–19,
220–21, 228–32, 252–57, 258–60, 264;
trees in, 182–83, 250–51, 255, 259–60,
263–64; Xingu River, 193–94
Britain, x–xi, 12–14. *See also* Europe
British Consul, 33
Brown, Jerry, 10
Brown, K. S., 102
Bursera, 36, 44
Bush, Mark: "Amazonian and Neotropical
Plant Communities on Glacial Time

Scale," 267, 268–69; "Amazonian Paleoecological Histories: One Hill, Three
Watersheds," 218, 306n. 9; Ecuador research by, 97; El Valle lake (Panama)
and, 190, 191; Lagoa Pata (Brazil) and,
242–43; Lagoa Verde (Brazil) and, 232–
33, 243–44; La Yeguada lake (Panama)
and, 188, 191; "A Long Pollen Record
from Lowland Amazonia: Forest and
Cooling in Glacial Times," 218, 241;
Manaus (Brazil) and, 229; "Orbital Forcing Signal in Sediments of Two Amazonian Lakes," 306n. 8; "Paleoenvironments and Human Occupation in
Late-glacial Panama," 305n. 3; "A Paradigm to Be Discarded," 268, 276–77,
283–84, 287–88, 290–91; "A Pollen
Record of a Complete Glacial Cycle
from Lowland Panama," 305n. 3; pollen
trapping and, xiv, 159–60; research
teams served on generally, 162, 185, 198,
211; San Juan Bosco section (Ecuador)
and, 164, 166; Six Lakes (Brazil) and,
224–25, 232–33, 242–43, 243–44; "Tropical Forest Disturbance: Paleoecological
Records from Darien, Panama," 187;
"Two Histories of Environmental
Change and Human Disturbance in
Eastern Lowland Amazonia," 305n. 1;
Wodehouse lake (Panama) and, 186, 187
butterflies, 102–4
Byrsonima, 258, 289

caatinga, 215, 258
cacti, 35–36, 64, 82, 263–64
caimans, 260, 261
campos rupestre, 177, 257, 290
Cana swamp (Panama), 185, 188
capybara, 187
Carajas lake (Brazil): age of, 177–78; core
samples from, 176–81; grasses in, 178–
80, 226; ice ages and, 178, 179–80, 183,
226; plants in, 178–80; pollen analysis
for, 178–80, 183, 197, 222, 226, 239, 250,
257–58, 267, 286, 289–90; radiocarbon
dating for, 177–78; rain and, 178, 179–